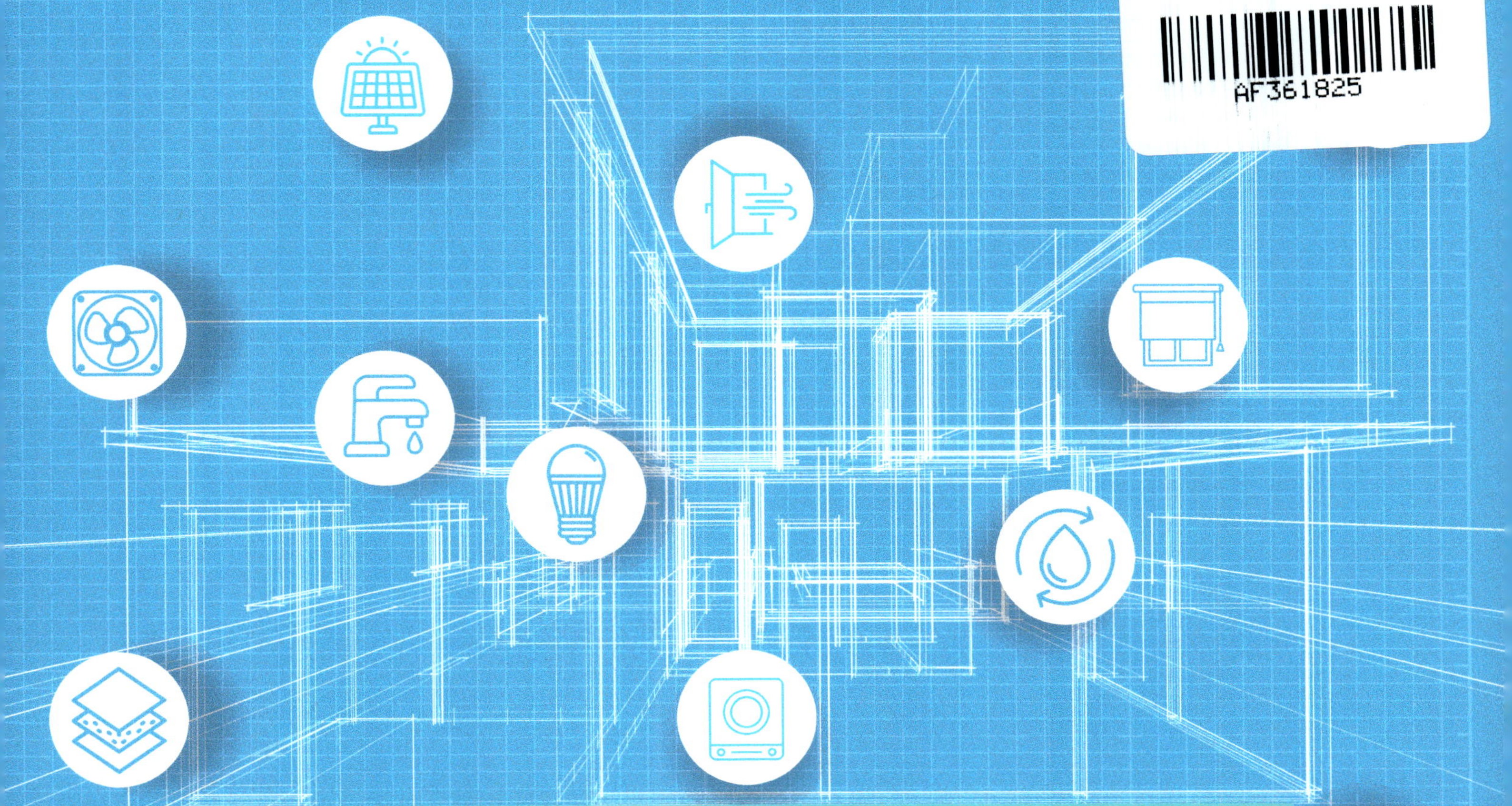

HANDBOOK ON ENERGY EFFICIENCY IN BUILDINGS

OCTOBER 2024

ASIAN DEVELOPMENT BANK

ADB

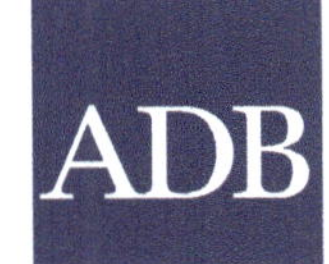

HANDBOOK ON ENERGY EFFICIENCY IN BUILDINGS

OCTOBER 2024

ASIAN DEVELOPMENT BANK

CONTENTS

Contents

TABLES, FIGURES, AND BOXES

Tables

Figures

Boxes

FOREWORD

The Asian Development Bank (ADB) has committed to becoming the climate finance bank for Asia and the Pacific. We have increased our level of climate financing to $100 billion cumulatively from 2019 to 2030. Increasing our level of climate financing means that any building financed by ADB from now onward will have to be sustainable, climate-resilient, energy-efficient, and/or be a certified green building. This handbook on energy efficiency in buildings provides detailed guidance and tools for ADB staff, technical experts, and governments on best practices and solutions.

According to the International Energy Agency, the buildings and building construction sector alone is responsible directly and indirectly for around one-third of global energy- and process-related carbon dioxide emissions. This includes emissions from the manufacture of cement, steel, and aluminum for building construction. With increasing urbanization and demand for cooling, the amount of energy used by buildings and cities is expected to increase considerably in Asia and the Pacific, where most housing is yet to be built. To meet the goals of the Paris Agreement, buildings across the globe will need to reduce energy consumption by 30%-50% per square meter. This represents a tremendous effort.

In December 2023, governments agreed at the Conference of the Parties of the UNFCCC to double the global average annual rate of energy efficiency improvements by 2030. Following this pledge, ADB has established a OneADB Taskforce on Energy Efficiency in Buildings as well as a Special Initiative on Accelerating Energy Efficiency. The aim is to embed energy efficiency in all projects across all sectors, particularly buildings.

ADB is supporting its developing members in several areas of energy efficiency, including clean and efficient cooling solutions, supply and demand-side energy efficiency, energy efficiency policies and regulations across all sectors including building energy codes, innovative financing instruments, and mobilization of private sector resources. In 2021, ADB launched its new energy policy, which has energy efficiency at its core. We believe that accelerating progress on energy efficiency gains across the region can contribute to the agendas of energy security and energy access while also delivering cost savings and environmental benefits.

We look forward to working closely with both the public and private sectors to implement the necessary policies and technologies highlighted in the handbook to accelerate investment in low-carbon, climate-resilient, efficient, and affordable buildings across Asia and the Pacific.

Priyantha Wijayatunga
Senior Director, Energy Sector Office

Ramesh Subramaniam
Director General and Chief of Sectors Group

ACKNOWLEDGMENTS

This handbook was prepared by the Asian Development Bank Energy Sector Office, Sectors Group, led by Kee-Yung Nam, principal energy economist, with guidance of Yongping Zhai, former chief of Energy Sector Group and Priyantha Wijayatunga, senior director, Energy Sector Office, Sectors Group. The team comprised David Morgado, senior energy specialist, for technical review; and Ute Zimmerman, Wei Zhou, and X Suliya, consultants, for analysis and technical writing.

The handbook was enhanced by the feedback from International Energy Agency colleagues: Ksenia Petrichenko, energy efficiency policy analyst; Clara Camarasa, energy efficiency policy analyst; and Cornelia Schenk, energy efficiency policy analyst.

Charity Torregosa, senior energy officer; Maria Dona Aliboso, associate project officer; Angelica Apilado, associate project analyst; and Marinette Glo, operations assistant, provided technical advisory and administrative support. Maria Theresa Mercado edited the report. Editha Creus, typesetter; Levi Lusterio, proofreader; and Lawrence Casiraya, page proof checker helped in the production of this publication. Insights and suggestions from colleagues in the Energy Sector Office was also instrumental in shaping the handbook.

ABBREVIATIONS

ABS	asset-backed security
ADB	Asian Development Bank
ASEAN	Association of Southeast Asian Nations
BEA	Berlin Energy Agency
BEE	building energy efficiency
BOT	build-own-operate
BSEEP	Building Sector Energy Efficiency Project
DMC	developing member country
EBRD	European Bank for Reconstruction and Development
EERF	Energy Efficiency Revolving Fund (Thailand)
EPC	energy performance contract
ESA	energy service agreement
ESCO	energy service company
ESP	Energy Savings Partnership
ETS	emission trading systems
EU	European Union
GDP	gross domestic product
HVAC	heating, ventilation, and air-conditioning
IDB	Inter-American Development Bank
IEA	International Energy Agency
IFC	International Finance Corporation
KeTTHA	Ministry for Energy, Green Technology and Water (Malaysia)
LABEEF	Latvian Baltic Energy Efficiency Facility
MDV	Malaysian Debt Ventures
MEPS	minimum energy performance standards

Mtoe	million tons of oil equivalent
NDC	nationally determined contribution
PACE	Property Assessed Clean Energy
PCG	partial credit guarantees
PPA	power purchase agreement
SMEs	small and medium-sized enterprises
SPV	special purpose vehicle
SUNShINE	Save your bUildiNg by SavINg Energy
TA	technical assistance
VAT	value-added tax

EXECUTIVE SUMMARY

The Asian Development Bank (ADB) recognizes the importance of supporting its developing member countries (DMCs) in scaling up energy efficiency investments in buildings. ADB has been providing financial and technical assistance to its DMCs to promote energy efficiency in buildings through various modalities, such as policy-based lending, project loans, results-based lending, guarantees, grants, and knowledge products.

This handbook provides practical guidance and tools for ADB staff and other stakeholders involved in designing, implementing, and evaluating energy efficiency projects in buildings. It covers the key stages of the project cycle, from project identification and preparation to implementation and monitoring. It provides a comprehensive overview of the range of actions and measures needed to help reduce emissions and ultimately decarbonize the sector such as the promotion of energy-efficient design strategies and technologies, the development of optimized financing instruments, and business models for future scale-up. It showcases good practices and lessons learned from ADB's experience and other international experiences. This handbook is a contribution to opening a sensible avenue for the transition to a zero-carbon, efficient, and resilient building stock.

The buildings and building construction sector is among the most emission-intensive sectors and offer considerable potential for carbon dioxide (CO_2) emission reduction. In 2020, the sector consumed about 36% of global total final energy consumption and generated 37% of energy-related CO_2 emissions. The global building stock is expected to increase by 75% from 2020 to 2050, 80% of which is projected to be in emerging and developing economies. Asia and the Pacific is the fastest-growing region in the world. It accounts for nearly 40% of global gross domestic product and generates about 50% of energy-related CO_2 emissions. The buildings and building construction sector in Asia is booming. In 2018, buildings in the Association of Southeast Asian Nations (ASEAN), the People's Republic of China (PRC), and India consumed 27% of these countries' total final energy consumption, slightly lower than the world average of 30%, and emitted 24% of energy-related CO_2 emissions, or 3.2 gigatons of CO_2, which accounted for nearly one-third of the global amount. The continuous economic growth, rapid urbanization as well as increasing population, are driving the fast expansion of buildings stocks, especially in developing and emerging economies. The International Energy Agency (IEA) projected that 65% of the new floor area to be constructed from 2017 to 2050, about 70 billion square meters, will be in the PRC, ASEAN, and India.

To decarbonize the buildings and building construction sector, electrification and energy efficiency are the two major measures, followed by fuel switching. Improving energy efficiency in buildings can reduce energy costs, enhance energy security, and mitigate greenhouse gas emissions. However, many barriers hinder the adoption of energy-efficient technologies and practices in the buildings and building construction sector, such as lack of information, awareness, and skills, high up-front costs, split incentives, and regulatory gaps. To accelerate the adoption of energy efficiency in the buildings and building construction sector (both new and existing buildings), first, it is imperative that governments reconsider the use of inefficient fossil fuels subsidies, improve availability and access to building energy use data, set national energy efficiency targets for buildings, and enable local governments to adopt and effectively enforce mandatory building energy codes. Second, green building certification and energy labeling can help building owners identify buildings that go beyond the national standards, and help the government provide incentives to cover the incremental costs or even expedite the permitting process of green buildings. Third, capacity building of government staff and the private sector will be important to ensure there is sufficient skilled workforce to deliver sustainable and efficient buildings in the country. Energy efficiency can contribute significantly to new job creation, particularly in building retrofit. Fourth, it is essential to have enhanced information and awareness on energy efficiency policies and technologies at different levels of government, building owners, developers, real estate companies, and even the banking and utility sectors.

The handbook is organized into four parts. Part I provides an overview of the global and regional trends and challenges in the buildings and building construction sector, the potential and benefits of energy efficiency, and the role of ADB in supporting DMCs. Part II presents the main elements and steps involved in developing energy efficiency projects in buildings, such as conducting market and policy assessments, identifying project opportunities, selecting appropriate financing and delivery mechanisms, and preparing project proposals. Part III discusses the key aspects of implementing and monitoring energy efficiency projects in buildings, including procurement, supervision, quality assurance, reporting, and evaluation. Part IV summarizes the main conclusions and recommendations of the handbook and provides some useful references and resources for further reading.

This technical handbook has been prepared for energy communities working in the buildings and building construction sector (e.g., engineers, architects, building developers) in members of the Asian Development Bank (ADB), as well as ADB operations teams to support demand-side investment projects and technical assistance.

There is growing consensus that over 50% of global emissions need to be reduced by 2030, from 2010 levels, to be on track for the 1.5-degree Celsius (°C) target by 2050.[1] The buildings and building construction sector is among the most emission-intensive sectors and offer considerable potential for carbon dioxide (CO_2) emission reduction. In 2020, this sector consumed about 36% of global total final energy consumption (i.e., residential, nonresidential, and buildings construction industry), as shown in Figure 1, and generated 37% of energy-related CO_2 emissions (i.e., construction and operation of buildings). Some progress has been made since 2015 in decoupling energy consumption from floor space expansion as well as the adoption of efficient heating technologies, particularly because of policy adoption in developed countries. In 2020, the buildings and building construction sector emitted 11.7 gigatons of CO_2, an unprecedented drop of emissions by 10% from 2019, largely due to the coronavirus disease

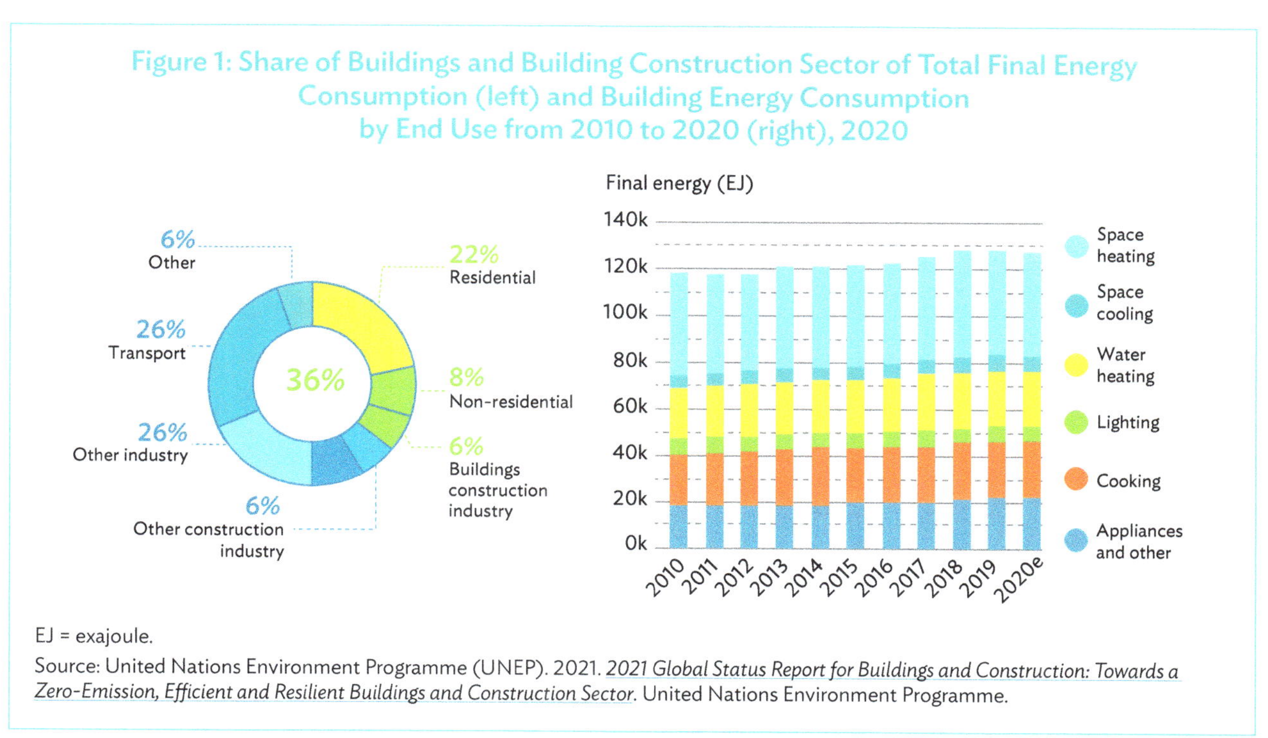

Figure 1: Share of Buildings and Building Construction Sector of Total Final Energy Consumption (left) and Building Energy Consumption by End Use from 2010 to 2020 (right), 2020

EJ = exajoule.

Source: United Nations Environment Programme (UNEP). 2021. *2021 Global Status Report for Buildings and Construction: Towards a Zero-Emission, Efficient and Resilient Buildings and Construction Sector*. United Nations Environment Programme.

---

[1]   United Nations, Net Zero Coalition.

(COVID-19) pandemic. However, the sector is not on track to meet the climate commitments by 2050, according to the International Energy Agency (IEA). In 2021, there was a sharp rebound of CO_2 emissions globally, nearly back to the pre-pandemic level, adding the urgency for concrete and immediate actions.[2] Energy efficiency offers the most cost-effective and one of the fastest measures to cut emissions. It is thus essential to improve energy efficiency and ultimately decarbonize the buildings and building construction sector.

This handbook covers in detail a range of actions and measures needed to help reduce emissions and ultimately decarbonize the sector, such as the promotion of energy-efficient design strategies and technologies, the development of optimized financing instruments, and business models for future scale-up. It is to develop capacity and knowledge of ADB operations, decision-makers, and professionals in developing member countries (DMCs) of ADB by providing a practical reference for the design of building energy efficiency projects. This handbook is a contribution to opening a sensible avenue for the transition to a zero-carbon, efficient, and resilient building stock.

Energy efficiency improvements typically refer to the utilization of less energy to provide the same services. Buildings, together with transport and industry, are among the key areas for energy efficiency improvement. In the global discussion, a distinction is made between supply-side energy efficiency (SSEE), aiming at decreasing energy losses in energy production and in the delivery of electricity or heat, while demand-side energy efficiency (DSEE) focuses on the consumption of less energy for the same level of service[3] in energy consuming sectors such as industry, buildings, and transport. Yet, buildings can be energy producers as well. Further, decarbonizing the buildings and building construction sector requires efforts to cut emissions throughout the entire life cycle, including land use and urban planning, design, construction and retrofits, manufacture of building materials, operation, decommissioning, and appliances. Energy efficiency in buildings usually covers building construction and operation, including building envelope, appliances, and equipment (e.g., lighting, space heating, cooling, ventilation, water heating, management system). This handbook looks at the efficiency at the point of consumption in buildings (DSEE) and focuses on heating, ventilation, and air-conditioning, which make up around 40% of energy use, in the technology chapters.

Energy-efficiency measures can be applied to all types of buildings. This handbook will differentiate between new construction and building retrofits where necessary. Most technologies considered apply to both levels irrespectively. As the first step, it is important to make an initial survey of the building types being considered, as shown in Figure 2, before formulating the energy efficiency project or program.

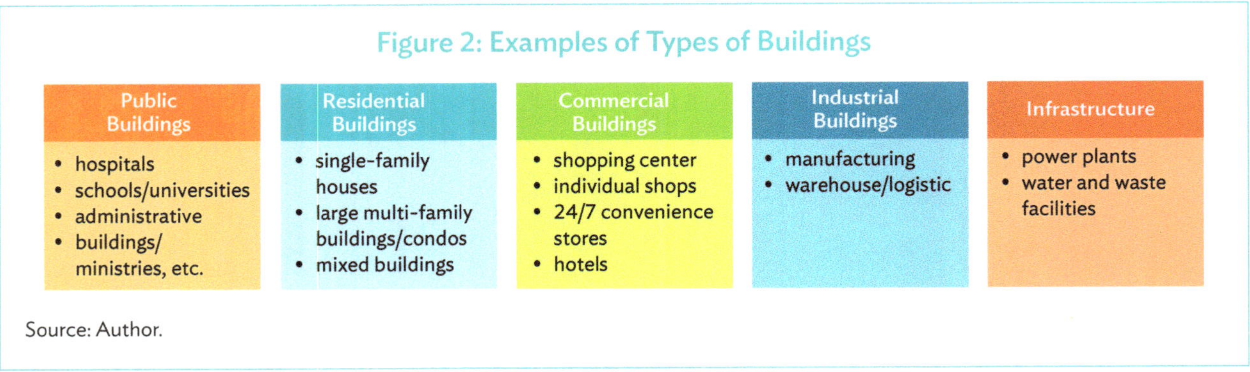

Figure 2: Examples of Types of Buildings

Source: Author.

---

[2]   International Energy Agency. 2021. Tracking Buildings.
[3]   See also 2017. Evaluation report of ADB. Asia Energy Efficiency Accelerator.

This handbook also aims to raise awareness of building energy efficiency measures and provide a better understanding of the potential, proven technologies, and financing instruments applicable to this sector to foster project opportunities within DMCs.

This handbook comprises the following chapters:

- Chapter 1 provides the objectives and scope of this handbook.
- Chapter 2 presents the latest building energy efficiency data in the world and in Asia and the Pacific, the key stakeholders of the buildings and building construction sector, four enabling factors of improving energy efficiency, the co-benefits, and ADB's efforts in this field.
- Chapter 3 focuses on green building[4] certification schemes, also known as rating tools, and provides an overview of international schemes and national schemes in Asia, and examines the energy criteria followed by cases applied in DMCs.
- Chapter 4 elaborates on selected sustainable cooling strategies and technologies throughout the design and operation phases. It looks broadly at the control of heat gains, ventilative cooling, cooling technologies and sources, and air distribution.
- Chapter 5 continues the technical discussions on heating and explores clean heating options, including heat pumps, solar space heating, and solar water heating.
- Chapter 6 provides an overview of investment in building energy efficiency and presents proven financial instruments, including traditional and specialized innovative instruments.
- Chapter 7 explores business models for building energy efficiency projects by presenting three successful examples from Europe and Asia.
- Chapter 8 concludes the handbook and provides policy recommendations.

Case studies and examples from developing and advanced economies from Asia and other regions are identified and illustrated in boxes.

---

[4] Green building is a holistic concept that starts with the understanding that the built environment can have profound effects, both positive and negative, on the natural environment, as well as the people who inhabit buildings every day. Green building is an effort to amplify the positive and mitigate the negative of these effects throughout the entire life cycle of a building. See more at https://www.usgbc.org/articles/what-green-building.

BACKGROUND

Key Stakeholders

Building energy efficiency (BEE) is not a single-party problem—it is cross-sectoral and involves different stakeholders. Collaboration among developers, users, owners, and others is essential. It applies to both public sector proponents as well as for the private sector ones.

Stakeholders within the public sector, policymakers and officials at all jurisdictional levels must cooperate. As BEE involves several ministries and public offices, it is critical that one of the ministries is mandated and responsible for BEE but respects the competence and professionalism of others.

The value chain for BEE is complex and varies according to the building type. Comprehensive support and multi-stakeholder analysis might be required. It may involve urban planners, architects, developers, investors, construction companies, utilities, banks, and financial institutions as well as retailers—e.g., for household appliances, installers, technical consultants, energy auditors, the owners, the lessors, and facility managers, and finally specialized companies such as energy service companies (ESCOs). It is not a homogeneous group. They all have individual goals, levels of expertise, knowledge, and needs that may differ within the stages of the process and the countries in question. They may all need initial support. Programs must be comprehensive and inclusive and consider the market as a whole instead of singling out one party. The gaps between them need to be mended.[5] Careful analysis of the drivers of each stakeholder is required to ensure a common objective and deliver a positive outcome. Figure 3 illustrates the positions of various stakeholders.

Sector Overview

Globally, buildings and building construction sector consumed 36% of the global total final energy consumption in 2020 as shown in Figure 1, among which operational energy use accounted for 31% or around 127 exajoule (EJ), and constructions accounted for 5% or around 22 EJ. In terms of global buildings, energy consumption by end use, spacing heating, and cooling together make up about 40%. Space cooling has been the fastest-growing use of energy, particularly in countries that are not members of the Organisation for Economic Co-operation and Development (OECD). In 2020, the CO_2 emissions of the buildings and building construction sector amounted to 11.7 gigatons, around 10% down from 2019, mainly due to the slowdown of the global economy and construction activities. Countries like the United States (US) already saw a sharp bounce-back of emissions in 2021, increasing by 6% from the 2020 level (Plumer 2022). The 11.7 gigatons of CO_2 emissions accounted for 37% of energy-related

---

[5] Further reading: Zedan, S. and W. Miller. 2018. Quantifying Stakeholders' Influence on Energy Efficiency of Housing: Development and Application of a Four-Step Methodology. *Construction Management and Economics*. 36 (7). pp. 375–393.

Figure 3: Stakeholders' Roles Throughout the Entire Life of Buildings

NEW BUILDINGS			EXISTING BUILDINGS				
Land Use/ Planning	Design	Construction	Sale or Lease	Tenant Build-Out	Operations and Maintenance	Retrofit	Demolition and Deconstruction
Local governments	Design and construction professionals	Design and construction professionals	Buildings owners and managers	Buildings owners and managers	Buildings owners and managers	Buildings owners and managers	Design and construction professionals
Developers and self-help builders	National and provincial governments	Building investors	Developers and self-help builders	Building occupants	Energy utilities	Building investors	Buildings owners and managers
	Local governments	Suppliers and manufacturers	Building occupants	Design and construction professionals	Building occupants	Building occupants	
						Design and construction professionals	

Source: World Resources Institute (WRI). 2016. *Accelerating Building Efficiency: Eight Actions for Urban Leaders*.

emissions: 27% from building operations, and 10% from construction and materials, i.e., embodied emissions[6] considering the carbon footprint across the entire life cycle of buildings. Embodied emissions have gained increasing attention over the recent years, as the sector is moving toward zero-carbon emissions, and a significant number of carbon-intensive buildings are still expected to be constructed. Several European countries have taken early actions to regulate embodied emissions that need to have comprehensive regulations in place.

The global building stock is expected to increase by 75% from 2020 to 2050, 80% of which is projected to be in emerging and developing economies (IEA 2021d). Asia and the Pacific is the fastest-growing region in the world and accounts for nearly 40% of the global gross domestic product (GDP) (IMF 2021) and generates about 50% of emissions. The buildings and building construction sector in Asia is booming. In 2018, buildings in the Association of Southeast Asian Nations (ASEAN), the People's Republic of China (PRC), and India consumed 27% of these countries' total final energy consumption, slightly lower than the world average at 30% and generated 24% of energy-related CO_2 emissions, or 3.2 gigatons of CO_2, which accounted for nearly one-third of the global amount (Figure 4).

Continuous economic growth, rapid urbanization, and increasing population are driving the fast expansion of building stocks, especially in developing and emerging economies. IEA projected that 65% of the new floor area to be constructed from 2017 to 2050, about 70 billion square meters (m^2), will be in the PRC, ASEAN, and India, as shown in Figure 5. This coincides with GlobalABC's study showing that Asia will accommodate nearly half of all new constructions (UNEP 2021). The Asia Pacific Energy Research Centre (2019) estimated that Southeast Asia will lead the growth of building energy demand in Asia and the Pacific. This will likely lead to significant increases in fossil fuel consumption and CO_2 emissions if the transition toward efficient and low-carbon buildings is slow. Taking India as an example, two-thirds of the buildings that will exist in 2040 in the country are yet to be built, based on the current policy settings (IEA 2021c). The buildings and building construction sector, together with industry and transport, will largely shape the energy demand of the country in the coming decades. In addition, the higher airflow rates for ventilation and filtration and changes in working patterns may increase energy consumption in the post-pandemic era.

---

[6] Embodied emissions or embodied carbon refer to the emissions across the entire life cycle of buildings, because of material manufacturing, construction, maintenance, renovation as well as deconstruction. The major emissions come from the production stage.

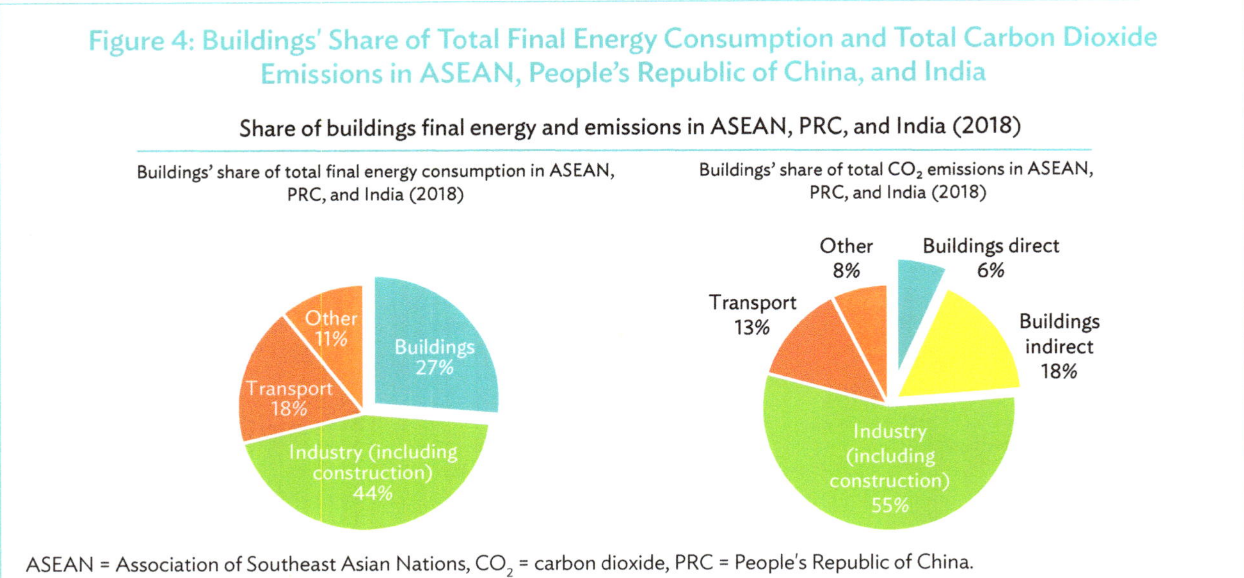

Figure 4: Buildings' Share of Total Final Energy Consumption and Total Carbon Dioxide Emissions in ASEAN, People's Republic of China, and India

ASEAN = Association of Southeast Asian Nations, CO$_2$ = carbon dioxide, PRC = People's Republic of China.
Sources: GlobalADB/IEA/UNEP. 2020. *GlobalABC Regional Roadmap for Buildings and Construction in Asia 2020-2050: Towards a Zero-Emission, Efficient, and Resilient Buildings and Construction Sector*.

Figure 5: Total Building Stock in 2017 and Expected Growth to 2060 in Key Regions

ASEAN = Association of Southeast Asian Nations, OECD = Organisation for Economic Co-operation and Development, PRC = People's Republic of China.
Source: International Energy Agency (IEA). 2017. Energy Technology Perspective 2017.

To decarbonize the buildings and building construction sector, electrification and energy efficiency are two major measures followed by fuel switching. According to British Petroleum's projection, all the growth of primary energy demand will come from the developing world (Figure 6). Under British Petroleum's net-zero scenario, which aligns with the 1.5°C target, the total primary energy demand of buildings is expected to rise only around 4.7% from 2018–2050, driven by energy efficiency improvement as well as behavioral changes, in comparison to a growth of 38% under the business-as-usual scenario, assuming the building stock grows by more than 75% over the same period.

Overall, the primary energy demand of buildings has to decrease by around 20%–30% by 2050 from the current level to achieve the 1.5-degree target. Meanwhile, electricity will become the dominant energy use, and on-site fossil fuels will be nearly eliminated, as shown in Figure 7. At the country level, fuel use varies significantly in Asia. For instance, electricity accounted for 35% of building energy consumption in the PRC in 2019 but only 19% in India.

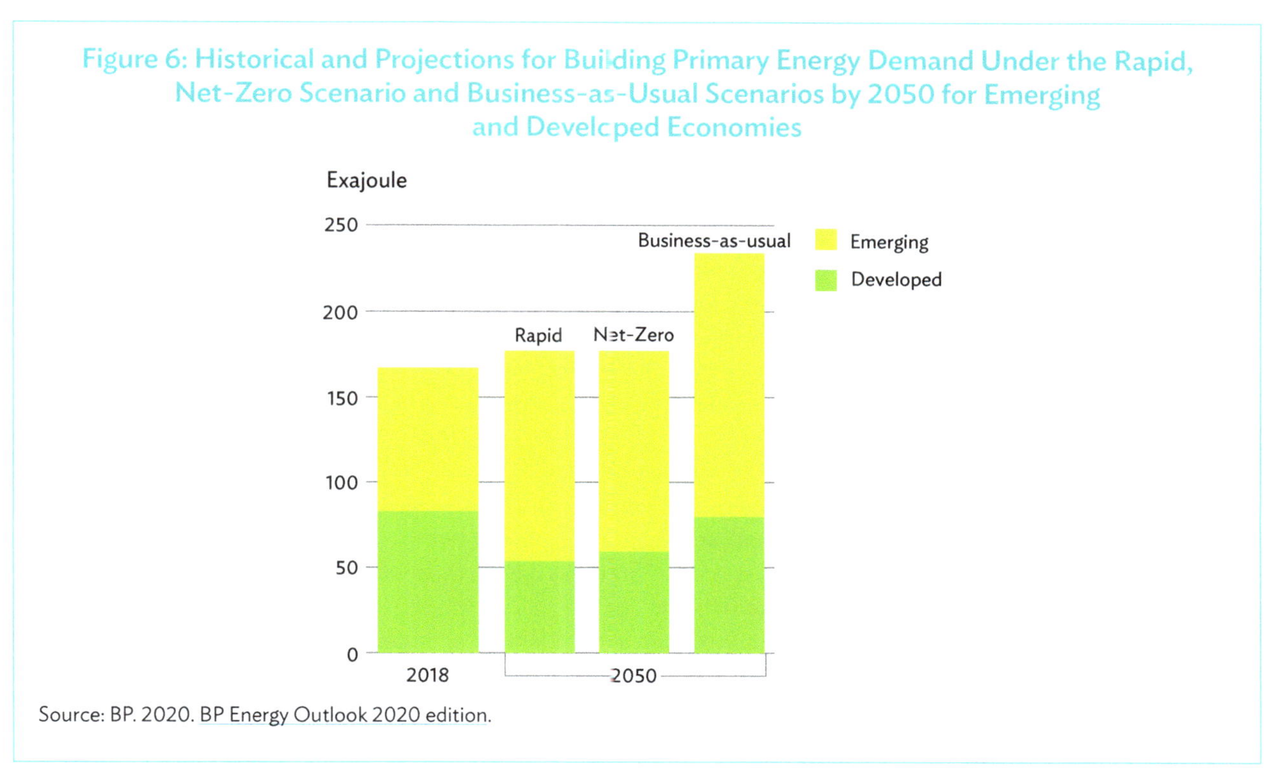

Figure 6: Historical and Projections for Building Primary Energy Demand Under the Rapid, Net-Zero Scenario and Business-as-Usual Scenarios by 2050 for Emerging and Developed Economies

Source: BP. 2020. BP Energy Outlook 2020 edition.

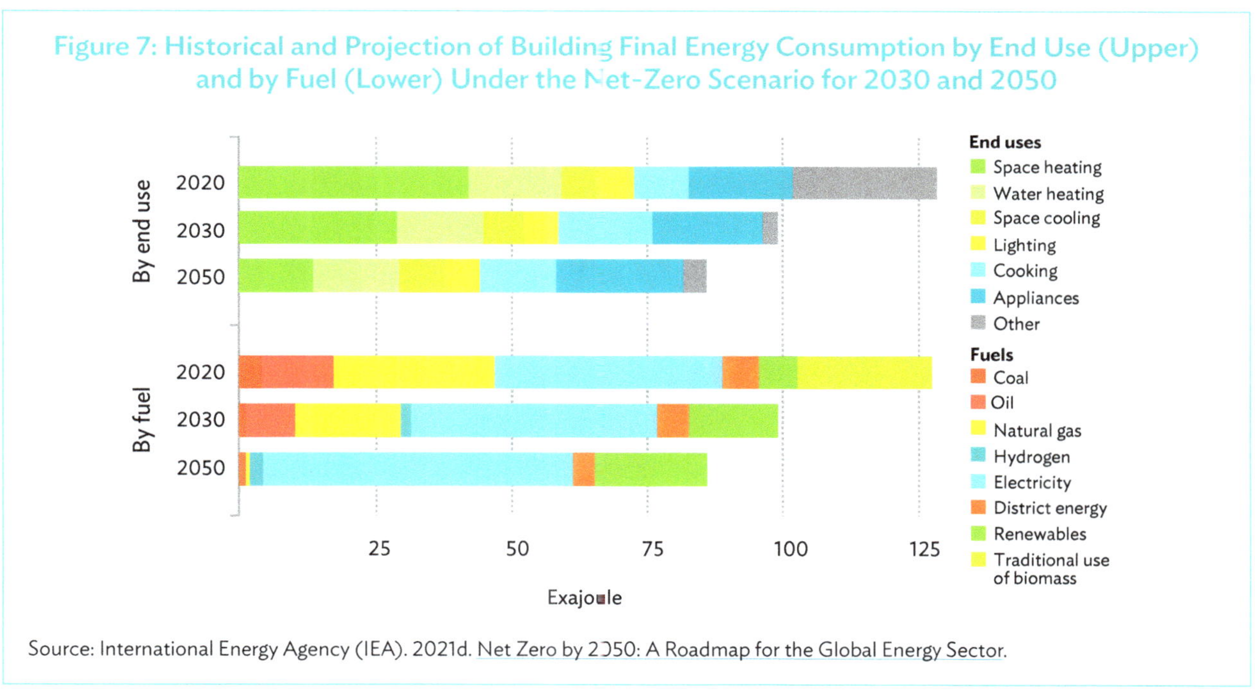

Figure 7: Historical and Projection of Building Final Energy Consumption by End Use (Upper) and by Fuel (Lower) Under the Net-Zero Scenario for 2030 and 2050

Source: International Energy Agency (IEA). 2021d. Net Zero by 2050: A Roadmap for the Global Energy Sector.

In terms of end use, the share of space heating is likely to decrease significantly, contributed by improved efficiency of buildings and appliances, as well as a warmer climate in the future. The heating market is transitioning slowly from inefficient carbon intensity status toward low-carbon solutions. Besides heat pumps, solutions like solar thermal heating and biomass boilers still need to be further and faster scaled up. In contrast, spacing cooling demand will see substantial growth. In particular, India and ASEAN countries, among the hottest regions in the world, are expected to have a great rise in cooling demand as living conditions improve from their existing low access rate to cooling facilities. Therefore, cooling design and efficient technology have been identified as the priority areas for action (GlobalADB/IEA/UNEP 2020). Chapter 4 will elaborate more on this topic.

Since the Paris Agreement, progress has been made in energy efficiency for the buildings and building construction sector in terms of energy consumption, enabling policies as well as investment (Figure 8).

Over the last 2 decades, the global building energy intensity improved by roughly 0.5%–1% annually. However, this is not sufficient to offset the growth of building stock. The global building energy intensity will need to drop at least 2.5% per year, i.e., 31% down in 2030 from 2019 level, to achieve the 2°C scenario. Europe is expected to achieve the intensity reduction of 23%; other developing Asia, 15%; and the PRC, 45% by 2030, as shown in Figure 9. Without accelerated actions, the global emissions from the buildings and building construction sector can be doubled or even tripled by midcentury (Broadwater 2016).

Energy efficiency is more than simply saving energy and mitigating emissions. It is a paradigm shift. First, policymakers need to understand the wider implications to be able to put out relevant policies, laws, and regulations. Second, the transition toward zero-carbon emissions will primarily rely on technological advancement, including digital or smart solutions and behavioral change. Third, investment and financing in zero-carbon buildings will be crucial. Last, the availability and the quality of data are worth mentioning separately concerning the lack of data and information at both global and regional levels, in particular in developing economies, which has become one major challenge. The following section will elaborate more on the four enabling factors for BEE.

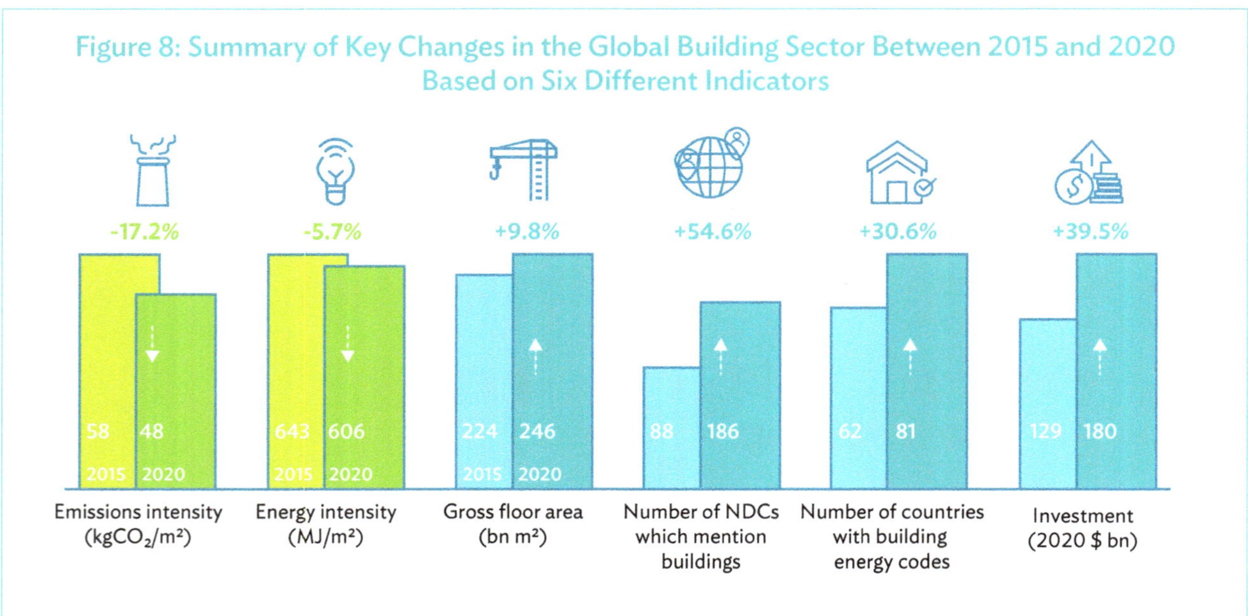

Figure 8: Summary of Key Changes in the Global Building Sector Between 2015 and 2020 Based on Six Different Indicators

bn = billion, CO$_2$ = carbon dioxide, kg = kilogram, m^2 = square meter, MJ = megajoule, NDC = nationally determined contribution.
Source: UNEP. 2021. *2021 Global Status Report for Buildings and Construction: Towards a Zero-emission, Efficient and Resilient Buildings and Construction Sector*. United Nations Environment Programme. .

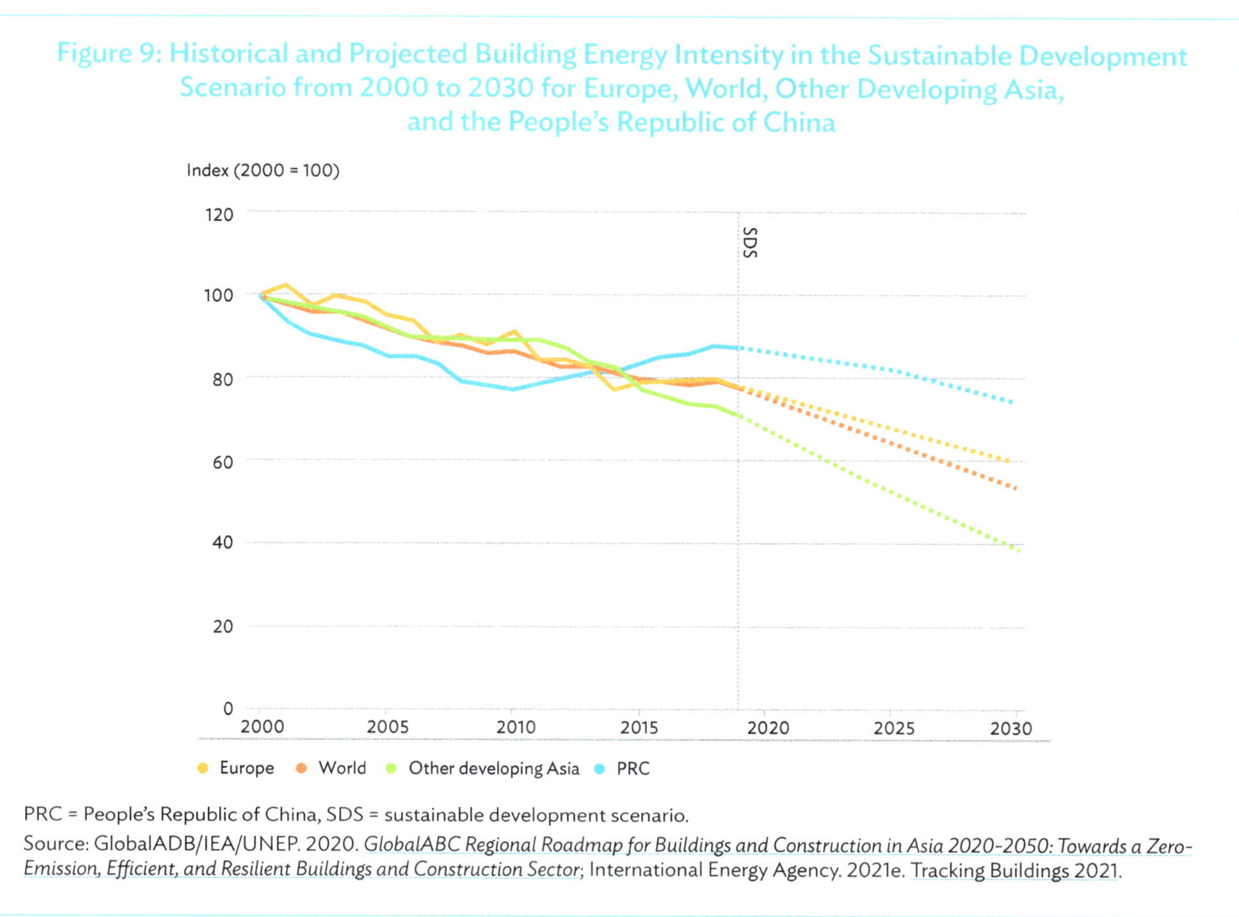

Figure 9: Historical and Projected Building Energy Intensity in the Sustainable Development Scenario from 2000 to 2030 for Europe, World, Other Developing Asia, and the People's Republic of China

PRC = People's Republic of China, SDS = sustainable development scenario.
Source: GlobalADB/IEA/UNEP. 2020. *GlobalABC Regional Roadmap for Buildings and Construction in Asia 2020-2050: Towards a Zero-Emission, Efficient, and Resilient Buildings and Construction Sector*; International Energy Agency. 2021e. Tracking Buildings 2021.

Enabling Factors

Political Framework and Regulations

Energy efficiency needs to be treated as a first fuel and by that receives the same attention and position as renewable energy. A successful policy should cover new and existing buildings of any type and the entire life cycle from planning to decommissioning phases.

Clear Target

It is vital to set up clear and specific targets for building energy efficiency in climate strategies and plans at the country level. From 2020, economies were requested to communicate their new or updated nationally determined contributions (NDCs) to the Secretariat of the United Nations Framework Convention on Climate Change (UNFCCC) as part of the 5-year cycle of reviewing their climate targets and commitments under the Paris Agreement. As of September 2021, 192 economies had submitted their first NDC, 113 of which were updated, and 11 economies had submitted their second NDC. Across the communicated NDCs, improvement in building energy efficiency has been explicitly mentioned in 63% of the NDCs, becoming the second most frequently indicated mitigation option following renewable energy generation (UNEP 2021).

About 72% of those economies in Central Asia, South Asia, East Asia, and Southeast Asia mentioned buildings in their NDCs by 2019. However, many still did not include specific actions to address buildings and building construction sector energy use and emissions (GlobalABC, IEA, and UNEP 2020). Following the new reporting phase beginning in 2020, various policy actions to support improvements in building energy performance and climate resilience have been increasingly mentioned in updated first or second NDCs submitted by many economies in Asia.

However, as shown in Table 1, the progress is far less than sufficient. The breadth and depth of the mitigation (and adaptation) measures for buildings and construction as laid out in most NDCs are not adequate to put the countries in the region on track with the envisaged net-zero emissions scenario by 2050. The increasing population, rapid urbanization, and growth demand for floor area and space conditioning are among the key factors that present a serious challenge to transforming the buildings and building construction sector in Asia toward energy efficiency, low-carbon emissions, and climate resilience. Based on the United Nations population growth projections, it has been expected that 73% of the 294.3 million population to be added in Asia from 2021 to 2030 will live in countries that have NDCs in which building codes and/or building energy efficiency measures are not mentioned (UNEP 2021). This indicates a significant gap in ambition and commitment, calling for increasing recognition of the strategic importance of buildings in NDCs and large-scale adoption of more explicit and ambitious actions and pathways to decarbonize buildings in line with the Paris Agreement.

Table 1: Building-Related Climate Actions in Nationally Determined Contributions of Selected Developing Member Countries of ADB

Country	Description	Building Standards	Appliance
Central and West Asia	Azerbaijan	Massive use of control and measurement devices in electrical, heat energy, and natural gas systems; application of energy-efficient bulbs; use of modern energy-saving technologies in heating systems; as well organization of public awareness programs on energy use.	First NDC (January 2017)
	Pakistan	Green building codes and certification for new and refurbished buildings, including revolving guarantee mechanism for energy-efficient appliance.	Updated first NDC (Oct 2021)
	Uzbekistan	Further introduced energy-saving technologies in construction.	Updated first NDC (October 2021)
East Asia	China, People's Republic of	Improved energy efficiency standards for new buildings and accelerate the development of ultra-low energy-consuming, near-zero energy-consuming, and low-carbon buildings on a large scale. Supported the energy-saving renovation of existing buildings in cities and towns as well as municipal infrastructures, improving their energy-saving and low-carbon level. Applied green building standards to all new urban buildings by 2025. Gradually launched limits management of energy consumption in buildings, practiced building energy efficiency labeling, and conducted performance assessment of low-carbon development in the building sector.	Updated first NDC (October 2021)
	Mongolia	Insulated old precast panel buildings in Ulaanbaatar City. Limited the use of raw coal in Ulaanbaatar City and switch to the use of improved fuel.	Updated first NDC (October 2020)
South Asia	Bhutan	Launched Energy Efficiency Roadmap 2030 covering buildings in 2019 covering 155 gigawatt-hours annually in materials, appliances, and construction. Solar photovoltaics on buildings. Energy-efficient building design.	Second NDC (June 2021)
	Maldives	Built labeling and building standards to improve energy efficiency. Improve building code for climate resilience.	Second NDC (December 2020)
	Nepal	Adopted national building code that emphasize low-carbon and climate-resilient urban settlements.	Second NDC (December 2020)
	Sri Lanka	Introduced mandatory building energy efficiency code in 2021–2022. Established sectoral databases for eco-certification system, minimum performance and energy efficiency labeling programs, green building, and building management system.	Updated first NDC (July 2021)

continued on next page

Table 1 *continued*

Country	Description	Building Standards	Appliance
Southeast Asia	Cambodia	Building codes, enforcement/certification for new buildings to reduce electricity consumption by 10% in 2030. Improved cooling in public sector buildings to reduce 43,000 tCO_2 per year. Passive cooling in buildings to reduce 74.5 tCO_2e.	Updated first NDC (December 2020)
	Thailand	Relies on the implementation of the National Energy Efficiency Plan B.E. 2558–2579 (2015–2036), which aims to reduce energy demand by 30% in 2036 compared to a business-as-usual trajectory	Updated first NDC (October 2020)
	Viet Nam	Reduced greenhouse gas emissions by replacing construction materials and improving the cement and chemical production processes together with reducing the consumption of Hydrofluorocarbons (HFCs). Improved, developed, and applied technology in manufacturing construction materials. Reduced clinker content and implemented other measures to reduce greenhouse gas emissions in cement production. Developed and used energy-saving construction materials and green materials in housing and commercial sectors	Updated first NDC (September 2020)
Pacific	Papua New Guinea	Improved building insulation and energy efficiency. Introduced building codes to mitigate impacts of heat waves and cyclones and for adaption.	Second NDC (December 2020)
	Tonga	Introduced energy efficiency standards for buildings and energy performance audits and minimum energy performance standards for appliances in buildings.	Second NDC (December 2020)

NDC = nationally determined contribution, tCO_2e = ton of carbon dioxide equivalent.
Source: UNFCCC. 2022. NDC Registry. United Nations Framework Convention on Climate Change.

Regulations

There is no single pathway to implement energy efficiency policies for buildings but a mix of various measures. They are broadly set along with the following categories.

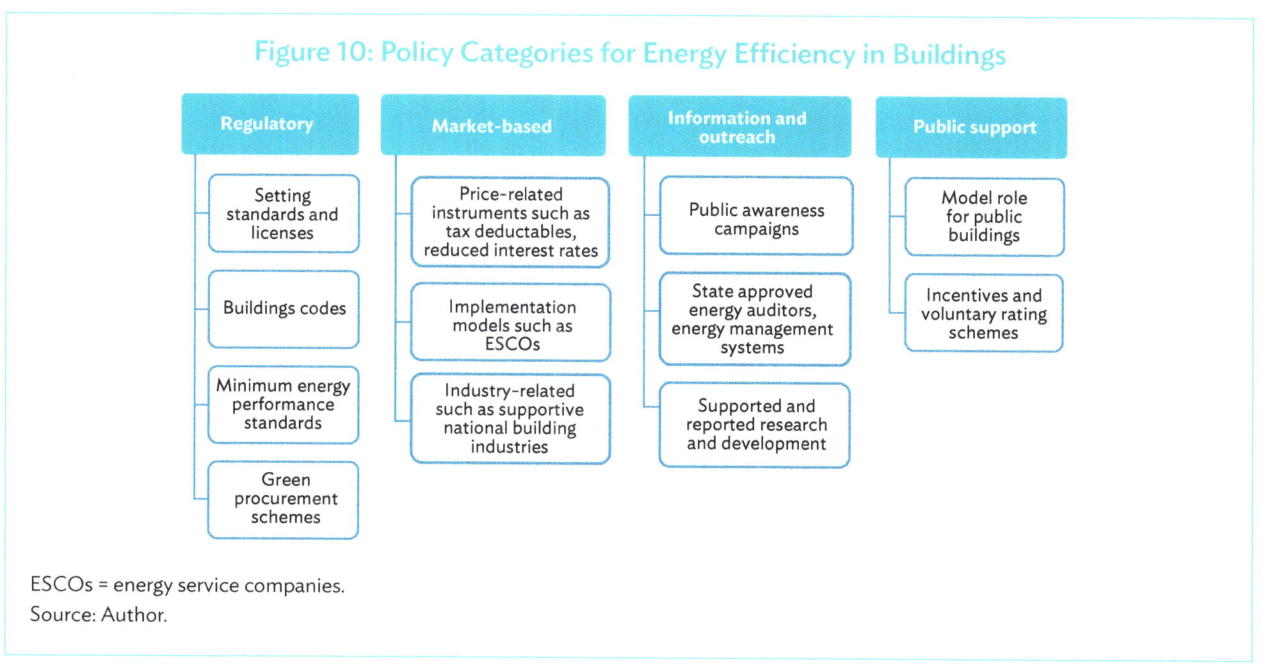

ESCOs = energy service companies.
Source: Author.

Building codes are probably the most important individual measure in promoting energy efficiency. Especially in countries with a fast-growing real estate sector, such as in developing and emerging Asia, the decisions laid out by the building codes today have far-reaching consequences in the future.

Building energy codes, a subset of building codes, set up baseline requirements for buildings' energy performance and building construction. Codes are either prescriptive, meaning that they describe in detail the technical standard or performance base, where the energy level of a building must meet a certain level, or even outcome-based, requiring a specified performance to be achieved over a certain period (C40 2019). They can be mandatory or voluntary, issued at national or subnational level. Usually, they apply to both old and new buildings as well as to public and private ones. Codes usually regulate areas of construction such as insulation; window and door specifications; heating, ventilation, and air-conditioning (HVAC) equipment efficiency; and lighting fixtures (WBDG 2016). Often, public buildings are used to break reservations and act as best practice examples. This can be a successful way if the example is easily replicable, and the proponent has smooth access to all necessary permits and administrative clearances.[7]

To meet the 1.5°C target by 2050, all countries must have comprehensive mandatory building energy codes implemented by 2030. All new constructions need to be compliant with stringent building codes by 2030, while the majority of retrofits need to be compliant by 2050. IEA's tracking data as of 2021 shows that only 80 countries globally, most of which are advanced economies, have launched mandatory or voluntary building codes (IEA 2021e). Around two-thirds of countries, where floor areas grow at a fast pace, are yet to launch mandatory building energy codes. Codes should be developed to govern the building retrofit as well. Furthermore, the building retrofit rate is less than 1% per year in most key countries, which is far below the requirement of 2.5% till 2030 for the 1.5°C target. The rate shall increase to 1.5% by 2025, at least in Asia (GlobalABC/IEA/UNEP 2020). This requires immediate action from policymakers. Concerning the long life span of buildings (30–100 years) compared to other energy-consuming assets, there are risks of lock-in effects of inefficient buildings, in particular in developing Asia, if the renovation rate of existing buildings remains low or new constructions are poorly designed and operated.

In Asia and the Pacific, 24 economies have mandatory and 6 have voluntary building energy codes out of 62 tracked economies, while 9 economies are developing the codes, and 23 do not have building energy codes as of the end of 2021 (IEA 2021b). As shown in Figure 11, many economies in Asia have or are in the process of developing building energy codes. However, some of them were not properly aligned with supporting policies such as energy performance certificates, incentives, and voluntary rating schemes, and they lack enforcement and compliance as well as monitoring and data evidence. Enforcement is one of the main stumbling blocks in energy efficiency in buildings. It requires designated authorities and an independent evaluation system.

Besides building energy codes, building performance standards, or minimum energy performance standards (MEPS, for buildings) is another important policy approach that is relatively new (BECWG 2021). Several economies regulate energy consumption in buildings by setting mandatory energy performance standards, such as performance rating and carbon intensity levels. Besides the disclosure of building energy consumption and regulations on green procurement, there are other proven energy efficiency policy options.[8] MEPS can be adjusted over time, starting with a modest but short implementation period to more stringent requirements with longer implementation periods.[9] It can allow building owners a longer time to bring buildings into compliance. They are closely linked to energy performance certification for buildings, like rating systems already well-known for appliances, such as Energy Star of the US. Building certification and rating systems can provide energy-related benchmarks and a basis for performance standards. However, it is worth mentioning that building certification schemes include more than energy criteria. Chapter 3 will elaborate more on popular international and national green building certification schemes.

---

[7] The Republic of Korea provides nonmonetary benefits of several energy-efficient investments in buildings by speeding up approvals.

[8] ISO 20400 is the norm for sustainable procurement and gives guidance on a comprehensive procurement scheme. ISO 20400. 2017. Sustainable Procurement - Guidance.

[9] Further reading: How to set MEPs: Minimum Energy Performance, various reports.

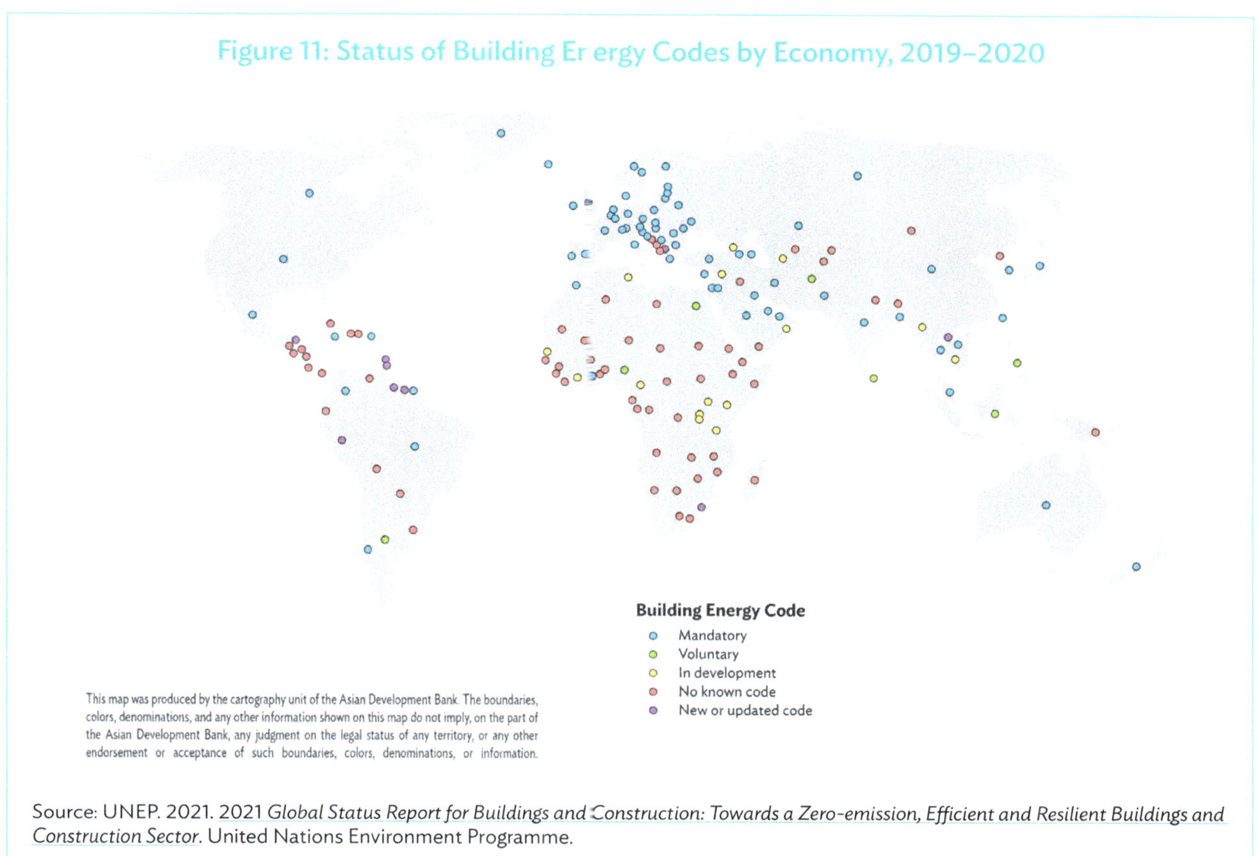

Source: UNEP. 2021. *2021 Global Status Report for Buildings and Construction: Towards a Zero-emission, Efficient and Resilient Buildings and Construction Sector*. United Nations Environment Programme.

In the PRC, the central government decided on comprehensive MEPS for buildings in different climate zones and introduced a BEE labeling scheme. In India, the Bureau of Energy Efficiency under the Ministry of Power set up the Energy Conservation Building Codes that provide minimum requirements for the energy-efficient design and construction of residential and commercial buildings. In the United Kingdom (UK), it has been unlawful since 2018, under the Energy Act 2011, for landlords to rent buildings that have an energy performance certificate of F or G until their energy efficiency deficits have been addressed.

Other common policy tools include MEPS and labeling for appliances and equipment, which are already mainstreamed in Asia and the Pacific, the requirement for energy audits, etc. These policies could help reduce the energy demand of buildings (APERC 2019). To reach 2°C scenario by 2050, three factors should be in place to drive down the building energy demand, which is further scaling up of efficiency technology, improvement in building envelops, and fuel switching. Besides building codes, the adoption of MEPS for appliances is vital as well.

BEE policies are usually part of an entire codex of energy efficiency measures. As for other policies, they must come along with awareness programs, technical and financial assistance, and other supporting mechanisms. Equally important is their long-term view, clear timelines, and proper enforcement measures.

Institutions

Only 20% of the global BEE potential has been leveraged. The buildings and building construction sector is highly fragmented. The lack of appropriate aggregators or dedicated institutions controlling the whole value chain and coordinating across national and local stakeholders is holding back the development. It is recognized that commitment, institutions, and supporting evidence are the major issues with split interest and diffused decision-making, while technologies and tools are available and cost-effective.

Most countries have a ministry for building and community and a ministry for energy. It is mandatory for energy efficiency in buildings that policies and implementation are coordinated.[10] This is best done by establishing an independent institution that oversees the policy measures and is responsible for its implementation. It should be well-financed and staffed with technical and financial experts and have sufficient authority to penalize non-implementation.

In the PRC, for example, all new urban residential and public buildings must meet energy-saving design standards established by the Ministry of Housing and Urban–Rural Development (MOHURD). MOHURD has also developed a Green Building Action Plan, with green building evaluation standards and a labeling program. Implementation is conducted by provinces and municipalities, which may adjust the standards but may not go below certain thresholds.[11]

Some countries have set standards and building codes at the national level, which are to be adjusted to reflect local conditions. In that case, a model building code is developed and modified accordingly to effectively cover different regional situations such as climate, building stock, and local conditions. Delegation of power is advisable in large countries and countries with very different climate zones. Implementation is often seen as a major issue because it requires vertical integration of central, federal, and municipal levels.[12] It is not uncommon to have mismatches between the central government's ambitious targets and the local's different regulation priorities and limited resources. The lack of implementation, enforcement, and monitoring mechanisms are pervasive issues. It is thus necessary to include inspection schemes, time frames for revision, and penalties in case of noncompliance with any of the regulations. In addition, cross-sector coordination is still weak, for instance, the integration of the buildings and building construction sector with urban planning. Capacity building and engaging local policymakers and practitioners will be necessary. It is often forgotten, but energy efficiency requires a skilled workforce. Various measures such as university programs and a state-certification system e.g., energy auditors have proven to be successful. These policy instruments should be also combined with incentives and financing programs to improve their effectiveness.

Technology Development

As noted before, technology development will play a major role in improving building efficiency and ultimately decarbonizing the sector. The existing buildings can leverage cutting-edge devices and technologies to mitigate emissions through retrofitting, and the new constructions shall employ sustainable and low-carbon materials, designs, and construction practices. Proven and cost-effective technologies are widely available in the market (WRI 2016), such as heat pumps, efficient appliances, advanced materials, etc. In addition, the large number of technologies for buildings at the demonstration and early adoption stages deserve attention, as shown in Figure 12. These two stages refer to the "valley of death." High costs and performance gaps exist. Support from the public sector, in particular financing, is critical to stimulate continuous innovation and help overcome the valley of death. Chapter 4 and Chapter 5 will discuss more on heating and cooling design technologies and appliances.

Moreover, the recent development of digital technologies creates chances to cater to diversified demands and accelerate the transition. Digital technologies have the potential to disrupt the whole-buildings and building construction sector, enable integration and information-sharing along the value chain, and unlock the improvement potential of the whole system. Concerning the complexity and diverse materials and structures, it is usually challenging to enforce building regulations on the ground. The smart meters and monitoring technologies, for instance, may help track the operational energy consumption, collect data for decision-making, and ultimately, optimize systematic performance and support behavioral changes.

---

[10]   The Republic of Korea has set national targets for energy efficiency as part of the countries "green growth" strategy, passed in 2009.

[11]   Further reading: World Green Building Council. China Green Building Council.

[12]   Further reading: APEC Energy Working Group, 2017 results of the Asia-Pacific Building Code Forum. Asia-Pacific Economic Cooperation. 2017. Asia-Pacific Building Codes Forum Workshop Outcomes. Singapore; Asian Development Bank Institute. 2020. Energy Efficiency in ASEAN: Trends and Financing Schemes. *ADBI Working Paper 1196*. Tokyo.

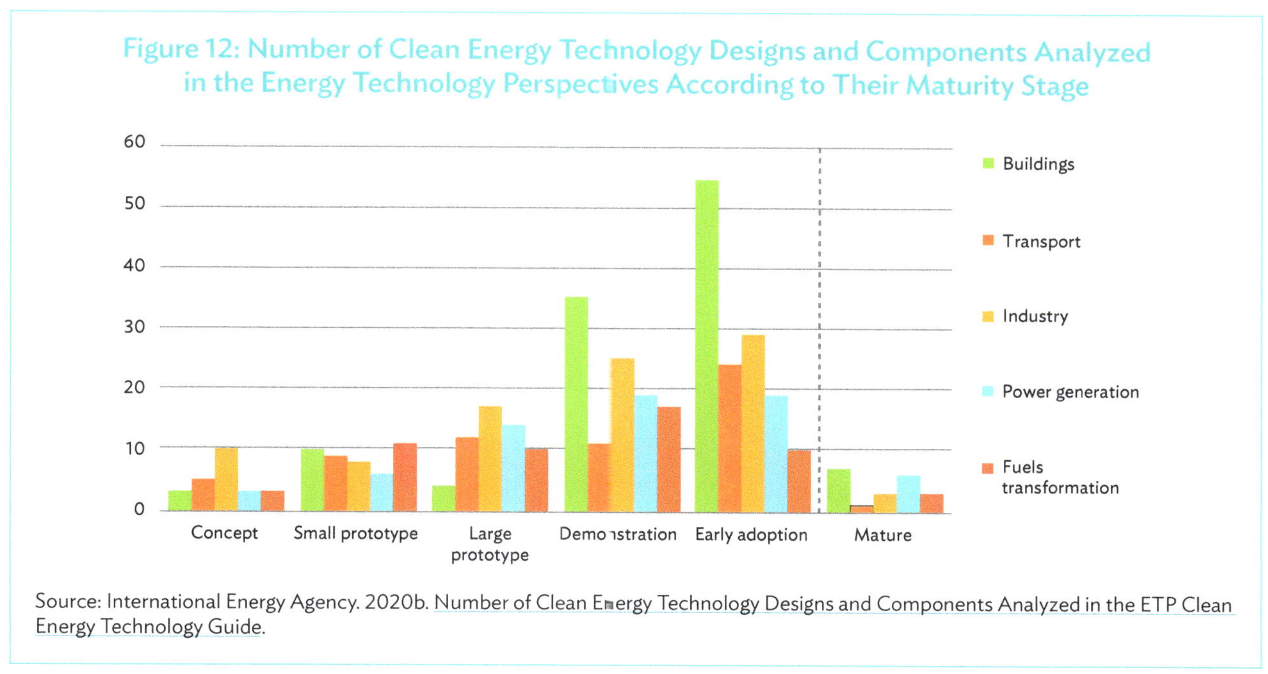

Source: International Energy Agency. 2020b. Number of Clean Energy Technology Designs and Components Analyzed in the ETP Clean Energy Technology Guide.

However, the adoption of digital technologies or so-called digitization requires regime changes, breaking silos and integrating the fragmented industry to share information, and certainly investment in research and development (R&D) as well as training and adaptation. The existing applications happen predominantly in advanced economies and create great potential to be scaled up in developing countries.

Investment Status

Governments often underestimate the economic and social benefits energy efficiency brings. The results are timid and unsuitable budgets. In supporting a strong policy framework, energy efficiency programs must have a consistently strong funding platform. Direct and indirect benefits need to be considered and valued, such as the reduction of CO_2 emissions, improved balance of trade, reduction in unemployment, and an increased net public budget.

In 2021, investments in energy efficiency were expected to reach a record high of $290 billion, increasing 10% from 2020, mainly due to the economic recovery. Among this, the buildings and building construction sector, the leading destination of efficiency spending, is likely to attract $190 billion, a spike of 11% from 2020, in contrast to the decline in industry and transport sectors (Figure 13). This strong growth in BEE is mainly driven by the increasing commitment of European countries but hides the sluggish development in other countries. In Asia, the growth has primarily concentrated in the PRC, the Republic of Korea, and Japan, while other countries, such as India, have seen contraction (IEA 2021b).

Europe continued leading the investment in BEE before the pandemic, while Asia and the Pacific has seen a high growth since 2015, thanks to the new constructions in the emerging economies (Figure 14). Meanwhile, it illustrates that Eurasia countries, including Russia and other eight DMCs in Central and West Asia, spent very limited in this sector.

The Asia Pacific Energy Research Centre (APERC) (2019) projected that the buildings and building construction sector in Asia-Pacific Economic Cooperation (APEC) member economies will need a total capital investment of $8.4 trillion and fuel costs of $15.3 trillion over 2017–2050 under the business-as-usual scenarios, and much higher capital investment at $19.1 trillion but 39% less fuel costs, which would be offset significantly by fuel saving from the reduced energy demand and improvided energy efficiency, under the 2°C scenario (Table 2).

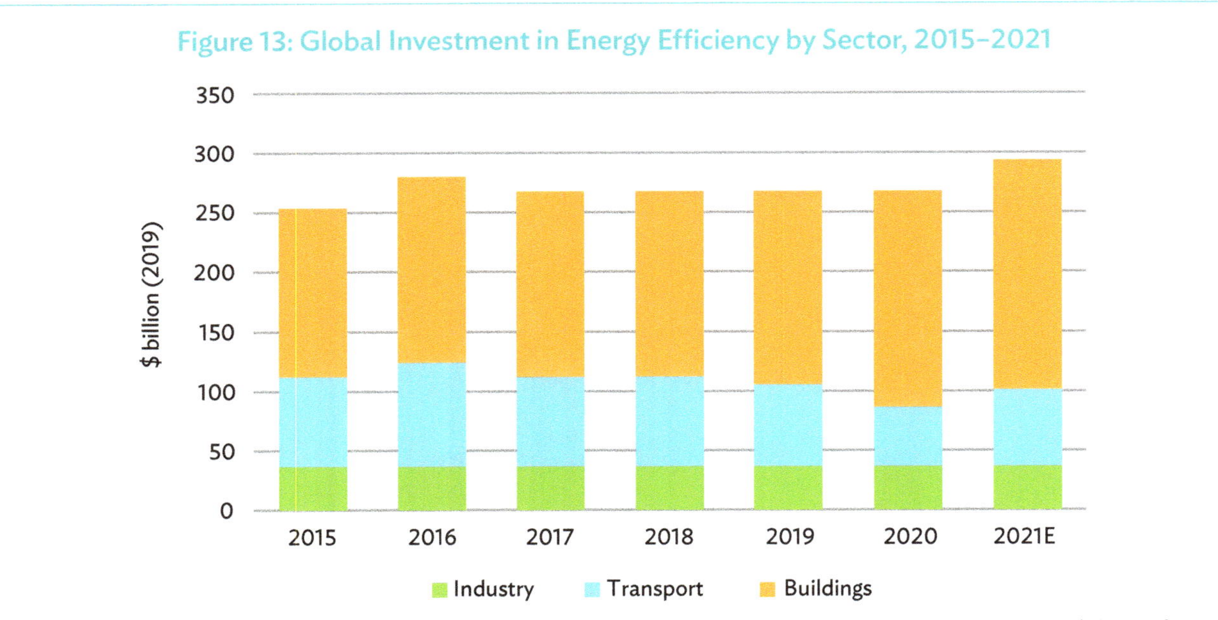

Note: An energy efficiency investment is defined as the incremental spending on new energy-efficient equipment or the full cost of refurbishments that reduce energy use. The intention is to capture spending that reduces energy consumption. Under conventional accounting, part of this is categorized as consumption rather than investment.

Source: International Energy Agency. 2021b. Energy Efficiency 2021.

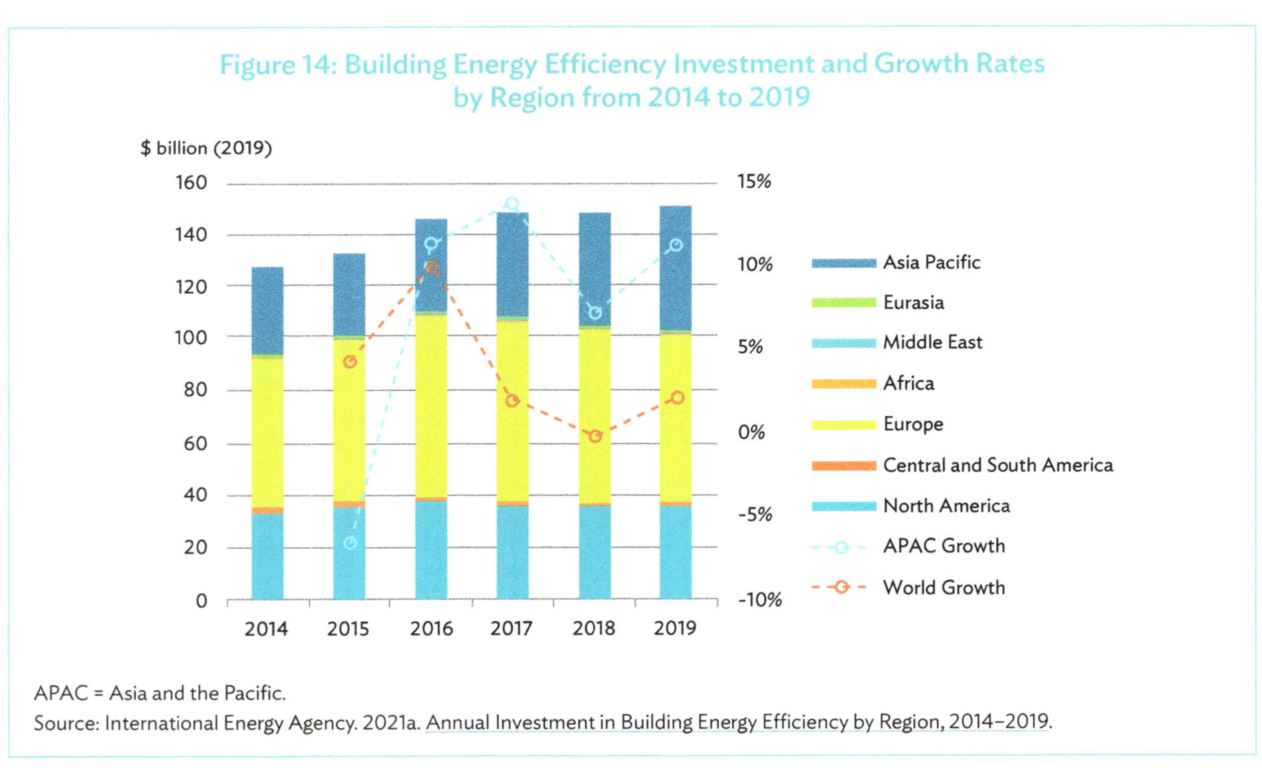

APAC = Asia and the Pacific.

Source: International Energy Agency. 2021a. Annual Investment in Building Energy Efficiency by Region, 2014–2019.

Table 2: Capital Investment and Fuel Costs in Business-as-Usual and 2°C Scenarios, 2017–2050 ($ trillion)

	Business-as-Usual Scenario			2°C Scenario		
	Capital	**Fuel Costs**	**Sum**	**Capital**	**Fuel Costs**	**Sum**
Total Energy Demand	13.6	49.2	62.8	25.5	32.3	57.8
Buildings Energy Demand	8.4	15.3	23.7	19.1	9.2	28.3

Source: APERC. 2019. *APEC Energy Demand and Supply Outlook 7th Edition 2019*. Vol. 1.

IEA projected that around $260 billion per year will be spent on energy efficiency improvement from 2021 to 2023, among which $70 billion will be from government direct investment, and the rest will need to be mobilized from public and private organizations (IEA 2021b). The current pledge on energy efficiency investment is insufficient toward the 1.5°C target. Moreover, the current investment growth is imbalanced among countries, and the largest gaps are in emerging and developing countries.

One promising trend is that energy-efficient buildings attract capital from green financing, such as green bonds. Green buildings rank second in terms of using green bond proceeds, following the energy sector, reaching around $76 billion by 2020. In ASEAN, buildings represented 48.6% of green debt investment (Climate Bonds Initiative 2021). Countries like India and Viet Nam are developing green taxonomy to quantify the building projects and attract international capital.

Technical development led by start-ups fosters new business models and financing measures. Building efficiency is one of the hot areas, which raised funding of $500 million in 2020, rising at a never-before-seen rate. Despite early-stage investment occurring in the US, Europe, and the PRC markets, there are still huge prospects for entrepreneurs to scale up these innovations aboard, particularly in developing economies. Energy-as-a-Service (EaaS) for buildings is a leading technology attracting investment (IEA 2021b).

Box 1: Energy-as-a-Service Business Model for Energy Efficiency in Buildings

Smart Joules Private Limited (SJPL) is cofinanced with the equivalent of ₹$2,000,000 by the Asian Development Bank (ADB) Ventures Investment Fund 1. The company is a provider of data-driven energy efficiency solutions for commercial buildings, targeting large hospitals as the beachhead market. SJPL delivers energy savings through a combination of (i) a proprietary internet of things-based sensing and analytics platform, and (ii) the application of an EaaS business model. The basic idea is that data are collected by the company and optimized to be able to react in real time and thus producing a continuously improved algorithm.

Source: Interview with project administrator Charles Cole Navarro on 21 April 2021.

Financing Sources

Investment in energy efficiency is not a one-size-fits-all matter. Equally, the starting position of energy efficiency considerations varies greatly among different building types. First, they do not have the same consumption patterns, and second, energy-consuming devices such as heating, cooling, installed equipment, and appliances vary due to the different usage of the buildings. It is important to survey which building types, as shown in Figure 2, are to be targeted and why, before formulating political programs establishing funding sources and creating financial support measures. Given the medium- to long-term effect BEE brings, it usually requires patient capital.

In addition to the usual taxed-based funding sources and external finance through multilateral development banks, creating national or international carbon pricing models must therefore play an important role. Currently, 64 carbon pricing instruments are operating worldwide, covering over 20% of global greenhouse gas emissions and generating $53 billion in revenue (World Bank 2021). Governments can take two ways to price carbon—through emissions trading systems (ETS) and carbon taxes. The PRC started its national ETS this year. It covers initially emissions in the power sector, meaning that 30% of national emissions are covered. Also, Germany and the UK start their national ETS this year, while the Netherlands and Luxembourg use the carbon tax system.[13] Income from CO_2 pricing is used to fund, among others, energy efficiency programs.

Another funding source is the private sector, which tends to be forgotten. It is estimated that half of the funding needed to refurbish buildings will have to come from the private sector. Hence, any funding platform needs to establish convincing business case situations for the private sector. A comprehensive approach covering the entire supply chain the private sector runs is useful.[14] However, emerging countries nowadays rely primarily on public sources for energy efficiency financing (IEA 2021b), which is not sustainable and hinders the scaling up of the market. High transaction costs, the uncertain future value of the building projects, and the asymmetry of information are among the major barriers to attracting private capital. The perceived costs of financing green buildings are higher, up to 30% more than the actual expense (Broadwater 2016). In addition, many financing bodies involved in BEE are small and medium-sized enterprises facing common challenges in accessing finance. Further, depending on the target the individual program has, key stakeholders may include technology suppliers, manufacturers, energy auditors, and banks. For all of them, the exercise may start with awareness-raising and educational measures. The private sector will buy into it and invest if risk and reward can be identified. It is, therefore, essential that de-risking instruments and the development of project pipelines are supported to create a market and the view of long-term profits. The funds spent on motivating the private sector can be leveraged at a large scale. Nonmonetary incentives such as expedited construction permits for green buildings, faster permits, and other administrative relaxations are also beneficial. Chapters 6 and 7 will provide a detailed analysis of BEE investment, available instruments, and best practices for innovative business models.

Data

In formulating a successful policy and developing useful financial supporting measures, a clear understanding of the building stock and the energy demand of different building types based on reliable data is required. It is, likewise, a prerequisite to analyze the situation, provide evidence-based information, decide, and communicate potential costs and benefits to stakeholders.

There are yet comprehensive data sources available for buildings globally in terms of energy performance and emissions (UNEP 2021). There is a lack of reliable and easily accessible building data at regional and country levels for developing Asia, for instance (Climate Bonds Initiative 2021). Also, project-level data and information are usually restricted to private access due to their confidential or sensitive nature, which presents serious challenges for (i) benchmarking the performance of site-specific buildings and setting up applicable targets and (ii) building owners and investors to evaluate the building asset portfolios. These data gaps may lead to a significant level of ambiguity for investment, research, policymaking, etc. Banks have been hesitant to invest in BEE projects partially due to the lack of data on loan performance (IFC 2019c). For instance, the operation data of existing buildings and retrofitted ones in Asia are missing, and so is the limited focus on building retrofits in the region (GlobalADB/IEA/ UNEP 2020).

---

[13]   Further reading: World Bank. 2022. State and Trends of Carbon Pricing 2022. State and Trends of Carbon Pricing. Washington, DC: World Bank.

[14]   In the past, pure on-lending programs for banks have proven not to be successful because of a lack of demand.

Moreover, the quality of data varies tremendously due to the difference in the age of the building stock. Many old buildings do not have any other information on their energy use than the utility bill at the end of the month. Other buildings can provide leading-edge information and information technology, such as the Internet of Things and others. However, building data are heterogeneous. They are still stored in different formats at different locations and do not communicate. It is, thus, necessary to create a so-called high-level data-driven architecture for building data exchange (Marinakis 2020). It allows a better understanding of energy consumption and may lead if not too different business models at least to better fine-tuned ones.

Data is needed throughout the use of energy-efficient measures. Once policies can be based on a realistic set of data, incentives can be better targeted, and the success of implementation is better monitored and constantly controlled. Even for the end consumer, a set of reliable and easy-to-use data can herald consumption behavior, which could become a self-controlled and cost-effective strategy to reduce energy waste in buildings.[15]

Impacts of Building Energy Efficiency

Buildings have a huge influence on our well-being and environment. The German Sustainable Building Council (DGNB 2020) identified that sustainable buildings can address in particular five Sustainable Development Goals (SDGs): good health and well-being (SDG 3), affordable and clean energy (SDG 7), sustainable cities and communities (SDG 11), sustainable consumption and production (SDG 12), and climate action (SDG 13).

Economic and Financial Relevance

At the heart of energy efficiency is the saving of energy.[16] In addition to a reduction in greenhouse gases, energy efficiency has a lot of other benefits (non-energy impacts [NEI] or non-energy benefits[NEB]), such as economic, social, and environmental improvements, which in many countries are not accurately shown in the national account system. The core elements of such cost-effectiveness calculations are challenging to identify, and it is hard to quantify a common denominator to monetize them, e.g., the impact on public health and welfare effects. A general add-on of 15% to the costs of avoided energy had been used as a proxy in some US federal states in the past to make up for it.[17]

Macroeconomic

Macroeconomically speaking, the benefits are tremendous and have an impact—directly and indirectly—on GDP.[18] Saving energy reduces the need to import primary energy, which has a direct effect on the national balance of payment, as well as an increase in public budgets reinforced by the fact that fewer power plants need to be financed. Less consumption also leads to reduced investment needs on the side of (public) utilities and may even result in a decline in energy prices as demand drops when less energy is used. The cut in dependency on fuel imports stabilized in addition to energy security in the future. The IEA estimated that worldwide energy efficiency improved by an estimated 13% between 2000 and 2017, most of which was achieved in the industry and the buildings and building construction sector (IEA 2019a). The amounts that can be saved are shown in Figure 15 and give an idea of how much public funding can be made available to incentivize building owners. Taking a life cycle perspective,

---

[15] Further reading: Himeur, Y., A. Alsalemi, A. Al-Kababji, et al. 2020 *Data Fusion Strategies for Energy Efficiency in Buildings: Overview, Challenges and Novel Orientations. Science Direct.* 64. pp. 99–120; and Batra, N., A. Singh, P. Singh, et al. n.d. *Data Driven Energy Efficiency in Buildings.*

[16] Further reading: European Commission. Energy; European Commission. 2016. *The Macroeconomic and Other Benefits of Energy Efficiency, Scenario in the EU with a 2030 perspective.* The EU Report. 2017.

[17] Further reading: Northeast Energy Efficiency Partnerships, Inc. 2017. Non-Energy Impacts Approaches and Values: An Examination of the Northeast, Mid-Atlantic, and Beyond. United States.

[18] See footnote 11, where calculations have been made for different scenarios, ranging from a 0.1% increase in GDP to up to 2.0%.

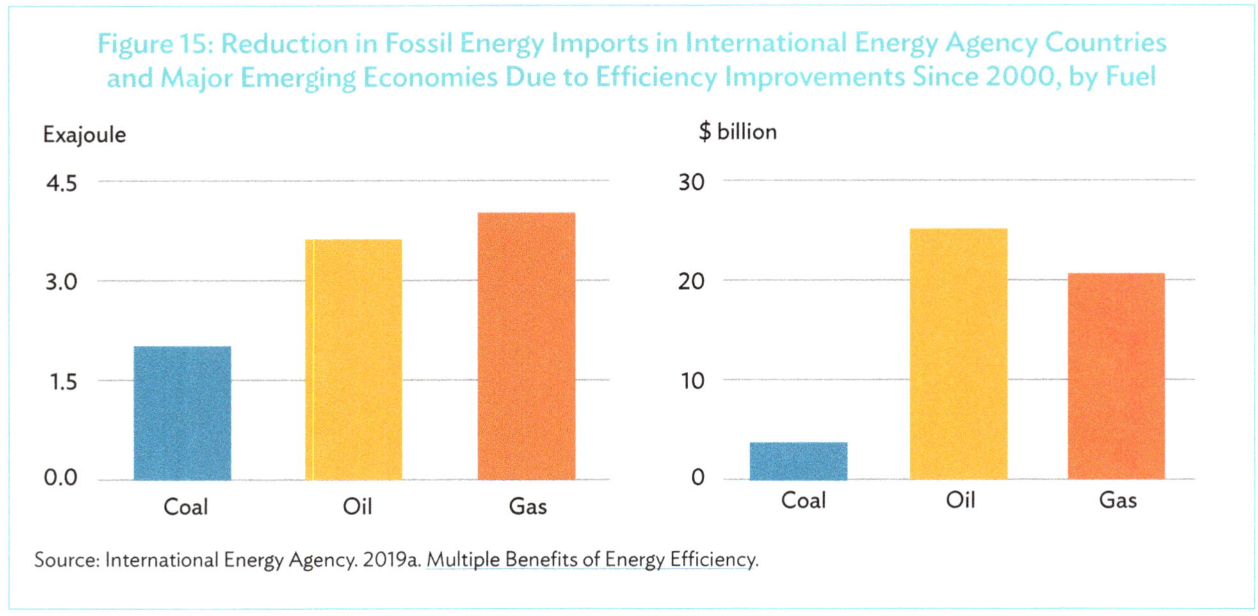

Source: International Energy Agency. 2019a. Multiple Benefits of Energy Efficiency.

efficient design and construction approaches can contribute up to 20% demand reduction for steel and over a 30% reduction for cement. Moreover, green buildings including energy efficiency improvement, feature a long value chain, involving multiple market players such as material manufacturers, planning, and design, and provide huge low-carbon investment opportunities in emerging countries, with a market size of $24.7 trillion by 2030 (IFC 2019c).

Cost-effectiveness. Energy efficiency, also called the "first fuel," results in a reduction in energy demand and greenhouse gas emissions. The monetary value of each unit of energy that can be saved needs to be seen against the expenses caused by generating, delivering, and storing the same amount of energy. Savings can be substantial if considering what it costs to generate energy versus how much energy efficiency measures cost. It seems that considerably less finance is needed to save energy by changing technologies than to generate energy. It supports the current discussion on the usefulness of DSEE. Generating electricity at the building site of larger residential areas and business centers and improving the energy efficiency standards of these buildings may even have a double effect when looking at network losses from source to delivery stations.

Employment effects. Employment effects have been noticed in the number of jobs created and in the level of productivity in different industries. Productivity also increased due to improvements in building quality, such as adequate temperature, air quality, and lighting. Further analysis must be made to calculate the indirect financial impact of decreased reported sick leaves of employees in office and commercial buildings related to energy efficiency upgrades.

Demand for more labor stems from conventional industries, but especially important for the buildings and building construction sector are new industries such as advanced manufacturing for new materials, data collection devices, innovative heat pumps, improved building design, and appliances. Locally made energy-efficiency equipment has the potential for high growth rates.

Employment opportunities in different industry sectors can lead to an average increase in available income at various levels of education. In some specific jobs, it stimulates further qualification and improves the overall skill set. Energy efficiency is said to be an employment machine. In the US, over 2 million jobs had been created in different sectors in 2016.

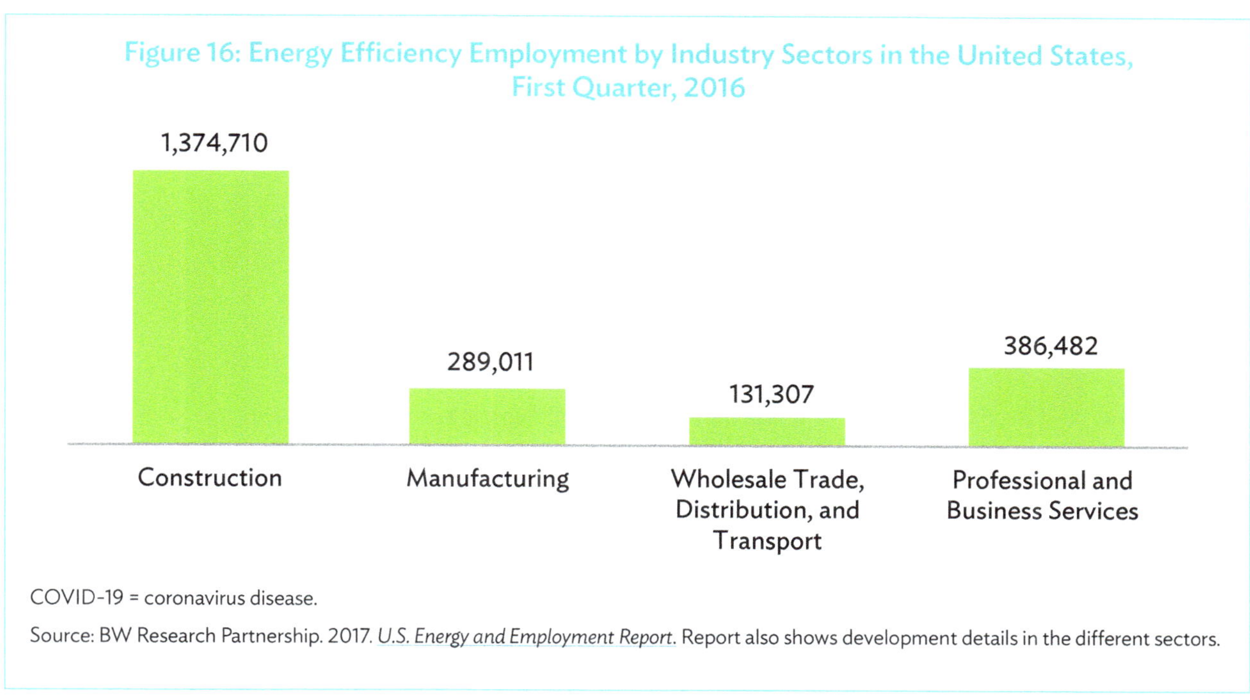

Figure 16: Energy Efficiency Employment by Industry Sectors in the United States, First Quarter, 2016

COVID-19 = coronavirus disease.

Source: BW Research Partnership. 2017. *U.S. Energy and Employment Report*. Report also shows development details in the different sectors.

The European Union report, 2017 supports this finding and projected the creation of 0.8 million jobs in 2020 due to energy efficiency measures. It was emphasized that energy service companies (ESCOs) are the main driver. In some of the main markets, the development has been consistently strong.

Table 3: Estimated Energy Efficiency Jobs in Selected Countries and Regions, 2020

Country/Region	Efficiency-Related Employment Numbers (pre-COVID-19 crisis)
Australia	60,000–236,000
Brazil	33,000–62,000
Canada	472,000
People's Republic of China	729,000–730,000
Europe	1 million–3 million
United States	2.4 million

COVID-19 = Coronavirus disease.
Note: For further information on energy efficiency and job creation: International Energy Agency. Energy efficiency jobs and the recovery.
Source: The EU in 2017.

Asset Value

Energy efficiency measures can increase the asset as well as the property value of buildings (Jensen, Hansen, and Kragh 2016). As studies show, the process, however, can be slow. In the case of Denmark, it took 15 years to show significant changes. Precondition is a building code that allows a recognized energy efficiency rating system. As also known, in the case of energy labeling for white goods, the efficiency labeling of buildings must be well communicated and needs a certain penetration before the market reacts.

Direct financial effects for building owners who undertake energy efficiency measures, including deep energy retrofits, can

- lower long-term operating and maintenance costs;
- lessen possibilities of becoming a stranded asset due to government regulations;
- potentially lead to higher returns on investment;
- potentially lead to higher asset value—lower risk—lower financing costs;
- potentially lead to higher sales price; and
- potentially lead to longer-term contracts with tenants as their utility costs are lower and quality of living standards higher, guaranteeing a more stable revenue stream.

An International Finance Corporation (IFC) report on green buildings states that there is evidence green buildings (new buildings) "can decrease operational costs by up to 37%, achieve higher sale premiums of up to 31% and faster sale times, have up to 23% higher occupancy rates and have higher rental income of up to 8%." (IFC 2019c, vi).

To transform technically doable and economically desirable energy efficiency measures into day-to-day reality experience shows that three preconditions should be in place: policies and implementation instruments, database, and a funding strategy.

SUMMARY BOX

Impacts of energy efficiency are felt macro-economically as well as in an increase in asset value. Energy efficiency has proven to be a source of employment and new types of jobs.

Social and Environmental Impact

Environmental impacts

Energy efficiency delivers a number of environmental benefits, the largest of which is the reduction of greenhouse gases and other pollutant emissions, e.g., from the power stations that generate electricity. Addressing BEE is the most cost-effective approach to mitigate emissions (WRI 2019). IEA (2019b) calculated that in the ASEAN countries, the PRC, and India, the buildings and building construction sector accounted for 27% of energy use and 24% of CO_2 emissions in 2018 (Figure 4). These figures do not include emissions from the manufacturing of building materials.

In the same report, the annual emissions from buildings could be reduced by nearly 3,000 metric tons of carbon dioxide ($MtCO_2$) by 2040 if the current Stated Policy Scenario were replaced by the Sustainable Development Scenario.[19] In case the material life cycle of buildings was included, the reduction of greenhouse gases could even be greater, e.g., 80% in the PRC in residential buildings. A view at Figure 17 makes clear where the main greenhouse gas emission savings would come from. Coal, oil, and gas are the largest energy providers of electricity and the largest polluters. This the more if the entire logistics such as shipments, etc., are included in the calculation. A reduction in overall energy demand would also put a hold on the depletion of natural resources as would a shift from fossil to renewable energy.

---

[19] The Sustainable Development Scenario is fully aligned with the Paris Agreement, holding the global average temperature rise to below 1.8°C.

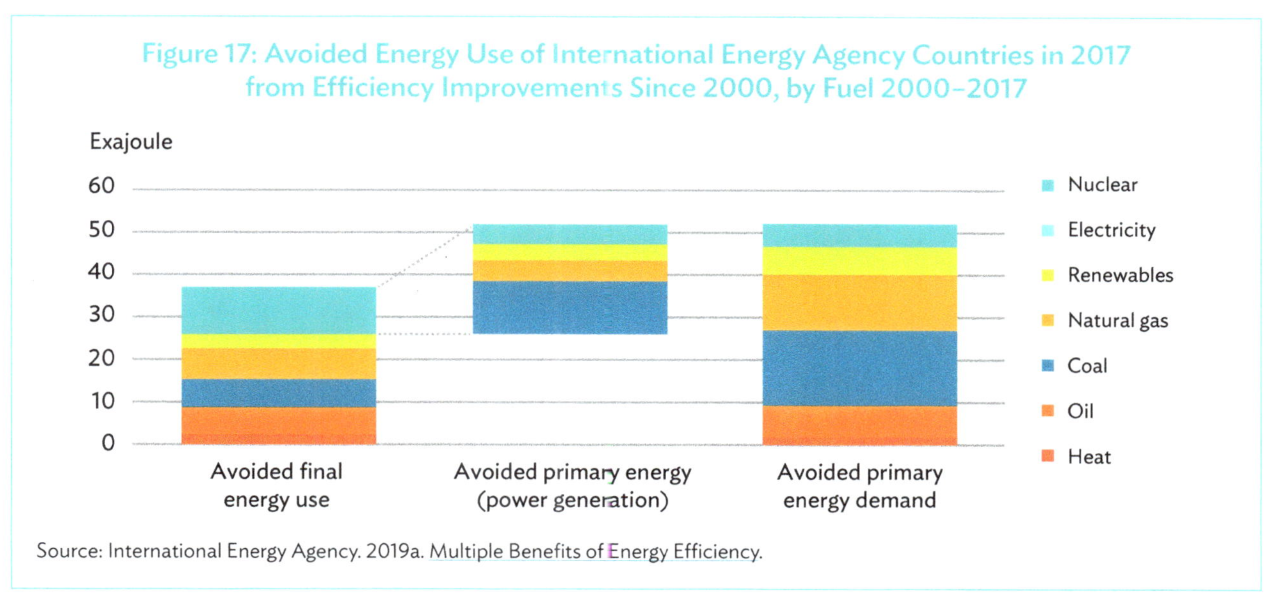

Source: International Energy Agency. 2019a. Multiple Benefits of Energy Efficiency.

An often-neglected environmental impact is the extraction of resources for the generation of power, such as coal in particular, but also petroleum. These can have devastating effects on the landscape and the entire ecosystem of the region. If not attended to properly, the detrimental effects on water and air quality of that region are felt for ages. Indian Green Building Council's data show that green buildings consume 20%–30% less energy and save around 30%–50% water (Dhir 2021). Similarly, nuclear power generation has environmental impacts seldom accounted for. It is not only the risk that after the accident in Fukushima that has provided ample evidence of what can happen, but the severe issues nuclear waste is posing has so far not been solved by any of the nuclear power-producing nations and remains an environmental time bomb.

Social Impacts

At a social level, physical and mental health are one of the core benefits. Energy efficiency improves mainly indoor living quality and, to a certain extent, the outdoor environment by improving the air quality,[20] humidity, temperature, and noise levels. Residential homes, as well as workplaces, benefit from it.

Especially for low-income groups who suffer from fuel poverty,[21] energy efficiency measures have a huge impact. Typically, this group lives in urban dwellings with poor building quality and relatively high energy costs. Poverty alleviation programs that encompass fully supported retrofit packages, and direct and indirect subsidies have proven to have positive results, whereby the retrofitting part was most successful. The New Zealand program "Heat Smart" showed that the monetized benefit after retrofitting for families from low-income groups was $531 annually compared to a control group at a higher income level, where $183 in benefits was achieved (WUNZ:HS 2012). Higher disposable income releases more people from fuel poverty.

Public health benefits were also found in the US due to improvements in air quality after utilities participated in an energy efficiency program. Not to speak of a reduction of overall hospital costs, which still needs to be quantified.

---

[20]   Studies point mainly to various cardiovascular and respiratory diseases.
[21]   Fuel poverty is defined as spending more than 10% of your available income on energy. Further reading: Liddell, C., and C. Morris. 2010. Fuel poverty and human health: A review of recent evidence. *Science Direct*. 38 (6). pp. 2987–2997.

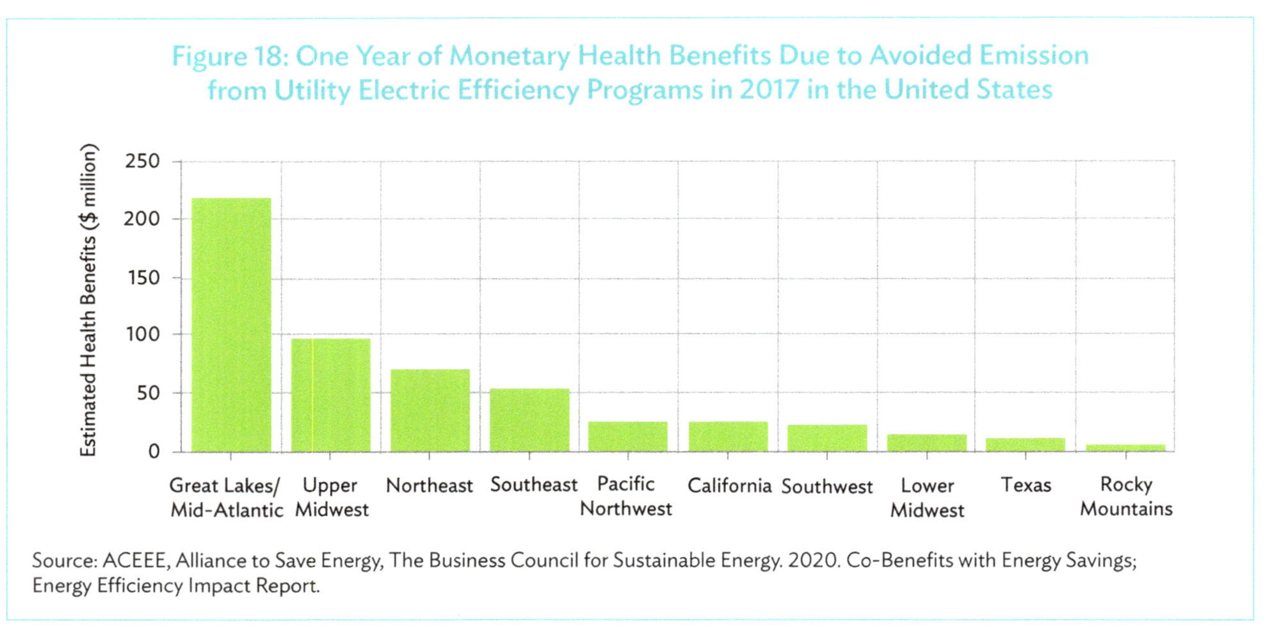

Source: ACEEE, Alliance to Save Energy, The Business Council for Sustainable Energy. 2020. Co-Benefits with Energy Savings; Energy Efficiency Impact Report.

Other social impacts of energy efficiency relate to the creation of new jobs and the demand for work also for non-skilled labor, for example, in construction. Most important, however, is the formation of a new industry that generates jobs at all levels. It offers opportunities to upgrade skills and technical knowledge as well as venture into entirely new jobs, such as energy auditors. The social impact, especially for an often younger but unemployed generation, is high. A stable, highly digitally developed industry has many positive effects and may stimulate the economy.

A social as well as an economic benefit of energy efficiency is the reduction in mainly imported fuels. A lot of the national wealth is transferred to foreign countries. If these expenditures can be saved society would benefit as more and stable financial flow is available for national projects.

Actions from ADB

ADB has a long-standing commitment to increase investment in energy efficiency, mainly SSEE, and renewable energy in its DMCs. Since 2010, ADB has spent, on average, over 20% of its portfolio in the energy sector, which amounted to over $42 billion. Many DMCs have yet to prioritize DSEE despite its great potential and co-benefits for financing power grid constructions and renewables (IED, ADB 2020). The PRC has taken proactive actions on energy efficiency, which, however, was initially driven by the purpose of air pollution control. Having recognized that, the bank invested $3.6 billion in energy efficiency from 2016 to 2020, among which $2.7 billion was on the supply-side and $838.7 million on the demand-side such as smart meter systems (ADB 2021).

In 2018, ADB approved the new long-term Strategy 2030 to respond to the region's changing needs.[22] It increasingly prioritizes DSEE through operational priority 3 on tackling climate change and emphasizes BEE and access to renewable heating and cooling through operational priority 4 on making cities more livable. In its bid to maximize the development impact and align with the significant changes in the energy landscape in DMCs, an independent evaluation of ADB's energy sector strategies and operations was carried out by ADB's Independent Evaluation Department (IED). The evaluation findings identified DSEE as one of the five key significant areas to be prioritized and scaled up in more DMCs (IED, ADB 2020).

---

[22] ADB. 2018. *Strategy 2030: Achieving a Prosperous, Inclusive, Resilient, and Sustainable Asia and the Pacific*. Manila.

In ADB's new energy policy[23] issued in 2021, the expanding efforts from the supply-side to DSEE and cross-sector operations have been highlighted. In particular, promoting cleaner cooling and heating has been identified as one of the prioritized operational areas under the energy policy's first guiding principle to securing energy for a prosperous and inclusive Asia and the Pacific. Recognizing the rapidly mounting demand for cooling and indoor thermal comfort in tropical and subtropical climate zones, ADB will support its DMCs in devising necessary policies and investment programs to accelerate the introduction and adoption of state-of-the-art sustainable cooling solutions. Renewable energy-based, energy-efficient, and thermally driven technologies and systems are among those considered highly relevant, applicable, and scalable in the context of fast-developing Asian cities. For heating, the energy policy has highlighted the competitive advantages of heat pumps and the strategic role they can play in decarbonizing heat generation and supply, which is still dominated by fossil fuels in many Asian countries. There is a large potential for various types of heat pumps, e.g., air-source, water-source, or ground-source, to be applied in these countries. As stated in the energy policy, ADB will help DMCs avail themselves of the benefits of heat pumps, through knowledge generation and demonstration projects that will help leverage project financing. In parallel with specific technologies, promoting DSEE through policy support, use of innovative financing instruments, and mobilization of private sector resources has also been identified by the energy policy as a priority for ADB operations. Residential and commercial buildings are among the key target areas of these cross-sector and cross-thematic interventions.

---

[23] Source: ADB. 2021. *Energy Policy*. Manila.

GREEN BUILDING CERTIFICATION SCHEMES

Green building certification schemes, also known as rating tools, rating systems, or certification systems, are comprehensive evaluation frameworks used to assess and recognize the compliance and performance of buildings or infrastructure projects from the perspective of life cycle resource efficiency and environmental sustainability. Usually, a scheme includes a set of explicit performance criteria against which a building seeking certification is evaluated, as well as related guidelines for the building design, construction, operation, and renovation toward meeting or exceeding the criteria. Commonly, a certification scheme can cover different types of buildings, with a specific module or a subset of the scheme applicable to a specific building type.

Green building certification schemes are playing an increasingly important role in promoting and transforming the sectoral development of a building. Baselines and benchmarks are established for measurement and comparison. Methods are provided to quantify the environmental effects of a building, which help inform sustainable integrated design solutions and facilitate decision-making. A certified building can have better quality and performance and less adverse environmental impacts as compared to what otherwise would have been the most likely case. There are also possible benefits relating to better occupant health and higher productivity that can be attributed to better indoor environmental quality, more access to natural daylighting, and healthier materials within green buildings. Beyond these benefits, green certification can differentiate a building and create monetary value, thus providing additional incentives to investors, developers, and owners (IFC 2021). In the broader context of sustainable development, the World Green Building Council (WGBC) highlights that green buildings can make a significant contribution toward meeting 9 out of the 17 SDGs defined by the United Nations (WGBC 2021).

Overview of Prevailing Certification Schemes

Globally, several green building assessment tools and rating and certification schemes have been developed by independent bodies and government agencies over the past three decades. Among them, the Building Research Establishment Environmental Assessment Method (BREEAM) and the Leadership in Energy and Environmental Design (LEED) have emerged as the leading rating and certification schemes and have been the most popular and widely used worldwide. BREEAM was developed by the Building Research Establishment (BRE) in the UK in 1990, becoming the world's first sustainability rating scheme for the built environment. LEED was created in 1995 by the US Green Building Council in partnership with the Natural Resources Defence Council. Both BREEAM and LEED have been used as the basis for the development of many other schemes. For example, the Green Globes for existing buildings were developed by ECD Energy and Environmental Canada in 2000, based on the 1996 Canadian Standards Association publication of BREEAM Canada.

In Asia, the Hong Kong Green Building Council was a forerunner of green building certification. In 1996, it developed the Building Environmental Assessment Method (BEAM) for new office designs and existing office premises in the local context of Hong Kong, China. From 2001 onward, a series of green building certification schemes were launched in Asian countries and regions. In 2001, the Comprehensive Assessment System for Build Energy

Environment Efficiency (CASBEE) was launched by the Ministry of Land, Infrastructure, Transport and Tourism of Japan as a method to evaluate and rate the environmental performance of buildings and the built environment in Japan. In 2003, the Green Building Council of Australia launched the Green Star to set the standard for healthy, resilient, and positive buildings and places. In 2005, the Building and Construction Authority (BCA) of Singapore launched the Green Mark, a benchmarking scheme to drive Singapore's construction industry toward more environmentally friendly buildings. The PRC issued its Assessment Standard for Green Buildings (ASGB) in 2006. As a government-led program, ASGB aims to regulate the evaluation and promote the development of green buildings across the PRC. There were three schemes launched in 2009. Malaysia launched the Green Building Index (GBI) as its industry-recognized green rating system for buildings to promote sustainability in the built environment and raise awareness among developers, architects, engineers, planners, designers, contractors, and the public about environmental issues. Qatar launched the Global Sustainability Assessment System (GSAS), which was the first integrated and performance-based system in the Middle East and North Africa region, designed for assessing and rating buildings and infrastructure for sustainability impacts. In the same year, the Philippines' national voluntary green building rating system, Building for Ecologically Responsive Design Excellence (BERDE), was launched through the Philippine Energy Efficiency Project-Efficient Building Initiative. In 2010, the Abu Dhabi Urban Planning Council developed and launched the Pearl Building Rating System for Estidama (EPRS) as a rating framework for sustainable design, construction, and operation of communities, buildings, and villas. In the same year, the pilot of the voluntary green building rating system of Viet Nam, LOTUS, was released by Vietnam Green Building Council. In 2012, the GreenSL rating system for built environment was developed by the Green Building Council of Sri Lanka. A year later, India, through its Green Building Council, launched the Indian Green Building Council (IGBC) rating system. The system is a voluntary, consensus-based, and market-driven program aiming to facilitate the development of energy-efficient, water-efficient, healthy, more productive, environmentally friendly buildings in India. In 2014, the Excellence in Design for Greater Efficiencies (EDGE) as an innovation of IFC was launched, with the view of responding to the need for a measurable and credible solution to prove the business case for green buildings and to unlock financial investments, especially in emerging markets.

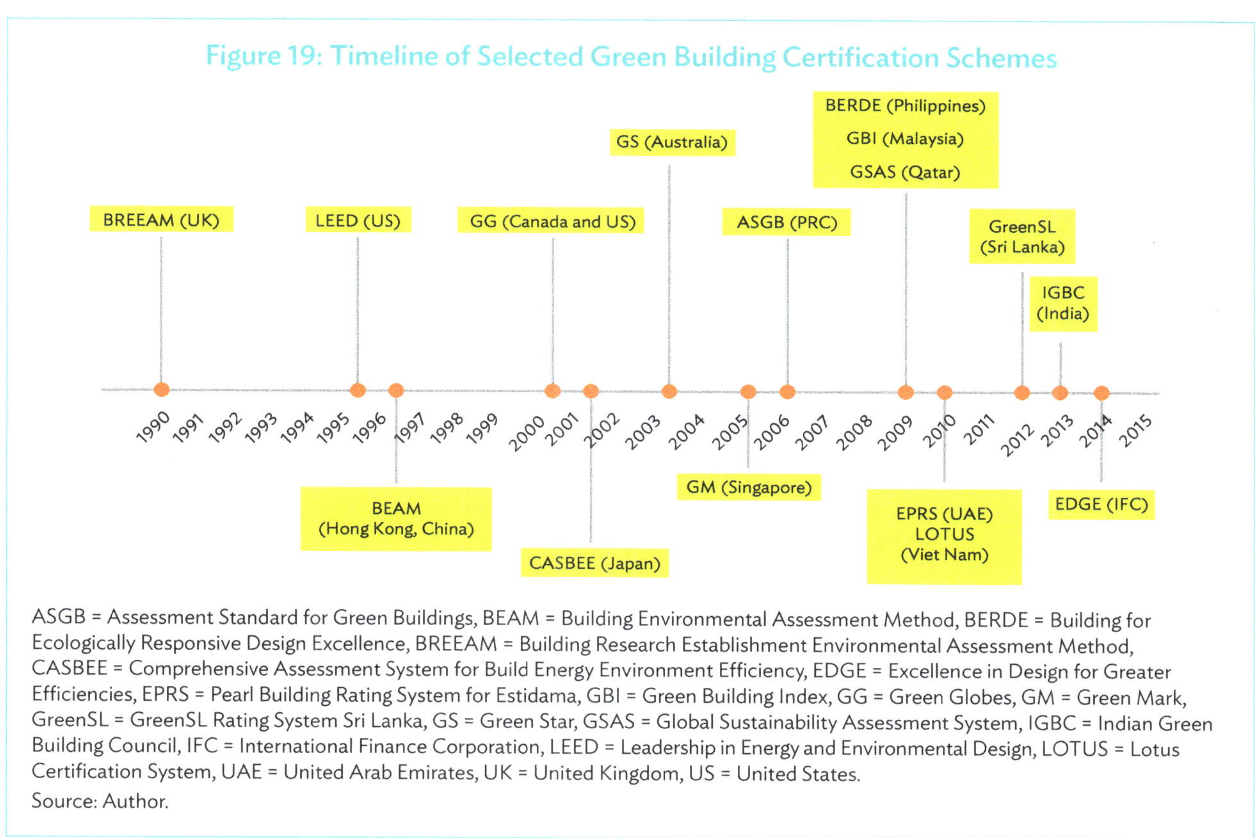

Figure 19: Timeline of Selected Green Building Certification Schemes

ASGB = Assessment Standard for Green Buildings, BEAM = Building Environmental Assessment Method, BERDE = Building for Ecologically Responsive Design Excellence, BREEAM = Building Research Establishment Environmental Assessment Method, CASBEE = Comprehensive Assessment System for Build Energy Environment Efficiency, EDGE = Excellence in Design for Greater Efficiencies, EPRS = Pearl Building Rating System for Estidama, GBI = Green Building Index, GG = Green Globes, GM = Green Mark, GreenSL = GreenSL Rating System Sri Lanka, GS = Green Star, GSAS = Global Sustainability Assessment System, IGBC = Indian Green Building Council, IFC = International Finance Corporation, LEED = Leadership in Energy and Environmental Design, LOTUS = Lotus Certification System, UAE = United Arab Emirates, UK = United Kingdom, US = United States.
Source: Author.

The timeline of the green building certification schemes is shown in Figure 20. The timeline is not intended to be an exhaustive coverage of all schemes in Asia, rather, it is a selective list covering some major economies to show the general evolution of green building certification schemes in the region. As can be seen, there has been a surge in certification schemes in Asia over the past 15 years. This can be largely ascribed to the good demonstration and successful implementation of BREEAM and LEED worldwide, including in Asia. In addition, what is not shown in the figure is the updating of these schemes from time to time to keep abreast of the development trend of green buildings against the backdrop of increasing emphasis on climate change mitigation and adaptation. Timely and appropriate updating of these schemes ensures their alignment with the continuously evolving building design standards and climate policies and enhances their applicability to the evaluation, rating, and certification of buildings in local contexts. For example, the BEAM in Hong Kong, China was updated to BEAM Plus in 2010. The ASGB in the PRC was revised twice: in 2014 and in 2019. The Green Mark of Singapore has undergone several rounds of updating since its launch in 2005, with the latest version being Green Mark 2021. The persistent development of new schemes and refinements of existing ones strongly demonstrate the increasing recognition and active pursuit of environmental, technological, economic, and social benefits of certified green buildings among governments, investors, planners, developers, architects, engineers, owners, and occupants.

The rest of this subsection provides an overview of these schemes. The subsequent two subsections focus on a brief discussion of key evaluation criteria of the schemes and case studies of certified green buildings in Asia, respectively.

Box 2: Overview of Selected Green Building Certification Schemes

BREEAM
(United Kingdom)

Building Research Establishment Environmental Assessment Method (BREEAM) is an international scheme that provides independent third-party certification of the assessment of the sustainability performance of individual buildings, communities, and infrastructure projects. Launched in 1990 by the Building Research Establishment (BRE), BREEAM was the world's first environmental assessment method for buildings and is defined by building science and research. It aims to deliver sustainable solutions, encourage a holistic approach to sustainability that is based on sound science and measures what is important, and improve building environmental performance. As the world's longest established rating scheme for building sustainability, BREEAM is internationally recognized and has been widely used in 80 countries worldwide, with more than 2,250,000 projects registered and more than 565,000 certificates issued. BREEAM has also served as the basis for many of the prevailing green building certification systems, including LEED and Green Globes.

BREEAM is highly flexible and can be applied to a wide range of building types and locations, with various technical standards for new construction, in-use, refurbishment and fitout, infrastructure, and master planning of communities. BREEAM measures the performance and sustainable value of a development across a broad set of indicators clustered in ten categories, covering energy, health and well-being, innovation, land use, materials, management, pollution, transport, waste, and water. Each of these categories addresses the most influential factors, including carbon emissions reduction, design durability and resilience, adaptation to climate change, and ecological value and biodiversity protection. Weighting is applied to each category to encourage projects to focus on the categories with the highest environmental impact. Minimum standards are set to ensure that key aspects of performance across a category are met to achieve potentially higher levels of certification. This provides a level of flexibility for use while maintaining the rigor of the standard. Where a project does not fit the scope of an existing standard, BREEAM offers a bespoke service that tailors the criteria in the existing standards to the development's specific use, sustainability opportunities and its location without compromising the rigor of the standard. The ratings of BREEAM range from acceptable (for in-use version only) to pass, good, very good, excellent to outstanding and it is reflected in a series of stars on the BREEAM certificate. The rating enables comparability between projects, provides reassurance to developers and users, and in turn underpins the quality and value of the asset.

Further information: BREEAM.

continued on next page

Box 2 *continued*

LEED
(United States)

Leadership in Energy and Environmental Design (LEED) is a voluntary green building rating and certification system developed by the U.S. Green Building Council in 1995. Available for virtually all building, community, and home-project types, LEED provides a holistic framework for healthy, highly efficient, and cost-saving green buildings. A LEED certification is a globally recognized symbol of sustainability achievement and leadership. LEED is one of the most widely used green building rating systems in the world, with more than 100,000 projects registered and 48,600 projects certified across 178 countries and territories as of 2020.

LEED is for all building types and all building phases including new construction, interior fit outs, operation and maintenance, and core and shell. It offers a range of specific rating systems, which are designed for building design and construction, interior design and construction, homes, building operation and maintenance, neighborhood development, and cities and communities. When several rating systems appear to be appropriate for a project that involves mixed construction or space usage type, the 40/60 rule is followed to make a decision. A particular system should be used if more than 60% of the project's gross floor area is appropriate for that system. If less than 40% is appropriate, then that system should not be used. If the appropriateness falls between 40% and 60%, the use of that system or a different one is at the discretion of the project team.

Projects pursuing LEED certification need to earn points for green building strategies across a range of categories, including location and transportation, sustainable sites, water efficiency, energy and atmosphere, materials and resources, indoor environmental quality, innovation in design, regional priority, and integrative processes. One hundred points are available across these categories. While project teams are free to go for any points under these categories within the chosen specific rating system, there are mandatory prerequisites that set the minimum requirements that all buildings need to meet to achieve LEED certification. Based on the compliance with prerequisites and the number of points achieved, a project is awarded one of four LEED certification levels: Certified, Silver, Gold, or Platinum.

Further information: LEED.

continued on next page

Box 2 continued

ASGB (People's Republic of China)

Assessment Standard for Green Buildings (ASGB) is a government-led green building certification program established by the Ministry of Housing Urban–Rural Development (MOHURD) of the People's Republic of China. The standard was initially issued in 2006, revised in 2014, and further revised in 2019. The purpose of formulating this standard is to regulate the evaluation and promote the development of green buildings. The assessment standard, covering both residential buildings and public buildings, consists of five main categories of indexes: safety and durability, health and comfort, occupant convenience, resources saving, and environment livability. Each category is assessed using a set of "prerequisite items," which are either pass or fail, and a set of "scoring" items, which are scores within predefined ranges. In addition, there is a separate score-based bonus category for improvement and innovation items, which aims to encourage the adoption of innovative technologies and products for higher building performance. The total score of assessment forms the basis of the grade of a building, i.e., basic, one-star, two-star, or three-star green building. When all prerequisite items are met, a building is rated as being of basic grade. On top of that, depending upon the scores of scoring items and the compliance with additional technical requirements (thermal performance of envelope, sound insulation, indoor air quality, etc.), a building may be rated as a one-star, two-star, or three-star green building. When an assessed building intends to apply for green financial services, it is required that the building's energy-saving measures, water saving measures, energy consumption and carbon emissions shall be calculated and reported in detail. This standard is supported by the Assessment Standard for Green Retrofitting of Existing Buildings, which was made available to the public for comments in early 2021 and is currently being finalized.

Further information: Green Building Network.

CASBEE (Japan)

Comprehensive Assessment System for Build Energy Environment Efficiency (CASBEE) is a method for evaluating and rating the environmental performance of buildings and the built environment. It was developed by a research committee established in 2001 through the collaboration of academia, industry, and national and local governments in Japan. CASBEE aims to both enhance the quality of life for building occupants and reduce the life cycle resource use and environmental loads associated with the built environment. Under CASBEE, there are a set of different tools, each of which is tailored to a certain type of buildings and/or a specific purpose of assessment. The latest versions of the tools cover buildings (new construction, existing buildings, renovation), commercial interiors, detached houses (new construction), dwelling unit (new construction), heat island, cities, and market promotion, among others. The assessment areas focus on energy efficiency, resource efficiency, local environment, and indoor environment. Both indoor and outdoor spaces are considered as part of the assessment but are assessed separately.

Further information: CASBEE.

continued on next page

Box 2 *continued*

BERDE (Philippines)

Building for Ecologically Responsive Design Excellence (BERDE) is the Philippines' national voluntary green building rating system recognized by the Government of the Philippines, through the Department of Energy's Philippine Energy Efficiency Project-Efficient Building Initiative. It is a tool to measure the resource efficiency and environmental performance of building projects and certify how a building performs above and beyond existing national and local building and environmental laws, regulations, and mandatory standards. The development process of BERDE was guided by internationally recognized methodologies for developing standards, and by the Quality Assurance Guide for Green Building Rating Tools issued by the World Green Building Council. The core framework of BERDE defines credits and requirements essential to a green building, across energy, water, waste, materials, emissions, land, and ecology, among others. The latest version of the BERDE (version 4.0.0) was launched in 2021. It introduced further socioeconomic priorities for green building projects, including the contribution of projects to health and well-being, community engagement, and economic opportunities. This version also considered the learnings from previous BERDE projects, feedback from industry and updates on laws, regulations, and standards.

Further information: BERDE.

BEAM Plus (Hong Kong, China)

Building Environmental Assessment Method (BEAM) Plus assessment is Hong Kong, China's leading initiative to offer independent assessments of building sustainability performance. The local BEAM scheme was established in 1996, largely based on the United Kingdom's BREEAM. In November 2009, the scheme was revamped and renamed BEAM Plus. Recognized and certified by the HK Green Building Council (HKGBC), BEAM Plus sets a comprehensive set of performance criteria for a wide range of sustainability issues relating to the planning, design, construction, commissioning, fitting out, management, and operation and maintenance of a building. There are four assessment tools under BEAM Plus: New Buildings, Existing Buildings, Interiors, and Neighborhood, covering the whole-building life cycle. New buildings cover the demolition, planning, design, construction, and commissioning of a new building project. It can also be applied to major renovations, alterations, and additions. Existing buildings evaluate the actual performance of existing building in energy efficiency and green management policies to ensure the building is operated in a sustainable manner. Interiors cover the design and construction of fitout, renovation and refurbishment works in nondomestic occupied spaces, focusing on energy use, sourcing of environmental-friendly building materials, and indoor environmental quality. The neighborhood adopts a holistic approach to assessing sustainability performance at the early or inception stage of a project. The assessment concerns the design of space between buildings and places the emphasis on socioeconomic elements.

Further information: HKGBC.

continued on next page

Box 2 continued

GM (Singapore)

The Green Mark certification scheme, launched by the Building and Construction Authority (BCA) of Singapore in January 2005, is a green building rating system designed to evaluate a building's environmental impact and performance. It provides a comprehensive framework for assessing the overall environmental performance of new and existing buildings to promote sustainable design, and best practices in construction and operations in buildings. Key criteria of the scheme cover climatic responsive design, building energy performance, resource stewardship, smart and healthy building, and advanced green efforts. For new buildings, the scheme encourages the design and construction of green, sustainable buildings that are more climate-responsive, energy- and resource-efficient, and have healthier indoor environments. For existing buildings, the scheme encourages the owners and operators to reduce the adverse impacts of their buildings on the environment and occupant health over the building's life cycle. The scheme also recognizes the role of building users in reducing energy consumption by having assessments specifically for office interior, retail outlets, restaurants, and supermarkets. To accelerate the adoption of environmentally friendly building technologies and building design practices, the Green Mark scheme has introduced a series of incentives in various forms. The currently available ones include the building retrofit energy efficiency financing scheme, which provides financing options to offset up-front costs of energy efficiency retrofits of existing building, and the built environment transformation gross floor area incentive scheme targeting private sector developments, which was recently introduced in November 2021.

Further information: Green Mark Certification Scheme.

Green Star (Australia)

Launched by Green Building Council of Australia (GBCA) in 2003, Green Star (GS) is Australia's only national, voluntary rating system for buildings, fitouts, and communities. It is an internationally recognized sustainability rating system setting the standard for healthy, resilient, and positive buildings and places. There are five GS rating tools, providing a means of certification for building design and construction, operation, fitouts, and communities. These tools have been developed by GBCA, in close consultation with industry and government, and continues to evolve. The GS Buildings guides the sustainable design and construction of schools, offices, universities, industrial facilities, public buildings, retail centers and hospitals. The GS Interiors rates the interior fitout works of buildings. The GS Communities assesses the planning, design, and construction of large-scale development projects at a precinct, neighborhood, and/or community scale. The GS Performance is dedicated to assessing the operational performance of existing buildings across various impact categories. The GS Homes, which was most recently launched in August 2021, is a standard assessing the health, resilience, and energy efficiency of homes. All GS rating tools include an "Innovation" category that rewards projects that implement new technologies and approaches or exceed current industry best practice.

Further information: Green Building Council Australia.

continued on next page

Box 2 continued

Pearl Building Rating System for Estidama (United Arab Emirates)

The Pearl Building Rating System (PBRS) for Estidama (EPRS) is the green building rating system developed by the Abu Dhabi Urban Planning Council as part of the sustainable development initiative, *Estidama* ("sustainability" in Arabic). PBRS encourages water, energy, and waste minimization, local material use, and aims to improve supply chains for sustainable and recycled materials and products. PBRS has established three rating stages: design, construction, and operational. The design rating rewards measures adopted during the design development of the project that meet the intent and requirements of each credit. The construction rating ensures that the commitments made for the design rating have been achieved. The operational rating assesses the built-in features and operational performance of an existing building and ensures the building is operating sustainably. The PBRS is applicable to all building typologies, their sites, and associated facilities. In essence, any building constructed for permanent use and that is air-conditioned must meet the PBRS requirements. An Executive Council Order of May 2010 stated all new buildings must meet the one Pearl requirements starting in September 2010, while all government funded buildings must achieve a minimum of two Pearls. Following this mandate, significant effort has been made to align the PBRS with the Abu Dhabi Development and Building Codes.

Further information: Pearl Building Rating System.

Green Building Index (Malaysia)

The Green Building Index (GBI) is Malaysia's industry-recognized green rating system for buildings to promote sustainability in the built environment and raise awareness among developers, architects, engineers, planners, designers, contractors, and the public about environmental issues. The tool provides an opportunity for developers and building owners to design and construct green, sustainable buildings that can provide energy savings, water savings, a healthier indoor environment, better connectivity to public transport and the adoption of recycling and greenery for their projects and reduce the impact on the environment. The evaluation process consists of a design assessment, leading to a provisional certification, and a completion and verification assessment, leading to a final certification. To maintain the certification, reassessments are required every 3 years. The assessment of commercial and residential properties under the GBI is based on six main criteria, including energy efficiency, indoor environment quality, sustainable site planning and management, materials and resources, water efficiency, and innovation.

Further information: Green Building Index.

continued on next page

Box 2 *continued*

IGBC (India)

The Indian Green Building Council (IGBC) rating system is a voluntary, consensus-based, and market-driven program aiming to facilitate the development of energy-efficient, water-efficient, healthy, more productive, environmentally friendly buildings. The system is applicable to all five climatic zones of India. It is designed to initiate a need to address national priorities and increase the quality of life for its occupants and it keeps pace with current standards and growing technology. The scope of the system covers new buildings and existing buildings across various types. The IGBC receives recognition from India's central and state government agencies which offers incentives to promote IGBC-related green building projects. For example, the Ministry of Environment, Forest and Climate Change offers fast track environmental clearance for green building projects which are pre-certified and/or provisionally certified by IGBC. Various state government agencies offer financial incentives in various forms, such as free-of-charge floor area ratio, exemption of building scrutiny fee, reduction on permit fees, reimbursement of certification fee, financial assistance at concessional rates, etc.

Further information: IGBC.

GSAS (Qatar)

The Global Sustainability Assessment System (GSAS) is the first integrated and performance-based system in the Middle East and North Africa region, designed for assessing and rating buildings and infrastructure for their sustainability impacts. It was developed by the Gulf Organisation for Research and Development and its collaborators. The primary objective of GSAS is to create a sustainable built environment that minimizes ecological impact and reduces resource use, while addressing the specific needs and environment of the Middle East and North Africa region. Representing green building regulations within Qatar construction specifications, GSAS takes an integrated life cycle approach to improve the design, construction, and operations of buildings. The system has three certifications, i.e., (1) GSAS Design and Build; (2) GSAS Construction Management; and (3) GSAS Operations. The sustainability assessment focuses on eight categories and criteria, including urban connectivity, site, energy, water, materials, indoor environment, cultural and economic values, and management and operations.

Further information: GSAS.

continued on next page

Box 2 *continued*

Green Globes (Canada and United States)

Green Globes is an online assessment protocol, rating system, and guidance for green building design, operation, and management. It is interactive, flexible, and affordable, and provides market recognition of a building's environmental attributes through third-party assessment. Green Globes originated in Canada and was brought to the United States (US) by the Green Building Initiative (GBI) in 2004. Now it is being used in both Canada and the US. The Green Globes brand and associated rating systems are administered in the US by GBI and in Canada by its wholly owned, nonprofit subsidiary, GB Initiative Canada. Green Globes have modules for new construction and significant renovations, sustainable interiors, and existing buildings. These modules can be used for a wide range of commercial, institutional, and multi-residential building types. A series of performance criteria are assessed, including project management, site, energy, water, resources, emissions, indoor environment, space, and amenities. The weighting of specific criterion varies depending upon the module. Different from LEED, there are no prerequisites under Green Globes. Based on assessment outcome, five levels of certification are offered, from 1 as the lowest to 5 as the highest.

Further information: Green Globes.

EDGE (International Finance Corporation)

The Excellence in Design for Greater Efficiencies (EDGE) is a green building certification system focused on making buildings more resource-efficient. Now covering more than 170 countries, EDGE was created and launched in 2014 by the International Finance Corporation (IFC), a member of the World Bank Group, to respond to the need for a measurable and credible solution to prove the business case for green buildings and to unlock financial investments. EDGE enables design teams and project owners to assess the most cost-effective ways to incorporate energy and water saving options into their buildings. EDGE includes a cloud-based platform to calculate the cost of going green and utility savings. The state-of-the-art engine has a sophisticated set of city-based climate and cost data, consumption patterns, and algorithms for predicting performance results. There are three levels of certification, applicable to new construction, existing buildings and major renovations of various types of buildings. On the condition that at least 20% savings in water and materials are achieved, a project may be awarded (1) EDGE Certified, if a minimum of 20% energy savings are achieved; (2) EDGE Advanced, if a minimum of 40% energy savings are achieved; or (3) Zero-Carbon, if 100% energy savings are achieved either through renewables or carbon offsets. The consortium of Sintali-SGS and Green Business Certification Inc. are the approved certifiers who operate in most countries where projects can be designed and certified with EDGE.

Further information: EDGE.

continued on next page

Box 2 continued

GreenSL (Sri Lanka)

The GreenSL rating system for built environment was developed by the Green Building Council of Sri Lanka as a "home-grown system" with all norms acceptable to leading rating systems. The formulation of GreenSL learned from the experiences and lessons of green building initiatives of India, the People's Republic of China, Indonesia, Malaysia, and Singapore. It followed an open, consensus-based process under the supervision of the Green Environmental Rating and Life Cycle Assessment Committee composed of a diverse group of practitioners and experts representing a cross section of the construction industry. GreenSL aims to encourage environmentally acceptable design, construction, and operation and maintenance of buildings. It consists of two major assessment categories, with one for new constructions and major renovations and the other for buildings in operation. A building is assessed across eight performance criteria, including management, sustainable sites, water efficiency, energy and atmosphere, materials and resources, indoor environmental quality, innovation and design process, and social and cultural awareness. Based on assessment outcome, a building is awarded one of the four levels of certification—certified, silver, gold, or platinum.

Further information: GreenSL.

LOTUS (Viet Nam)

LOTUS is a set of voluntary green building rating systems developed by Vietnam Green Building Council (VGBC) based on established building physics science and adapted to the conditions, climate, and regulations of the Viet Nam construction sector. The pilot program of LOTUS was released in 2010. With several updates and revisions, the latest version was released in 2019. LOTUS provides a holistic assessment framework for design, construction, and operation of buildings from a whole-building perspective. It assesses and recognizes performance in a series of key evaluation categories, including energy, water, material, indoor environment, and site. The set of evaluation categories and weighting of a specific category vary, depending upon building types and project scopes—new construction, major renovation, buildings in operation, single-dwelling homes, small non-housing projects, interior fitout projects, and small interior fitout projects. Points are awarded for the achievement of credit requirements in each category. In addition, there are certain prerequisites under each category that a building must meet to qualify for consideration of other credits. LOTUS certification has four levels of certification—certified, silver, gold, and platinum. To be certified, a project must earn a minimum of 40 points. The LOTUS assessment and training programs are managed by the Vietnam Green Building Social Enterprise Co. Ltd. registered by VGBC.

Further information: LOTUS Certification System.

Energy as a Key Evaluation Element of Green Building Certification Systems

A further review of the summarized green building certification schemes identifies a wide range of themes used to evaluate green building strategies and approaches across various aspects. Water, materials, and energy are all covered in the schemes. Subsequently, the themes covered in descending order include indoor environment, site, land and outdoor environment, and innovation. Associated with the seven themes are specific evaluation criteria for green buildings.

It is useful to further investigate the relative importance of the seven themes. Based on the manuals, guidelines, and instructions of the rating schemes, the respective weighting of the seven themes is calculated for each scheme. The average weighting of each theme across the examined schemes and the corresponding standard deviation are also calculated (Table 4).

Table 4: Weighting of 7 Essential Themes of 12 Rating Schemes

No.	Scheme	Energy	Site	Indoor Environment	Land and Outdoor Environment	Material	Water	Innovation
1	Assessment Standard for Green Buildings (ASGB)	19.0%	–	14.0%	17.0%	14.0%	16.0%	9.0%
2	Building Research Establishment Environmental Assessment Method (BREEAM)	20.0%	–	–	7.7%	12.5%	6.7%	10.0%
3	Comprehensive Assessment System for Build Energy Environment Efficiency (CASBEE)	20.0%	–	20.0%	15.0%	12.0%	3.0%	–
4	Pearl Building Rating System for Estidama (EPRS)	24.0%	–	8.5%	19.5%	16.0%	24.0%	2.0%
5	Green Building Index (GBI)	23.0%	33.0%	12.0%	–	8.0%	12.0%	8.0%
6	Green Globes (GG)	39.0%	12.0%	16.0%	–	13.0%	11.0%	–
7	Green Mark (GM)	18.0%	–	–	–	9.4%	9.4%	–
8	Global Sustainability Assessment System (GSAS)	24.0%	13.0%	16.0%	–	10.0%	16.0%	–
9	Green Star (GS)	25.0%	–	16.0%	10.0%	14.0%	10.0%	9.0%
10	Building Environmental Assessment Method Plus (BEAM Plus)	35.0%	25.0%	20.0%	–	8.0%	12.0%	–

continued on next page

Table 4 *continued*

No.	Scheme	Energy	Site	Indoor Environment	Land and Outdoor Environment	Material	Water	Innovation
11	Indian Green Building Council (IGBC)	28.0%	14.0%	12.0%	–	16.0%	18.0%	7.0%
12	Leadership in Energy and Environmental Design (LEED)	28.0%	12.0%	14.0%	–	16.0%	9.0%	4.0%
Average weighting of each theme		25.2%	18.2%	14.9%	13.8%	12.4%	12.3%	7.0%
Standard deviation		6.4%	8.8%	3.6%	4.9%	3.0%	5.6%	2.9%
Average weighting of all themes across all schemes		14.8%						

– = not applicable.

Source: Adapted from (Hwang, B.-G. 2018. Green Building Rating Systems: Practices and Research Efforts. In *Performance and Improvement of Green Construction Projects: Management Strategies and Innovations*. Elsevier Science and Technology. pp. 23–44; Shan, M. and B. Hwang. 2018. Green Building Rating Systems: Global Reviews of Practices and Research Efforts. *Sustainable Cities and Society*. 39. pp. 172–180.).

The examined schemes have their weighting in a relatively wide range of 18% and 39%, with an average value of 25.2% and a standard deviation of 6.4%. The average weighting of 25.2% substantially exceeds that of the other six themes, namely site (18.2%), indoor environment (14.9%), land and outdoor environment (13.8%), material (12.4%), water (12.3%), and innovation (7%). It is also much higher than the average weighting of all seven themes across all schemes (14.8%). Despite having a relatively large standard deviation, energy has the highest weighting among the seven themes under 9 out of the 12 examined schemes. For each of the nine schemes, the difference between energy and the specific themes having the second highest weighting varies considerably, ranging from 2% to 23%. The three exceptions are CASBEE of Japan, which allocates the highest weighting (20%) to both energy and indoor environment; EPRS of the United Arab Emirates, which allocates the highest weighting (24%) to both energy and water (24%); and GBI of Malaysia, which defines a site as the most important criterion (33%) and energy as the second most important one (23%). The energy weighting under all 12 schemes is higher than the cross-scheme average weighting of all the seven themes, i.e., 14.8%. These results suggest that energy carries the most importance in evaluating the performance of green buildings. The reduction of energy use, particularly the reduction of peak load, use of energy efficiency appliances, equipment and systems, use of renewable energy, and energy monitoring and reporting are among the main indicators to evaluate the energy criterion, as defined in the examined schemes.

Case Studies of Certified Green Buildings in Asia

This subsection provides seven case studies of certified green buildings by selected certification schemes introduced in the preceding subsections. The case studies have been purposively selected to cover diverse geographical locations and climatic conditions across Asia—Azerbaijan, Bangladesh, India, Kazakhstan, the PRC, the Philippines, and Singapore. Some of the cases have been certified by two or more schemes. As energy is the most important evaluation theme of green buildings and the focus of this handbook, these case studies are presented to highlight the design features of energy-related strategies and technologies of the buildings. Where possible, the actual energy performance of the buildings is discussed. Typical energy efficiency strategies, technologies, and practices adopted in these case studies are described in more detail in chapters 2 and 3 of this handbook.

Box 3: Case Studies

CASE 1: MEGA SILK WAY SHOPPING MALL, KAZAKHSTAN
(BUILDING RESEARCH ESTABLISHMENT ENVIRONMENTAL ASSESSMENT METHOD [BREEAM] 2020)

Summary of Certification

- Scheme and Version: BREEAM In-Use International 2015

- Certification Stage: In-Use

- Rating: Excellent

- Overall Score: 75.4%

Mega Silk Way, Kazakhstan
Source: Mega Kazakhstan.

Description of Building

MEGA Silk Way is the largest and most modern shopping and entertainment center in Kazakhstan, built in the capital of the country as a flagship cutting-edge facility among shopping and entertainment centers in Central Asia. It is an important project in terms of tourism and investment prospects of the capital, combining the best qualities of other MEGA network shopping malls. The mall was designed on the principle of a shopping street. It is a "city within a city" with its own streets and sheltered areas. The mall features a unique concept of social space. A total of 1.5 million cubic meters (m^3) of internal space is bright and airy. The building has three levels, including two elevated floors and one lower ground floor, and nine entrances, two of which are underground. It has parking capacity for 2,113 cars. The ratio of the commercial premises of the shopping and recreation center to the gross floor area is 53%. Monthly visit to the shopping and recreation center is about 1.1 million people.

Green Strategies

The architectural design with regard to orientation, facade, window, roof lights, and so on reduces the cooling and heating loads of the mall. The mall uses supply and exhaust units with heat recovery and frequency converters, as well as an intelligent Building Energy Management Systems. The building and the surrounding area are lit by 4,800 LED luminaries powered by renewable energy.

More than 18% of the building envelope is glazed and provides good access to daylight. Temperature zoning is provided in the mall, which meets design parameters throughout the year. The ventilation system mitigates air pollution by placing air intake grids of 74 air-handling units at the required distance from any sources of air pollution. The use of building materials and products with low emissivity of volatile organic compounds also contributes to keeping a high quality of indoor air.

The mall has more than 300 units of water-efficient sanitary equipment. A rainwater collection system is in place for irrigation of green areas. Local facilities for treatment of rainwater collected from the open car park and specialized well-insulated rooms for placing diesel generator sets and fuel tanks help to minimize the risks of contamination of soil and rainwater by oil products.

The facility is operated in compliance with the approved environmental policy, which contains performance targets for energy efficiency, water consumption reduction, minimization of general waste, and increasing the biodiversity of the facility.

continued on next page

Box 3 *continued*

CASE 2: BAKU WHITE CITY, AZERBAIJAN
(BREEAM 2016)

Summary of Certification

- Scheme and Version: BREEAM International New Construction 2013
- Stage: Final
- Location: Baku, Azerbaijan
- Size: 15,861 square meters (m²)
- Score and Rating: 48.5%, Good

Source: ArchDaily. 2013. Baku White City Office Building Proposal.

Project Background

The impetus for this redevelopment was provided in 2006 when the decree entitled "Comprehensive action plan for improving the ecological conditions in the Azerbaijan Republic during 2006–2010" was issued. The area formerly known as the Black City was being gradually transformed into a vibrant, comfortable business and residential destination, which was symbolic of Baku's development progress. The special meaning of the project for the country brought particular attention to environmental issues. In line with the ecological decree principles, the project team chose BREEAM as the international best practice standard practice for designing the buildings. The White City office building was the first project of the regeneration plan and aimed to provide class A offices with world-class specifications. Along with the building's accessibility and a host of local amenities, the state-of-the-art offices were designed to give its occupants a distinctive market presence.

Green Strategies

The building was developed by the architect to focus on reducing its impact on the local environment, as well as on improving the quality of the internal environment to provide a high standard of healthy office space. Each floor has its own balcony and the five top levels of the building feature extensive terraces. These provide occupants with access to fresh air and good views. Natural light and views create an inspiring environment for business on each floor. The facade of the building is designed as a "quiet" wall system that incorporates high-performance double-glazing to help minimize sound and provide a calm workplace. Modern and sustainable materials are used in the building structure and fitout. For cooling and heating, the building uses a high-efficiency fourth generation variable refrigerant volume system and high efficiency (92%) boilers. Fresh air is distributed with the use of mechanical ventilation, while air extract is performed in a natural manner. Daylight is combined with efficient LED and T5 artificial lighting, controlled through detection sensors. Energy use within the building is sub-metered to provide better building management. Water efficiency is achieved through measures to reduce water use (efficient sanitary fittings and irrigation) and avoidance of leakage. During the design stage a series of studies were conducted to make informed decisions on design alternatives, such as energy modeling in line with the American Society of Heating, Refrigerating and Air-Conditioning Engineers standards, or life cycle impact assessment of materials with the use of IMPACT tool.

continued on next page

Box 3 *continued*

CASE 3: SDE 4, NATIONAL UNIVERSITY OF SINGAPORE (BCA 2019; NUS 2020; NUS 2021A; NUS 2021B)

Summary of Certification

- Scheme and certification: Green Mark Platinum Zero Energy; WELL Gold
- Location: Singapore
- Occupancy type: Academic
- Typology: New construction
- Climate: Tropical
- Size: 8,588 m²

Source: National University of Singapore.

Description of Building

School of Design and Environment Building 4 (SDE 4) is a new addition to the existing three buildings of the School of Design and Environment at the National University of Singapore. The new building, which was completed in 2019, functions as a living laboratory to promote research collaboration with public agencies and industry partners. With a gross floor area of 8,514 m², the building houses a mix of research laboratories, test-bedding facade, design studios, as well as teaching and common learning spaces. It also includes a 3D scanning lab, green building technologies lab, urban greenery lab, and NUS-CDL Smart Green Home.

SDE is characterized by its contemporary architecture design which demonstrates a deep understanding of the tropical climate of Singapore. The design concept incorporates a large overhanging roof, which together with the double facades on the east and west of the building, shade it from solar radiation and provide a cooler interior environment. The building design also makes use of the architectural concept of "floating boxes," where its shallow plan depth and porous layout allows for cross-breezes, natural lighting, and views to the outdoors. Weather permitting, rooms can also be opened to natural breezes, and air-conditioning is used only where it is needed, thereby reducing the electricity consumption of the building.

Net-Zero Energy Strategies

SDE 4 utilizes solar energy by having more than 1,200 solar photovoltaic (PV) panels installed on the roof. On days when there is insufficient solar energy, the building draws electricity from the power grid. When the solar energy is more than enough, electricity generated by the PV panels is supplied to the grid. Throughout a year, the net amount taken from the grid is zero, or even negative, thereby achieving net-zero energy consumption.

Integral to SDE 4's concept of net-zero energy consumption is the need to rethink air-conditioning, which typically accounts for up to 60% of a building's total energy consumption in a tropical country like Singapore. This has resulted in the adoption of an innovative hybrid cooling strategy to avoid overcooling of rooms, which is a common issue in buildings in Southeast Asian countries. Rooms are supplied with 100% fresh precooled air at higher temperature and humidity levels than a conventional system and this is augmented with increased air speeds from ceiling ventilators. This strategy, coupled with the building's tropical architecture design that encourages natural ventilation, creates a thermally comfortable indoor environment while ensuring much less energy consumption than conventional air-conditioning systems. In addition, with daylight utilization maximized through the architectural design, energy savings are further enhanced through a network of photocell and occupancy sensors.

continued on next page

Box 3 continued

SDE 4 offers useful lessons to architects and engineers on how energy use and occupant comfort might be better balanced in future buildings. It also demonstrates that a building need not deliver the same conditions all day to everyone. Giving occupants the option to control indoor environment is a viable solution to achieve comfort.

Actual Energy Performance of SDE 4 (April 2019 to March 2020)

Actual energy use	470,750 kilowatt-hours (kWh)
Actual energy produced	619,345 kWh
Net energy use	(148,595 kWh)
Energy use intensity	55 kWh/square meter/year

Note: ( ) = negative

CASE 4: ARTHALAND CENTURY PACIFIC TOWER, PHILIPPINES (USGBC 2018A; IFC 2019A; WGBC 2019)

Summary of Certification

- Scheme and certification: Leadership in Energy and Environmental Design (LEED) Platinum; EDGE Zero Carbon; BERDE 5-stars
- Location: Taguig, Metro Manila
- Climate: Tropical
- Type: Commercial
- Size: 38,714 m^2

Source: Edge Buildings.

Project Background

Arthaland is a publicly listed company in the Philippines and the pioneer developer for premium green and sustainable projects. At the heart of every Arthaland project is sustainability, exceptional and innovative design, and high-quality construction standards. All residential and office projects of the company aim to adhere to global and national standards for green buildings, e.g., the LEED by the US Green Building Council and the BERDE by the Philippine Green Building Council. Arthaland is the world's first developer to earn the International Finance Corporation's EDGE Zero Carbon certification for its flagship office building, the ArthaLand Century Pacific Tower (ACPT). The building reduces energy use by 45% (with 100% hydroelectric energy supplied by the Pantabangan–Masiway Hydroelectric Plant), water use by 64% and embodied energy in materials by 34%. It is the only triple-certified project in the Philippines, having received the EDGE Zero Carbon certification, the LEED Platinum rating, and the BERDE 5-star certification. All these certifications are the highest and most prestigious awards in the mentioned three rating systems.

continued on next page

Box 3 *continued*

Energy and Carbon Strategies

- ACPT has been designed to achieve its exemplary sustainable energy performance by means of the following:
- The building is designed with triple- and double-glazed curtain wall systems strategically oriented to insulate the building from solar gain while allowing natural light to be transmitted in, thereby reducing the energy required for cooling and lighting.
- The mechanical systems optimize thermal zoning to consider the different cooling requirements of the interior spaces. This further minimizes the amount of energy required for air-conditioning.
- An energy recovery ventilation system is used to recover the cool air from the exhaust system. Through heat transfer, incoming hot outdoor air is precooled, thereby reducing the cooling load and energy consumption of the air-conditioning system.
- The ventilation system is designed with a variable speed drive to communicate with the carbon dioxide sensors located in regularly occupied spaces. The sensors identify the density of a space at a given time and allows the ventilation system to appropriately function only at the required speed, thereby avoiding unnecessary energy wastage when the space is not fully occupied.
- The lighting system of the building uses intelligent design such as daylight and occupancy sensors. The lamps can accommodate the needs of the space by adjusting the lux level of the lighting fixtures to complement available daylighting and shut off when the space is unoccupied.
- The energy required by the building, after the reduction and efficiency strategies, is supplied by renewable sources. ACPT is 100% powered by FirstGen's Pantabangan–Masiway Hydroelectric Plant located in Nueva Ecija, Philippines.

CASE 5: MOBIL HOUSE, DHAKA, BANGLADESH (USGBC 2019; NZEB ALLIANCE 2021F)

Summary of Certification

- Scheme and certification: LEED Platinum (score 85/110)
- Location: Dhaka, Bangladesh
- Occupancy type: Office
- Typology: New construction
- Climate: Tropical wet and dry
- Size: 6,673 m²

Source: MJL Bangladesh PLC.

Description of Building

The Mobil House building is the head office of MJL Bangladesh Limited (formerly Mobil Jamuna Lubricants Limited), a joint venture between the state-owned Jamuna Oil Company and EC Securities Limited (a subsidiary of the East Coast Group) in Bangladesh. The building consists of office spaces, meeting areas, and conference halls. The building achieved a platinum LEED certification by following an integrative approach that engaged the clients, architects, engineers, energy

continued on next page

Box 3 continued

specialists, and green building consultants from the early design stage until post-occupancy. Their inputs at different stages were incorporated into the building design, construction, and operation to help ensure the cost-effectiveness and operational efficiency of the building. Compared to conventional office buildings in the region, the building consumes 30% less energy and 50% less water.

Green Strategies

The architectural design of the building features sizable courtyards and building cutouts that are populated with foliage and vertical gardens. They act as thermal buffers for the interior spaces of the building. This design compensates the building's lack of surface green cover due to site constraints.

The building envelope made of 300-millimeter (mm)thick concrete walls with high thermal mass and the windows using double-glazed panels with low emissivity and a U-value of 1.1 W/m2K reduce heat gain during the daytime. The building is oriented such that circulation elements like lift core and staircases are situated along the west facade. This shields the regularly occupied spaces, e.g., offices and reception, from the solar gains through the west facade. The northeast facade is exposed to less solar gain and therefore is designed to have large windows to allow daylight and outdoor views.

An efficient chilled water system is incorporated in the building's HVAC design to reduce energy consumption by cooling. The key technical features of this system include a chiller coefficient of performance of 6.3, a low approach cooling tower with variable-frequency drive, and an energy recovery wheel.

The lighting system of the building integrates daylight and occupancy sensors to provide the optimum levels of light during daytime. The sensors detect the ambient daylight levels as well as number of occupants to determine the light output. This strategy effectively reduces lighting electricity consumption.

The building utilizes renewable energy by installing a solar PV system on its rooftop. The PV system comprises of an array of 60 PV modules with a total capacity of 18 kilowatt-peak. Based on the average solar irradiance in Bangladesh, the system has the potential to generate 24,000 kWh per year.

CASE 6: UNNATI OFFICE, GREATER NOIDA, UTTAR PRADESH, INDIA (USGBC 2018B; NZEB ALLIANCE 2021I; NZEB ALLIANCE 2019; GAINWELL 2021)

Summary of Certification

- Scheme and certification: LEED Platinum (score 89/110)
- Location: Greater Noida, Uttar Pradesh, India
- Occupancy type: Office
- Typology: New construction
- Climate: Composite
- Size: 3,740 m²

Source: One Click LCA.

continued on next page

Box 3 *continued*

Description of Building

The Unnati Office Building is the regional headquarter (north) for the heavy machineries and construction equipment company Gainwell Commosales Pvt. Ltd., located in its industrial campus in Greater Noida. With an energy performance index of 60 kWh/m²/year, the building design of Unnati demonstrates how climate-responsive architecture is integral to achieving energy efficiency in a composite climate. It is the first building in India (and second in Asia) to achieve Platinum certification under LEED v4 BD+C.

Green Strategies

The three-story building is a cuboid with a central courtyard. It is oriented northeast-southwest, with the core areas distributed in the east and the west orientations. Passive design strategies have been integrated with the building design. With the design of form, central courtyard, shallow floor plates, appropriate sizing and distribution of openings, 90% of the office spaces, including the core and service areas, receive uniformly distributed daylight, thereby reducing electricity use for artificial lighting.

The building adopts climate-responsive facade concepts, including green wall and shading on all windows, to protect the interiors from direct sun and reduce heat gain. All external surfaces, including the walls, roof, and foundation, are insulated using polystyrene panels. The roof insulation also includes a 300 mm layer of green roof soil substrate. The window-to-wall ratio is 30%. High performance double-glazed windows provide improved protection against sunlight with integrated motorized blinds and shading.

Passive design features reduce the total cooling load to 208 kilowatts (kW) for 3,740 m². This load is met by a hybrid heating, ventilation, and air-conditioning (HVAC) system, which is a combination of water-cooled air-handling units and ceiling-embedded radiant cooling system. The water-cooled air-handling units meet 45% of the cooling load, whereas the ceiling-embedded radiant cooling system meets 55%. Based on indoor design conditions of 24°C and 55% relative humidity, the room dew-point temperature is 12°C and chilled water is supplied at 7°C to avoid any condensation on surfaces. Dry outdoor ventilation air is supplied through an externally mounted unit that dehumidifies the air before it is supplied to occupied space. This dry outdoor air acts as primary air to the chilled beams.

An energy-efficient lighting system with daylighting controls is used by the building. Occupancy sensors minimize the need of lighting. Lighting controls reduce internal heat gain and air-conditioning load. Access to daylight for 90% of regularly occupied spaces, e.g., offices, helps to substantially lower the lighting power density. The building utilizes renewable energy by installing a 100-kW solar PV system on its rooftop, which generates 146 megawatt-hours (MWh)/year meeting 40% of its energy demand per year.

continued on next page

Box 3 continued

CASE 7: CONSTRUCTION INDUSTRY COUNCIL– ZERO CARBON PARK, HONG KONG, CHINA (ARUP 2021; HKGBC 2021; ZCP 2018; WGBC 2015)

Summary of Certification

- Scheme and certification: BEAM Plus Platinum
- Location: Kowloon East, Hong Kong, China
- Occupancy type: Multipurpose (office, exhibition, ecological park)
- Typology: New construction
- Climate: Subtropical
- Size: 14,700 m²

Source: Meetings and Exhibitions Hong Kong.

Description of Building

Opened in June 2012, the Construction Industry Council's (CIC) Zero Carbon Park (ZCP) is home to the first zero-carbon building in Hong Kong, China. It acts as a test bed for state-of-the-art eco-building design and technologies and aims to promote a low-carbon mentality both locally and internationally. CIC-ZCP also serves as an exhibition, education, and information center to raise awareness on the importance of green building design, beyond the industry, and into the community. CIC-ZCP is designed to be energy-positive over the course of its life cycle. This exemplar project has achieved BEAM Plus Platinum rating, the highest rating for building environmental performance in Hong Kong, China. It has also received a series of awards relating to green buildings, such as the Green Building Awards 2012 - Hong Kong, Australian Institute of Quantity Surveyors 2013 International Project of the Year, World GBC's Asia-Pacific Leadership in Green Buildings Awards, among others.

Energy and Carbon Strategies

The ZCP building adopts a series of passive design strategies which collectively contribute to reducing energy consumption by 20%. The roof is covered with cellular glass insulation, screeding, and a protective surface membrane. About 85% of the roof area is covered with solar PV panels while the rest is green roof. The building's elongated form enables a suitable space depth between the northwest and southeast facade for daylighting. Light shelves on its southeast and northwest facades shade the periphery and distribute daylight further away from the windows and deeper into the space. A large glass wall on the northwest facade is positioned to face the relatively unobstructed sky to provide ample daylight, while being sheltered from direct sunlight or glare. The glazing around the model exhibition area on the ground floor is fitted with movable heat reflecting shades behind the glazing. The shade with an aluminum sheet on one side emits radiant heat that is absorbed while allowing the diffusion of light. Light pipes are installed to capture light from domes on the roof to provide light in windowless interior space of the building.

Cross-ventilation is a key strategy adopted by the building to address specific environmental challenges in the climatic conditions of Hong Kong, China. The building has an open-plan cross-ventilated layout, and its main facade faces southeast to optimize the use of prevailing summer breezes coming from that direction. There are also two wind catchers installed on the roof to supply the high-velocity outdoor air through the shaft down to the building. These design features, coupled with the use of high-volume-low-speed ceiling fans, promote a gentle and uniform air velocity throughout the building that can effectively counter the effects of the usually humid weather. This provides free cooling

continued on next page

Box 3 *continued*

equivalent to approximately 30%–40% of the building's cooling load during a year. A number of high-level windows are centrally controlled by coordinating their operation with the air-conditioning strategy. At the same time, there are a number of low-level windows that can be controlled by the user to tailor the amount of ventilation and air velocity at the occupant zone.

The building is equipped with an earth cooling tube, which provides naturally precooled air to reduce the cooling load and energy use in the hotter months when the mechanical system is in operation. Stormwater (at a temperature of about 20°C) inside the box culvert running underneath is used as the condensing medium for the air-conditioning system. Chilled water is circulated to chilled beams to provide radiant cooling effect. To avoid having to overcool outdoor air for dehumidification as in the case of conventional systems, a desiccant dehumidification unit is used to pre-treat the humid outdoor fresh air. This reduces the load of the chillers. The separately cooled and dehumidified air is supplied to the room at a low velocity, through an underfloor displacement ventilation system. With thermal stratification of indoor air, only the occupied zone at low level in the room needs to be cooled. This low velocity and relatively high temperature of the supply air help to reduce the energy consumption of air-conditioning system.

ZCP utilizes solar energy by installing PV panels at rooftop. Estimated annual power generation is 87 MWh/year. It also operates a tri-generation system which is powered by biofuel made of waste cooking oil. Thermal energy from combusting the biofuel is first used for electricity, which is estimated to be 143 MWh/year. Then the thermal energy at a lower grade is used for the adsorption chiller and the desiccant dehumidification process. This energy use in cascade yields an energy efficiency of 70%. Overall, ZCP produces more energy than its own needs, with a surplus of about 99 MWh/year.

Source: Compiled by authors.

SUSTAINABLE COOLING STRATEGIES AND TECHNOLOGIES

Integral to developing energy-efficient and low-carbon buildings, sustainable cooling strategies and technologies are expected to be able to reduce the use of mechanical ventilation and cooling systems while avoiding the occurrence of overheating and thermal discomfort in the summertime and ensuring reasonably good indoor air quality. Cooling load, cooling source, and air distribution are key metrics in designing and operating a cooling system aiming to achieve this overarching objective. In general, a sound sustainable cooling strategy should, to the extent realistically possible, explore the possibilities of controlling solar heat gains, choosing appropriate ventilative cooling techniques, applying low-energy cooling technologies, harnessing cooling energy from renewable sources, and using sustainable air distribution systems (Figure 20).

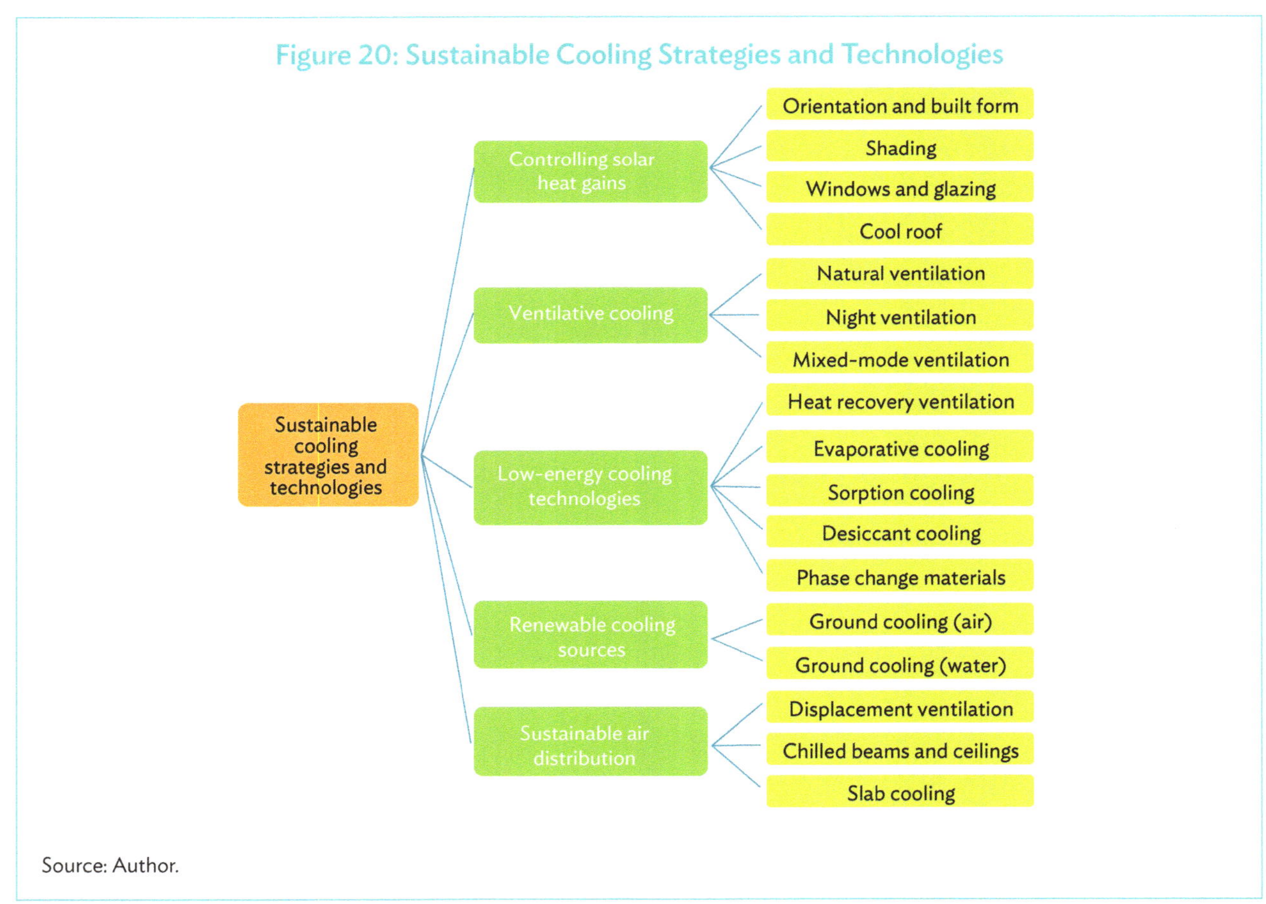

Source: Author.

Controlling Solar Heat Gains

Orientation and Built Form

The form and orientation of a building have direct and substantial impacts on the building's exposure to solar radiation and heat gain and, therefore, are among the most important passive design strategies to reduce building energy consumption and ensure indoor thermal comfort. Different from individual building systems, equipment, and materials that can be improved in later stages through renovation or retrofit, the core characteristics of the built form are difficult and costly to alter to increase the building's thermal and energy performance once its geometry and orientation have been fixed. In other words, design strategies relating to form and orientation are largely "one-off" interventions at the design stage before a building is built, and can substantially affect the actual energy performance during the operational stage of the building. If the potential benefits of these interventions are missed, more complex and costly solutions would likely have to be devised and implemented at later stages to achieve similar benefits of energy savings and thermal comfort.

Solar geometry is the determining factor of heat gain, shading, and the potential daylight penetration of a building. Thus, the design of a building must be responsive to solar orientation on the site where the building is built. For a building in the northern hemisphere, the sun is at a low angle during winter and to the south of the east–west axis. During summer, the sun path is at a high angle and, depending on the latitude of the building, may be entirely north to the east–west axis (e.g., in a tropical climate) or partially north to the east–west axis (e.g., in a temperate climate). The change in path affects solar radiation penetration patterns during different seasons and the resultant heat gain and loss of a building (Figure 21).

While the design of a building varies with its climate, location, and site conditions, the underlying principle to achieve energy efficiency remains largely the same—choosing a practically optimum orientation to minimize solar heat gain in summer, maximize daylight harvesting, and minimize heat loss in winter. In hot climates where no heating is required and other climates where cooling demand far exceeds heating demand, the priority is to minimize solar heat gain. It is desired that a building is shaped such that it is elongated in the east–west direction, i.e., the longer or larger facades of the building are oriented toward north–south. This is because glazing areas on east or west facades, which receive higher intensities of solar radiation than those on south or north facades, are more difficult to shade from direct sunlight due to the low angles of the sun. Where possible, surface areas and openings on east or west facades should be minimized.

In addition to exposure to solar radiation, prevailing wind direction is another important factor to consider when orienting a building. This is particularly relevant to hot-humid and warm-humid climates where the full potential of natural ventilation and cooling should be availed of. To do so, a building needs to be oriented at an angle to the prevailing wind direction to facilitate maximum airflow and cross-ventilation through the building. Often this direction does not necessarily coincide with the preferred sun orientation. Thus, it is not possible to simultaneously optimize both design parameters, which means a trade-off between them. In this case, a reasonable compromise should be made based on a detailed analysis of the situation. Possibilities for diverting the wind direction using vegetation and/or structural arrangements should be explored, such as the use of parapet walls within the external adjoining space. As a rule of thumb, usually, low-rise buildings would not receive much solar radiation and, therefore, could be oriented according to the prevailing wind direction to maximize the benefit of natural ventilation. For high-rise buildings, on the contrary, the priority should be given to best orienting the building to minimize solar radiation and heat gain. When there is a group of buildings on a site, they should be properly laid out such that minimal wind shadows would be created, and wind movements could be accentuated (Figure 22).

Figure 21: Illustration of a Building's Solar Exposure in Subtropical Climate in Northern Hemisphere and Sun Path Diagrams of Singapore, Manila, New Delhi, and Beijing

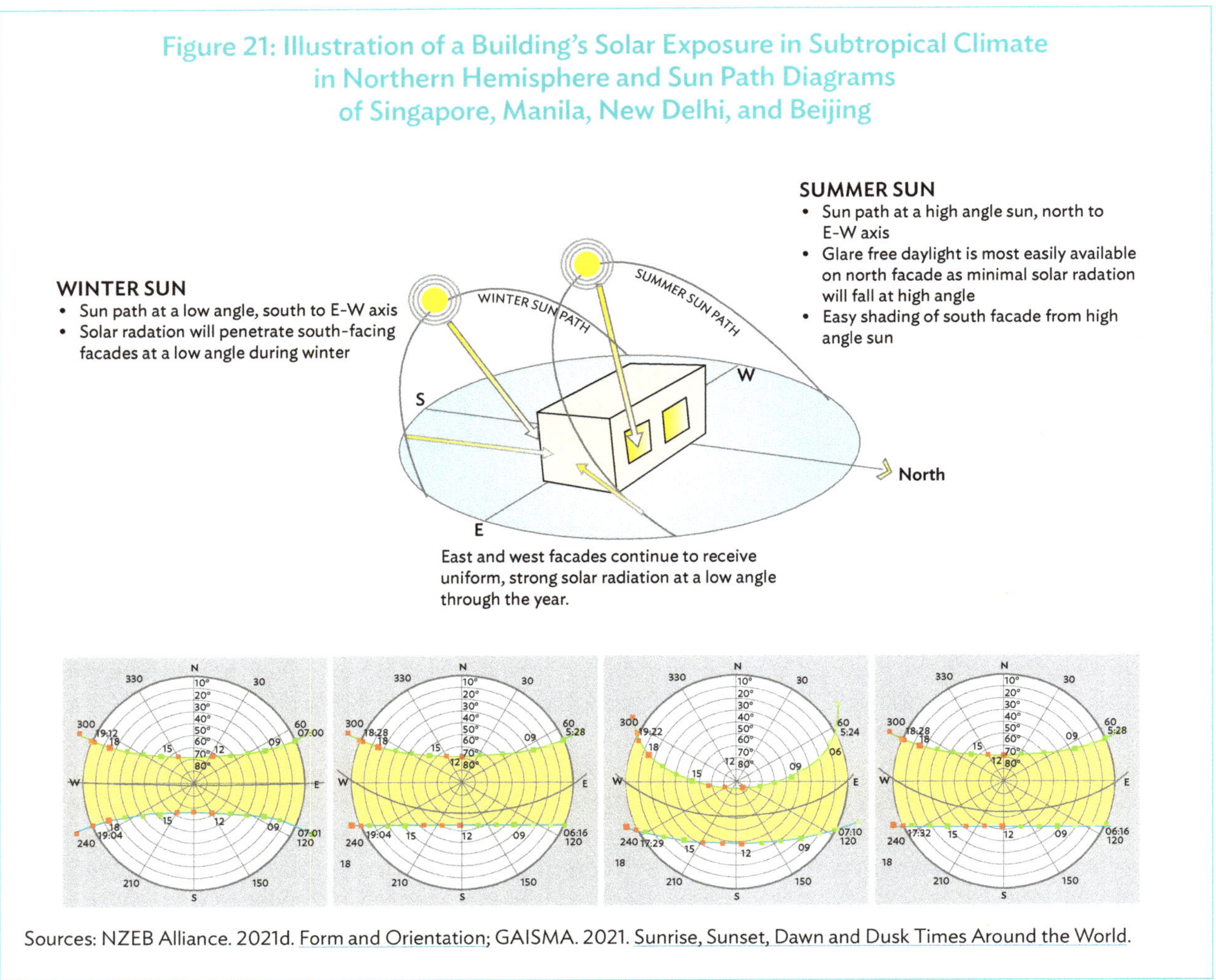

Sources: NZEB Alliance. 2021d. Form and Orientation; GAISMA. 2021. Sunrise, Sunset, Dawn and Dusk Times Around the World.

From the perspective of reducing solar heat gain through building envelope, there are two important built form design parameters to consider: (i) the ratio of perimeter-to-area, and (ii) compactness (Lim and Kim 2018) surface-to-floor ratio, area-to-perimeter ratio, and volume ratio to evaluate building energy performance. Also, the paper focused on the relation between the air-conditioned room and non-air-conditioned room. This approach affects both the design stages of the floor plan and the main designing factors that decide which spaces would become air-conditioned spaces such as those mostly occupied by residents or non-air-conditioned space such as staircases and elevators. The heating load and cooling load were calculated using the new equation based on the location of non-air-conditioned spaces and envelope ratio facing the outdoor. Both the width-depth ratio and envelope ratio were analyzed using the IES-V.E (Integrated Environmental Solutions Virtue Environment. The perimeter-to-area ratio can be an indicator of radiative heat gain. For a given area, a higher value of the ratio corresponds with a longer boundary and increased exposure to solar radiation. Thus, a low ratio is desired for hot climates. The compactness is the three-dimensional extrapolation of the perimeter-to-area ratio. It is measured using the ratio of surface area to volume. For a given volume, the greater the surface area, the greater the potential solar radiation and heat gain through the surface. Therefore, it is desirable to design a compact shape to minimize unwanted solar heat gain (CLEAR 2004; Fairconditioning 2021a). For office buildings, a compact design may also bring the benefit of having less space requirements on distribution of horizontal and vertical building services, particularly for air-conditioning ductwork (CIBSE 2012).

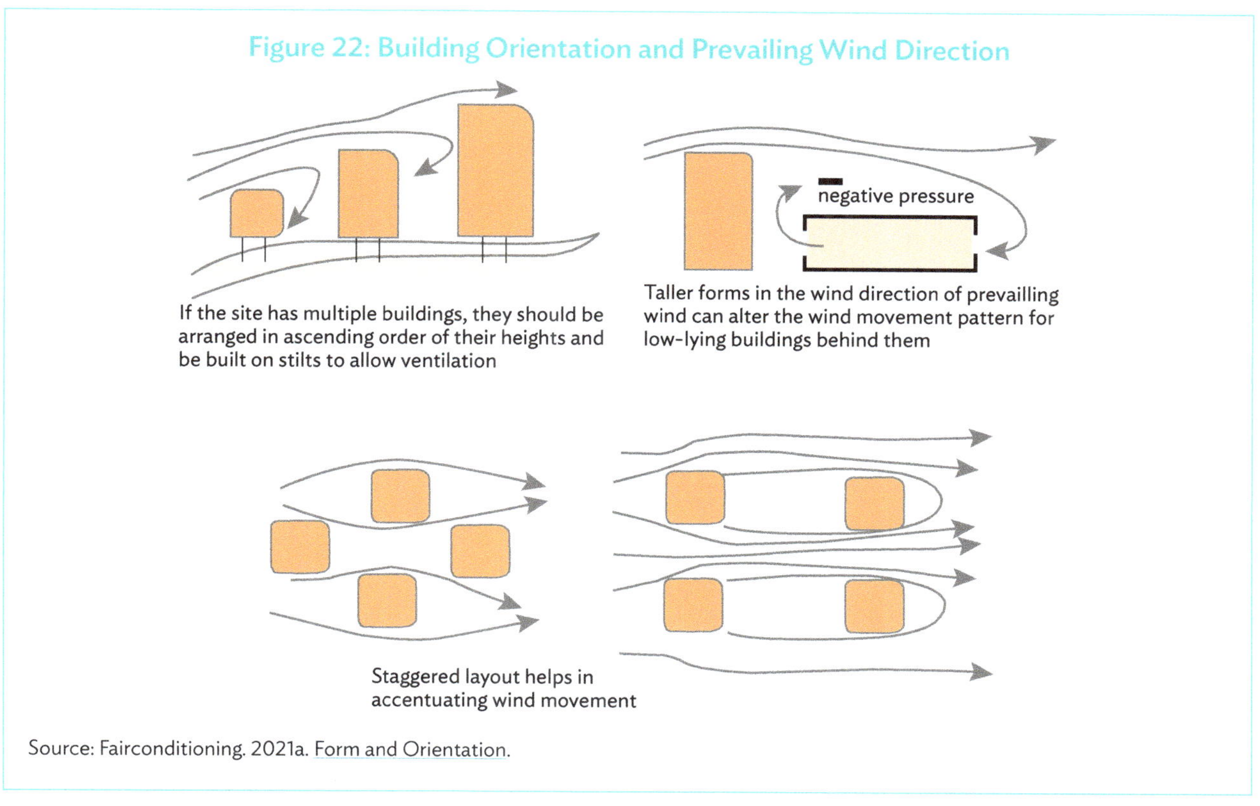

Source: Fairconditioning. 2021a. Form and Orientation.

In theory, the most compact building shape would be a cube. In practice, this configuration is often not realistic
or acceptable due to various reasons, such as planning regulations, functional requirements, site availability,
topographical conditions, neighboring buildings, daylight access, or simply architectural aesthetics. From the
perspective of energy use, a form design of high compactness may result in a deep plan, which may lead to a
greater complexity of configuring building services. A deep plan also means a large portion of the floor area is
placed far from the perimeter daylighting and thus may require continuous electric lighting (CIBSE 2012). Also,
internal activities in this inner area may require mechanical ventilation and/or cooling, which otherwise would not
necessarily be necessary if the inner area can be naturally ventilated. In contrast, a less compact design of built form
may prioritize access to daylighting and natural ventilation by placing more portion of the floor area closer to the
perimeter. While this may result in a suboptimal arrangement for solar exposure, there is the possibility that the
reduced lighting load and cooling load may well offset the increased solar radiation and heat gain (CLEAR 2004).
In addition, for office buildings, the view out and the location of the core (building services such as elevators,
toilets, air-handling units, staircases, risers, etc.) are also important aspects that can have a substantial impact
on form design and the energy performance of a building (Box 4). Therefore, similar to building orientation as
earlier discussed, a trade-off across a range of design factors should be adequately analyzed to make an informed
design decision.

Box 4: Impact of Built Form, Orientation, and Core Location on Energy Performance of Multistory Office Building in Malaysian Climate Zone

The selection of built form often decides where the location of service core can be placed, while the core location decides whether a mechanical system is required to ventilate spaces such as pantries and toilets. It is also possible to take advantage of the core location by using the core as a buffer zone to reduce the impact of solar radiation in the air-conditioned space of the building. Therefore, the selection of the core location would have a substantial impact on the main orientation of the building and the extent to which the air-conditioned spaces are exposed to direct solar heat gain.

Against this backdrop, a simulation-based study was conducted to investigate the interplay between built form, core location and orientation and its implication for energy performance of a multistory office building in Malaysia. The results showed that it was possible to save up to 7.2% energy based on the selection of built form, core location, and orientation (Box Figure 4.1). The most significant factor for lowering building energy consumption was the glazing area. Although the external facade's window-to-wall ratio of the office space was fixed at 70% for every scenario simulated, the absolute external glazing area differed by up to 48% between models because of different core locations.

Box Figure 4.1: Building Energy Index as a Function of Built Form, Core Location, and Orientation

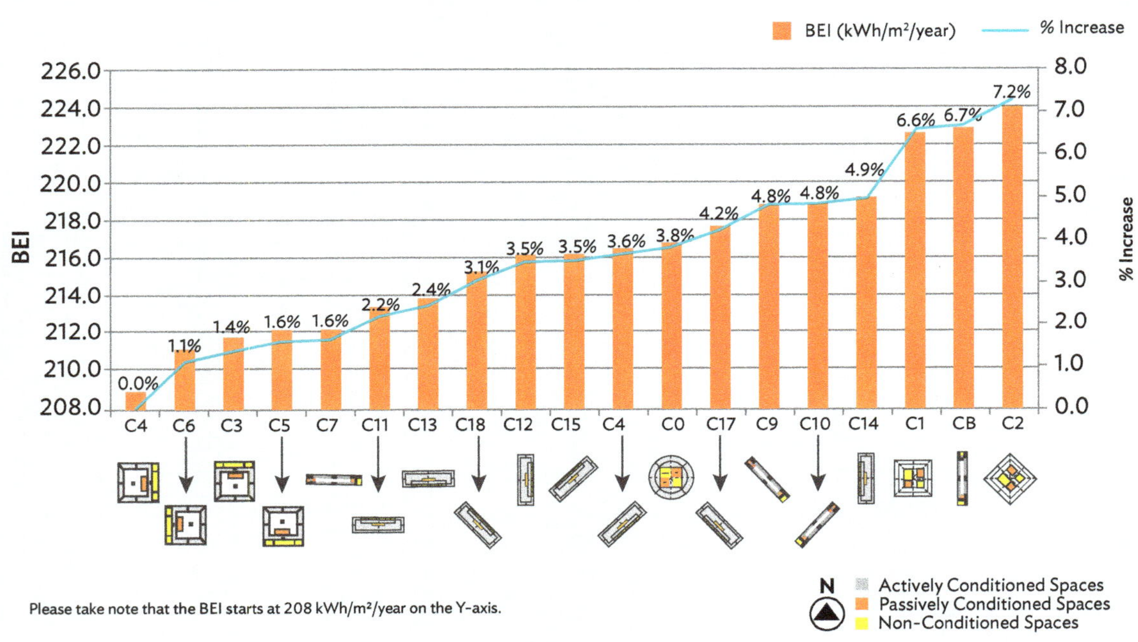

kWh = kilowatt-hour, m² = square meter.
Source: BSEEP Malaysia. 2017a. Building Energy Efficiency Technical Guideline for Passive Design.

The second significant factor was the ability to view out of the building. The view out was computed using the assumption of a person standing in the middle of the building looking out through the external facade. Internal partitions were assumed to be invisible or transparent, while external facade was opaque. A ratio of building energy index (BEI) in kilowatt-hour/square meter/year to view out in degrees was computed as the proxy of the energy cost for the view out of different built forms, core locations and orientations. This ratio indicated the amount of energy used for each degree of view out for each building model scenario. The lower the ratio, the less energy required for each degree of view out. The simulation results revealed that the energy penalty for increasing the view out reduced as the view out was increased (Box Figure 4.2).

Box Figure 4.2: Ratio of Building Energy Index to View Out as a Function of Built Form, Core Location, and Orientation

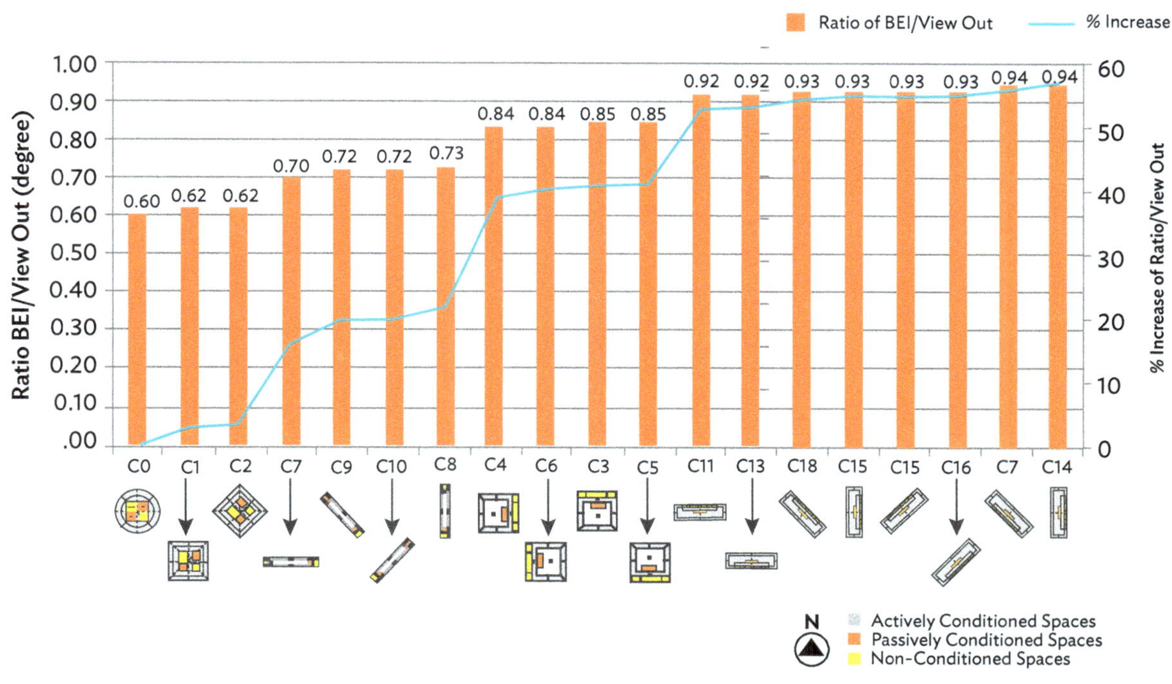

BEI = building energy index.
Source: BSEEP Malaysia. 2017a. Building Energy Efficiency Technical Guideline for Passive Design.

The orientation of the building was also shown to have a significant impact on the energy performance. The benefit of having a good orientation allowed for a higher glazing area and a better view out while reducing the BEI of the building in certain scenario. In short, having the glazing facing north and south resulted in a lower energy consumption than having the glazing facing east and west.

As key recommendations from the study, the selection of built form, core location and orientation for improved energy performance should have the following priorities:

- Choose a built form and core location that provides the least amount of glazing but maintains the necessary aesthetic appeal of the building and the view out.
- If the building is in a place that will benefit from having views in all directions, selection of a building with a 360° view out is an acceptable option due to the low ratio of BEI/view out as compared to other built forms, core locations and orientations.
- Expose glazing to the north and south but limit the exposure to the east and west. Whenever possible, locate the service core in the following order of preference: east, north–east, south–east, west, north–west, south–west, north, and south.

Source: BSEEP Malaysia. 2017a. Building Energy Efficiency Technical Guideline for Passive Design.

For building blocks such as buildings in a residential compound, there is a possibility to benefit from mutual shading to minimize the solar exposure of building elevations during summer (Figure 23). The effectiveness of mutual shading depends on a building's latitude and location with respect to the other buildings, the height of the context buildings, and the distance between the buildings (Bureau of Energy Efficiency 2014). The benefits of mutual shading in reducing solar exposure are possible if the buildings are closely placed to the east and west of the reference building. There is less shading effect from the buildings located south of the reference building, with minimal shading effect during peak summer. Except in tropical climates where the latitude is low, there is negligible shading from the buildings located north of a reference building.

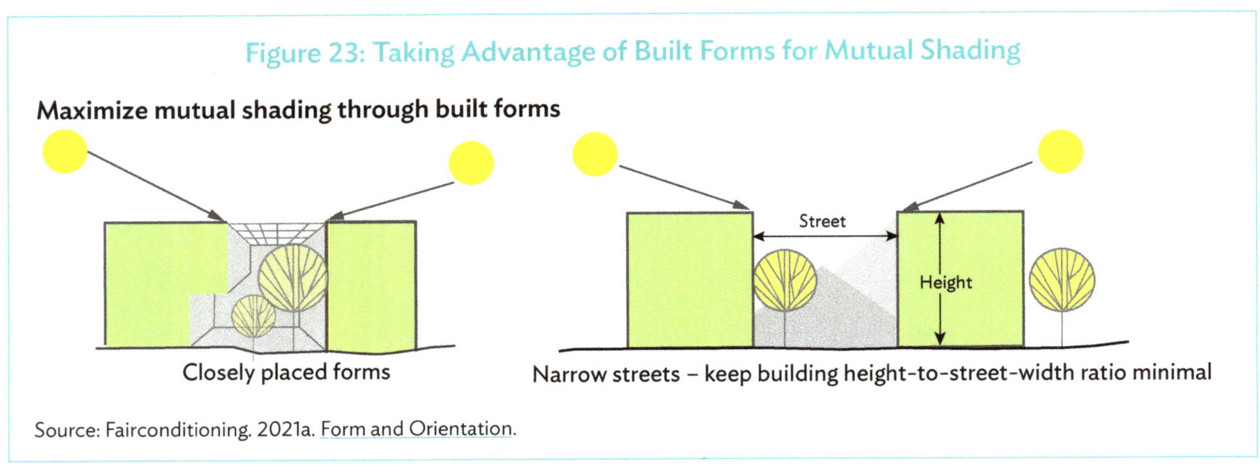

Figure 23: Taking Advantage of Built Forms for Mutual Shading

Source: Fairconditioning. 2021a. Form and Orientation.

Shading

External shading, as an integral part of building envelope design, is a simple but effective method to intercept excessive solar radiation falling on the glazed surfaces and entering the building. With reduced solar heat gain, the need for mechanical ventilation and cooling to maintain a thermally comfortable indoor environment can be reduced, and considerable energy savings can be achieved. External shading plays a particularly critical role in tropical and subtropical climates where buildings are exposed to strong solar radiation for long periods during a year and the cooling load and associated energy cost are high.

Well-designed external shading devices have the potential to reduce solar heat gain through glazing by up to 80% (2030 PALETTE 2021c; Bureau of Energy Efficiency 2021b). This brings the potential benefit of cost-effectiveness. With well-designed external shading devices, low-cost single-glazing with a high solar heat gain coefficient (SHGC) may be used in lieu of high-cost, low-SHGC double-glazing, which otherwise would have been used.

External shading is much more effective than, and therefore greatly preferred over, internal shading, because the latter allows substantially more solar radiation and heat to be transmitted through the window into the building in spite of its role in blocking the glare of the sun (NZEB Alliance 2021h; The Society of Building Science Educators 2012). Depending on the reflectivity of the internal shade, some of this transmitted heat is reflected straight back out of the window, but the rest is absorbed by the shade, which is then heated up. The shade releases the heat in the form of long-wave radiation, which is trapped inside the room (BSEEP Malaysia 2017a). As shown in Figure 24, the SHGC of an internal movable blind can be as high as more than 40%, whereas that of an external movable blind can be as low as less than 12% (BEEP 2020b).

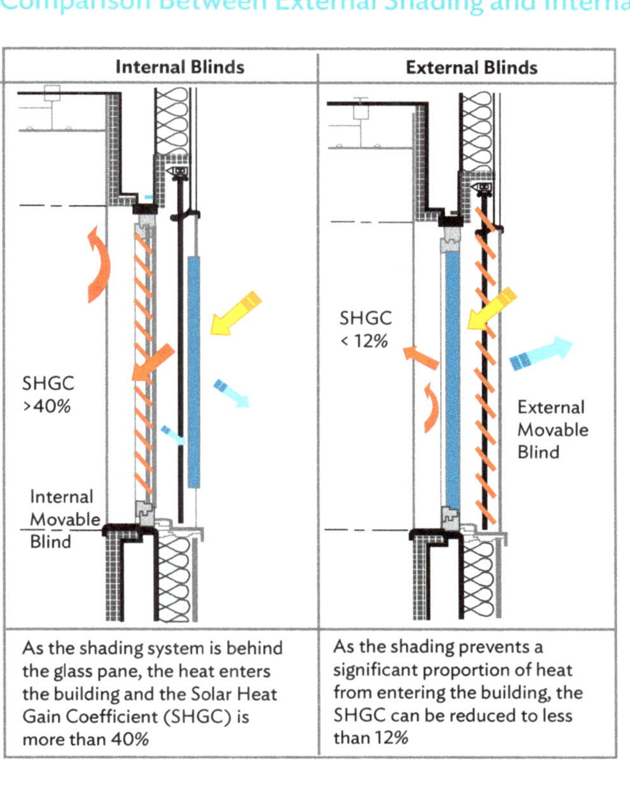

Figure 24: Comparison Between External Shading and Internal Shading

Source: Fairconditioning. 2021e. Shading.

The orientation of an opening on a building facade is a key factor in the design of external shading devices. The impact of daily and seasonal variations in the sun path and the incident solar radiation is closely linked to the orientation. As the exposure of each facade of a building to the sun is different and varies with the sun's position (its altitude and azimuth), it is necessary to take a different approach to the design of shading for each orientation of a building. For a building located in the northern hemisphere, the south elevation receives both direct and diffused solar radiation and can be relatively easy to control. Shading devices are normally designed as horizontal projections above the windows, e.g., horizontal overhangs. The length of the projection is determined as a geometric function of the height of the window and the angle of elevation of the sun at solar noon. Such shading devices can be designed to eliminate sun penetration in the summer. For example, a vertical side fin of a triangular or rectangular shape on either side of the window can be used in combination with an overhang to cut off the direct solar radiation of the early morning (as the sun rises) and late evening (as the sun sets). Other typical shading devices for south elevation also include horizontal louvers in the horizontal plane, horizontal louvers in the vertical plane, a vertical panel parallel to the wall and hung from a horizontal overhang, large roof overhangs, deeply recessed windows, etc.

The necessity of shading for the north elevation depends largely upon the latitude of the building. When the latitude is low, e.g., in tropical Asian cities such as Singapore, Ho Chi Minh City, or Manila, the north elevation of the building is exposed to strong direct solar radiation for a long period throughout a year. Thus, proper shading is important, and the mentioned shading devices for the south elevation can also be used for the north elevation in these locations. With the increase in latitude, the north elevation's exposure to direct solar radiation reduces, and the angle of incidence of the sun's rays on the north elevation becomes less impactful. For example, the north elevation of a building in Hong Kong, China or Ha Noi is exposed to direct solar radiation for a much shorter period throughout the year, as compared to the exposure in Singapore, Ho Chi Minh City, or Manila. At further northern latitudes, such as Beijing or Tashkent, the north elevation of a building receives direct solar radiation only in the summer months during the early morning and late evening hours. During these hours, the sun angle is so low that horizontal projections would be of little use as shading devices. Vertical side fins on the eastern and western sides of the window on the north elevation can be used to shade the window. Alternatively, since the impact of solar radiation and heat gain is limited in this circumstance, windows on the north elevation can go without shading devices (Bureau of Energy Efficiency 2021a).

As for the east and west elevations, they are exposed directly to strong solar radiation for the entire morning on the east face and the entire afternoon on the west face. The sun has low angles and penetrates deeper directly from the front of the window. This precludes shading using overhangs, as the overhangs will have to be too deep to be architecturally and structurally reasonable (The Society of Building Science Educators 2012; UN-Habitat 2018). More practical solutions for these orientations include multiple louvers, vertical fins, slanted vertical fins, eggcrates, eggcrates with horizontal louvers, and their variants. These shading devices are also effective on the near east and near west exposures, e.g., east–northeast, east–southeast, west–northwest, and west–southwest. Figure 25 illustrates some of the typical external shading devices.

Figure 25: Schematics of Typical External Shading Devices

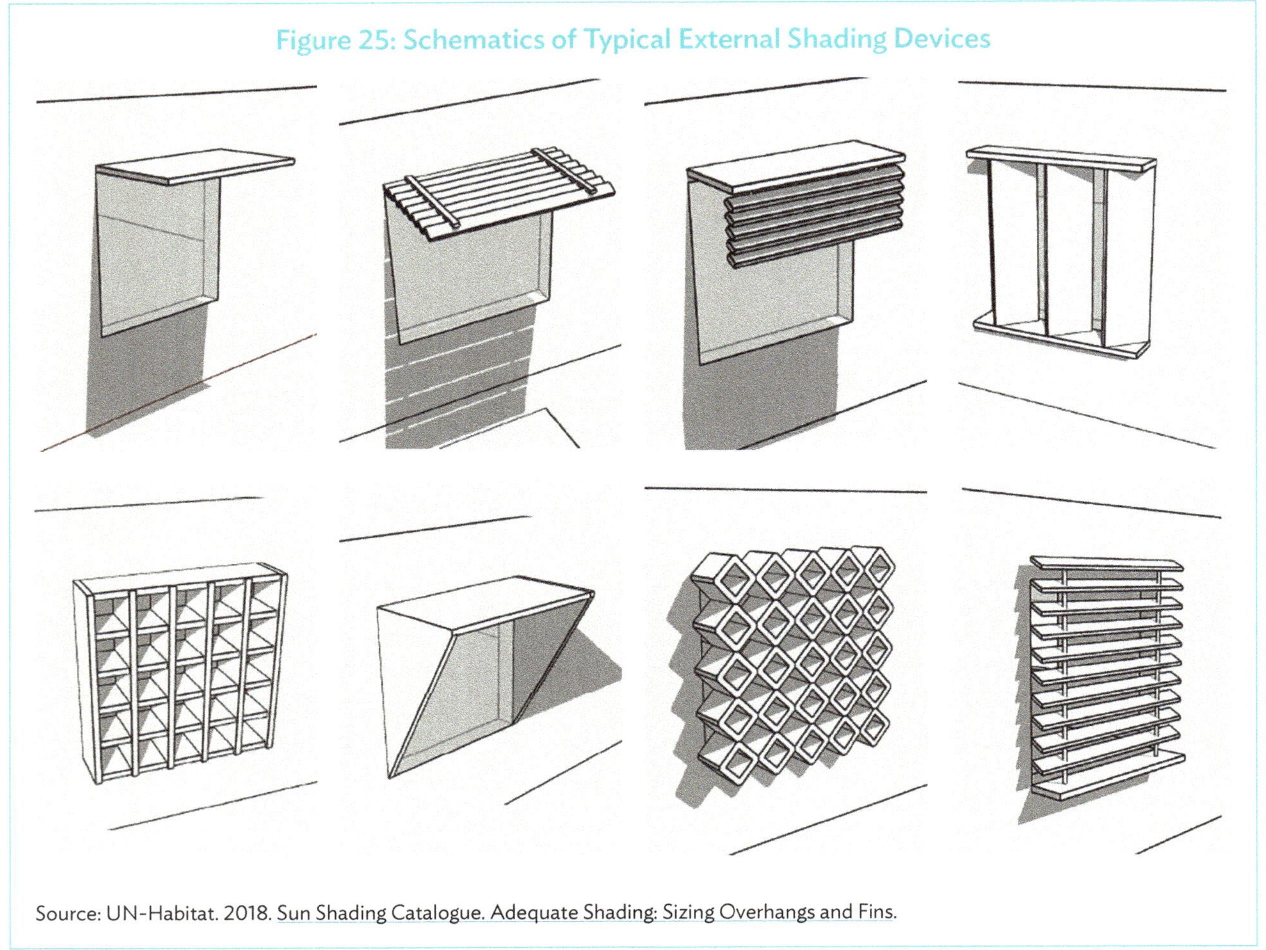

Source: UN-Habitat. 2018. Sun Shading Catalogue. Adequate Shading: Sizing Overhangs and Fins.

The shading devices illustrated are fixed ones, which have the advantages of simplicity, low cost, and low maintenance. A major disadvantage, however, is that the shading effect cannot be adjusted, and therefore the effectiveness can be limited. Movable shading devices are installed on the windows and glass facades of buildings to dynamically control the solar heat gain and the visual light transmission. Compared to fixed shading, movable shading devices have greater effectiveness against diffused radiation (BEEP 2020b). A typical example is rotating vertical fins, which are highly flexible and adjustable for daily and seasonal conditions and are most effective on east and west exposures (Figure 26). Other typical movable shading devices include rotating horizontal louvers, eggcrates with rotating horizontal louvers, shutters, solar screens, roller blinds, retractable awnings, etc. For movable external shading devices that are automatically controlled, it is desirable that manual adjustment by occupants be permitted. Also, it is important to have automatic retraction to prevent wind damage (CIBSE 2012).

Figure 26: (a) A Schematic of Rotating Vertical Fins; (b) Application of Rotating Vertical Fins at the Institute of Robotics and Mechatronics at German Aerospace Center

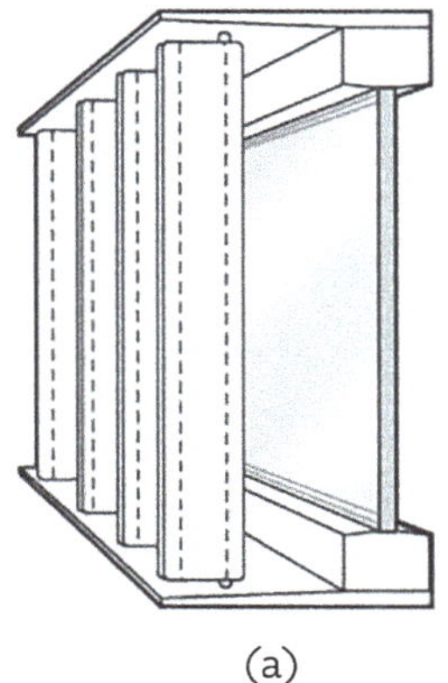

(a)

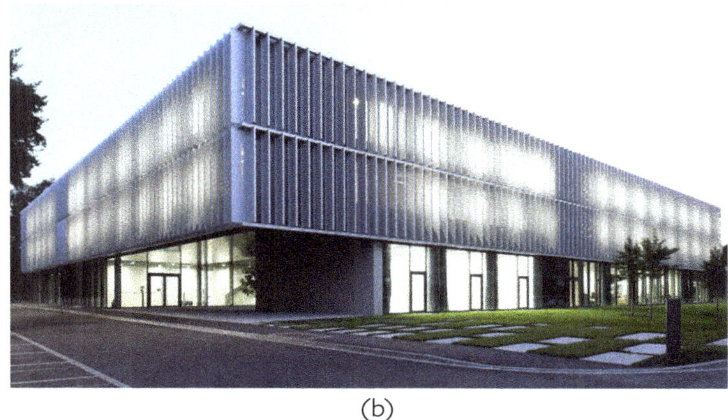

(b)

Sources: CIBSE. 2012. Guide F Energy Efficiency in Buildings; German Aerospace Center. 2021. Institute of Robotics and Mechatronics.

In designing a shading device for windows, it is important to understand the sun angle, which is critical to various aspects of design, including determining the basic building orientation and selecting appropriate shading devices. The key metrics are horizontal shadow angle (HSA) and vertical shadow angle (VSA). The HSA, which describes the performance of vertical shading devices such as vertical fins, is the angle between the solar azimuth and the normal of the windowpane, i.e., the orientation of the building facade under consideration. The VSA is required for designing horizontal shading devices such as overhangs. It is the angle between the horizontal plane of the building facade under consideration and a virtual tilted plane that contains the bottom edge of the window and the sun (Figure 27).

The latitude of a building and the orientation of its facade under consideration determine the period (both daily and annually) when shading is needed and the angle of solar radiation on the facade. A sun path diagram can be used to obtain the azimuth and altitude of the sun at each of the cut-off periods. And then the solar shading protractor can be used to determine the HSA and VSA, based on which a shading device satisfying performance specifications can be designed (Figure 28).

Figure 27: Horizontal Shadow Angle and Vertical Shadow Angle

VSA
ALT
HSA
VSA
ALT
HSA

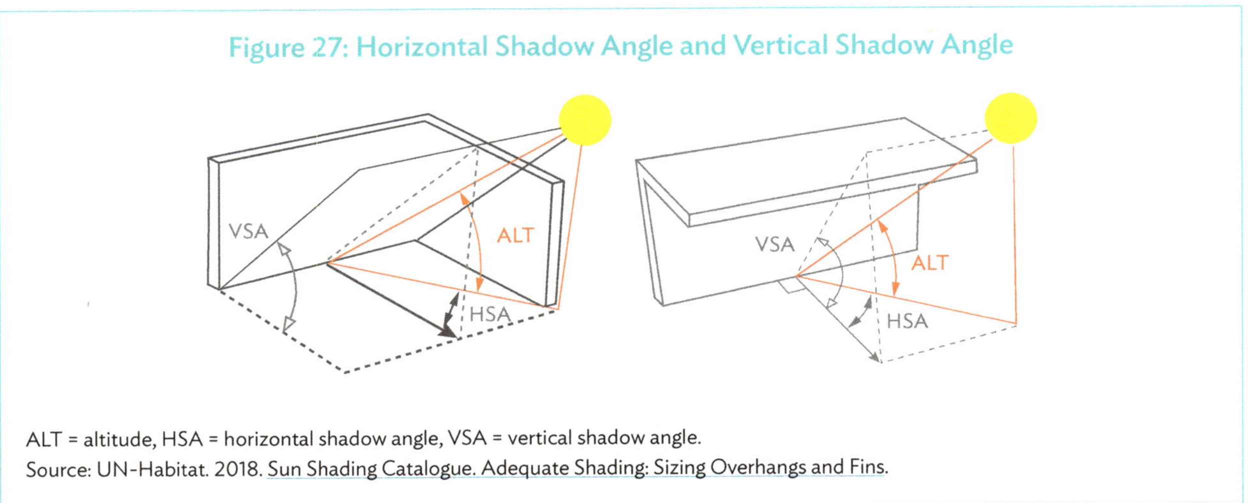

ALT = altitude, HSA = horizontal shadow angle, VSA = vertical shadow angle.
Source: UN-Habitat. 2018. Sun Shading Catalogue. Adequate Shading: Sizing Overhangs and Fins.

Figure 28: Design of External Shading Devices

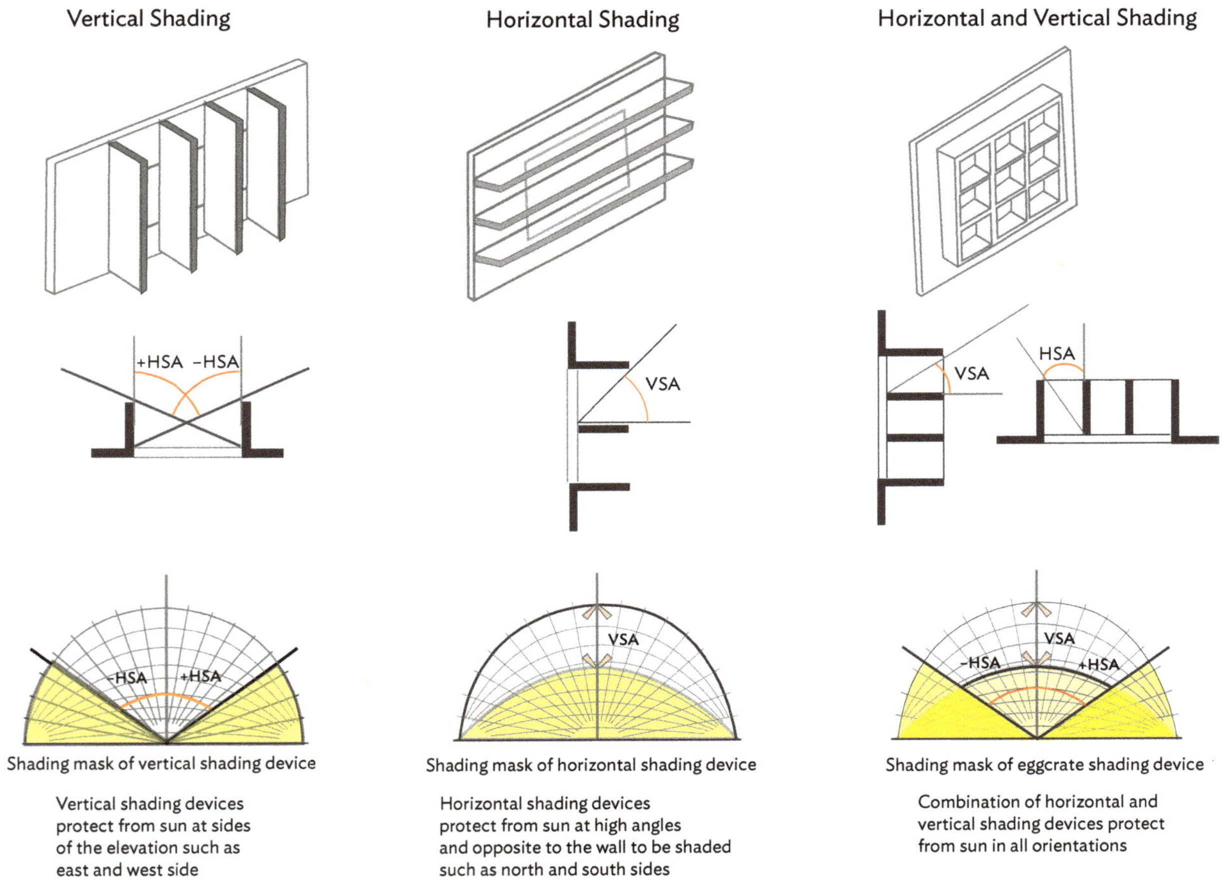

Vertical Shading
Horizontal Shading
Horizontal and Vertical Shading
+HSA -HSA
VSA
VSA
HSA
-HSA
+HSA
VSA
VSA
-HSA
+HSA
Shading mask of vertical shading device
Vertical shading devices protect from sun at sides of the elevation such as east and west side
Shading mask of horizontal shading device
Horizontal shading devices protect from sun at high angles and opposite to the wall to be shaded such as north and south sides
Shading mask of eggcrate shading device
Combination of horizontal and vertical shading devices protect from sun in all orientations

HSA = horizontal shadow angle, VSA = vertical shadow angle.
Source: NZEB Alliance. 2021h. Shading.

In general, it is recommended that the longer sides of a building be oriented north–south to minimize overall solar radiation and heat gain through the building envelope. The use of a large number of windows on the east and west facades should be avoided to the extent possible. East and west vertical fins, or eggcrates, should be angled to maximize the shading effect, which is important for buildings in tropical and subtropical climates. Exterior drop-down shades and adjustable horizontal louvers or blinds are also effective in blocking east–west sunlight. When external shading is not possible due to architectural, structural, aesthetic, or other reasons, it is necessary for the glazing to have a lower SHGC. Also, the building facade may be configured such that the windows are designed to avoid east and west orientation and thus solar heat gain (Figure 29).

Where possible, shading through landscaping and foliage should be availed of, especially for low buildings. To achieve desired shading effects, plant characteristics, such as foliage density, canopy height, and spread, need to be carefully considered (Fairconditioning 2021e). Deciduous trees or vines can be effective in screening the solar heat and glare in summer (and allowing solar penetration when passive solar gain is desired). High-branched canopy trees can be used to shade the roof, walls, and windows, whereas shrubs are appropriate for more localized shading of windows. Sometimes, using plants can also address the issue of reflected radiation from neighboring structures, water, or ground finishes (CIBSE 2012).

Other than using external shading devices, a building can be designed to provide shading for itself through different architectural forms. Typical examples include deep porches, covered verandas, recessed windows, and extended elements blended with envelope structural features (2030 PALETTE 2021a; Fairconditioning 2021e). In hot and humid climates, a good form of design is to both shed water and shade exterior walls and outdoor living spaces. In hot and dry climates, admitting cooler outdoor air through shaded outdoor living spaces or courtyards is a good strategy for natural ventilation (2030 PALETTE 2021b).

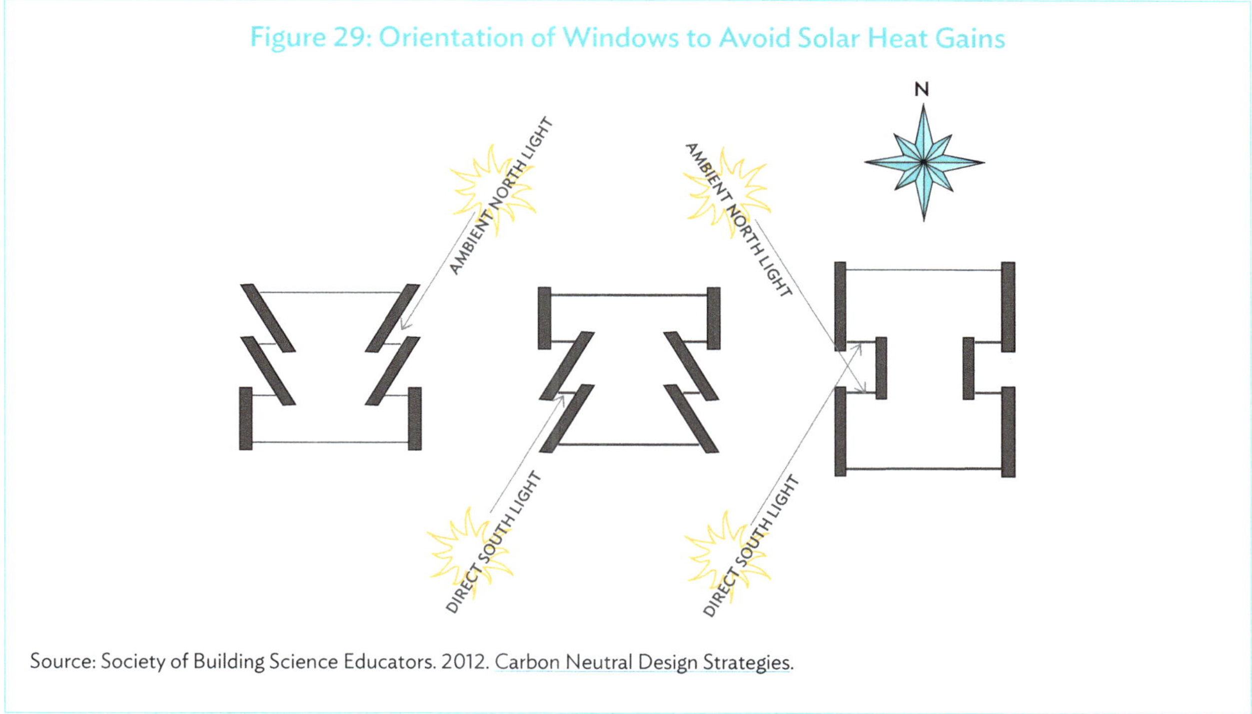

Figure 29: Orientation of Windows to Avoid Solar Heat Gains

Source: Society of Building Science Educators. 2012. Carbon Neutral Design Strategies.

Windows and Glazing

Windows comprise glass panes, structural frames, spacers, and sealants. With technological advances, a large variety of glass types, coatings, tints, and frames are available for use in window systems, providing opportunities to optimize selection and performance in different climates. Strategic decisions in building design require adequate interplay and collaboration between architects and building services engineers to appraise key factors that influence the design of the window and/or glazing system as an integral component of the building facade. Essentially, a good balance should be achieved across: (i) effective control of heat gains; (ii) providing daylight and avoiding the use of electric lighting at certain times; (iii) allowing for natural ventilation and avoiding the need for mechanical ventilation or cooling in some operating conditions; (iv) permitting some degree of occupant control over the local environment, which may reduce cooling demands; and (v) controlling glare and providing an adequate view out. For residential buildings, optimum window design and glazing specification can reduce energy consumption by 10%–50% below accepted practice in most climates. For commercial, institutional, and industrial buildings, well-designed window and/or glazing systems have the potential to reduce lighting and air-conditioning costs by 10%–40% (Whole-Building Design Guide 2016).

A window's ability to control solar heat gain and provide daylight to interior space depends mainly on its glass portion, namely glazing. Other than that, the type, shape, location, and functionality of a window are key factors in the utilization of natural ventilation, which reduces the need for mechanical ventilation or cooling. For example, long horizontal strip windows can ventilate a space more evenly, whereas tall windows with openings at the top and bottom can use induce local stack ventilation, with cool air supplied at the bottom and warm air exhausted at the top (Fairconditioning 2021f). Horizontal pivoting windows can produce effective ventilation because large open areas are created at a separation equivalent to the window height. By contrast, vertical pivoting windows are less efficient because the open area is uniformly distributed through the height of the window (CIBSE 2012). Coordination and integration with solar control strategies, particularly the use of shading devices, are important for the ventilation capacity and characteristics of windows. Further, the placement, orientation, and size of windows should be an integral part of the design of the built form and facade of a building.

Glazing plays a key role in determining a window's suitability in a specific climate and energy efficiency characteristics. A good glazing technology and its appropriate application may be more advantageous than the use of external shading devices. A good glazing technology can reduce solar heat gain from both direct and diffuse solar radiation, whereas external shading devices mainly block direct radiation with limited influence on heat gain from diffuse radiation (BSEEP Malaysia 2017a). Also, the control of solar heat gain is often coupled with the control of visible light transmittance, which, if well-designed and operated, may promote daylight harvesting. Apart from reducing the use of electric lighting, this may also generate the potential benefits of reducing the need for mechanical cooling and even allowing for a smaller air-conditioning system. As shown in Figure 31, the primary energy-related properties of glazing can be measured by the following metrics (Efficient Windows Collaborative 2021a; Efficient Windows Collaborative 2011; Efficient Windows Collaborative 2021b; Bureau of Energy Efficiency 2009; Whole-Building Design Guide 2016).

Solar heat gain coefficient. SHGC is the fraction of solar radiation admitted through a window or skylight, both directly transmitted, and absorbed and subsequently released inward. SHGC is expressed as a number between 0 and 1. The lower the SHGC value, the less solar heat is transmitted and the greater the shading ability. Key factors influencing SHGC include type of glass and number of panes, tints and coatings on the glazing, gas fill between glass panes, and solar shading (Bureau of Energy Efficiency 2013). SHGC can be expressed in terms of the glass alone or can refer to the entire window assembly. Whether a higher or lower SHGC is desirable depends on the climate, orientation, shading conditions, and other factors. In tropical and subtropical climates where cooling demand is high, the general preference is to use glazing with low SHGC values.

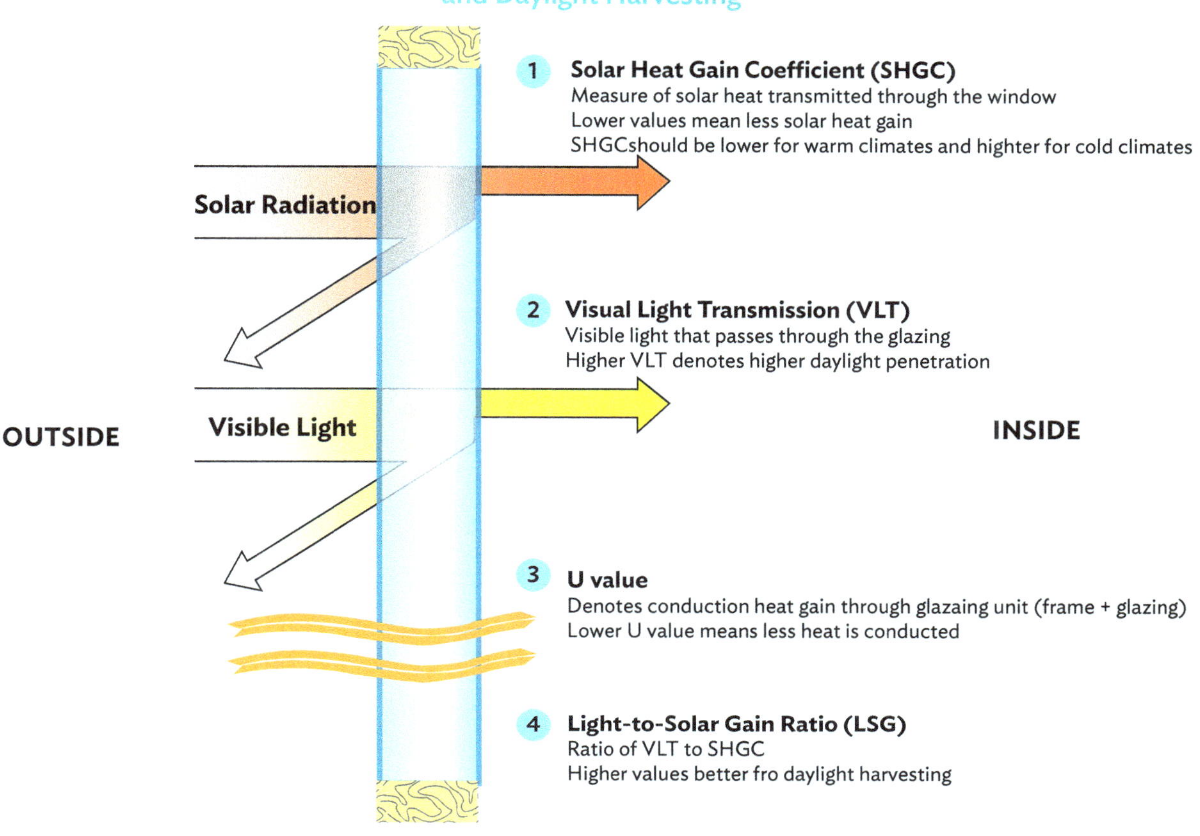

Source: NZEB Alliance.2021c. Fenestration.

Visible light transmittance, also known as visible transmittance. Visible light transmittance (VLT) is an optical property that indicates the portion of incoming visible light transmitted through a window. It is a whole window rating and includes the impact of the window frame that does not transmit any visible light. While the theoretical value of VLT ranges from 0 to 1, most values in practice are between 0.3 and 0.7. The higher the value, the more light is transmitted, suggesting a greater opportunity for harvesting daylight. Key factors influencing VLT include the color of the glass, tints and coatings on the glazing, and the number of glass panes (Bureau of Energy Efficiency 2013). VLT affects building energy use by providing the opportunity to harvest daylight and reduce electric lighting as an integral part of internal heat gain. Thus, it indirectly helps reduce cooling loads.

Light-to-solar gain ratio. The light-to-solar gain (LSG) ratio is a measure of the ability of glazing to provide light without excessive solar heat gain. It is the ratio between the VLT of a glass and its SHGC. The higher the LSG is, the better it is for daylight harvesting.

Insulation value. The insulation value (U-value) indicates the rate of heat flow due to conduction through a window as a result of a temperature difference between the inside and outside. Measuring non-solar heat gain or loss, the U-value may be expressed for the glazing alone or for the entire window assembly, which includes the effect of the frame and the spacer materials. Thus, key factors influencing U-value include the coatings on the glazing, the size of the air gap between glass panes, gas fill between glass panes, and window frame construction

(Bureau of Energy Efficiency 2013). The lower the U-value, the greater a glazing or window's resistance to heat flow, and the better the insulation value. U-value can be in the unit of British thermal units (Btu)/hour (hr)·feet (ft)2·°F (US) or Watt/square meter (W)/m²)·°K (European metric). The typical U-value ranges from a high of 1.25 for single-glazing in an aluminum frame to a low of 0.15 for low-emissivity-coated triple-glazing in an insulated frame. In a tropical climate, U-value has a relatively smaller influence on building energy consumption as compared to the VLT, SHGC, and LSG properties of glazing. For a tropical climate, it is recommended to select glazing based on VLT, SHGC, and LSG values, with the U-value as a secondary criterion of the selection process. This is because a good glazing selection based on VLT, SHGC, and LSG would typically come with an improved U-value, except for normal, single-tinted glazing. Thus, in a tropical climate, it is not economically justifiable to select glazing based solely on U-value (BSEEP Malaysia 2017a).

The properties of a given glass can be altered by tinting or by applying various coatings or films to it. Coatings, usually in the form of metal oxides, can be applied to glass during production. For example, reflective coating increases glass surface reflectivity, which can reduce solar heat gain substantially and control glare. This makes it suitable for large windows in hot climates. However, this is at the cost of a substantial reduction in visible transmittance, which disadvantages daylight harvesting (Efficient Windows Collaborative 2011). An increasingly popular type of coating is low-emittance (low-E) coatings, which are microscopically thin, virtually invisible, metal, or metallic oxide layers deposited on a window or skylight glazing surface to reduce radiant heat transfer through the glazing. Low-E coatings have been designed to allow for high solar gain, moderate solar gain, or low solar gain. The solar reflectance of low-E coatings can be manipulated to be spectrally selective so that the desirable wavelengths of the solar spectrum, such as the visible light spectrum, are transmitted and the others, such as short-wave and long-wave infrared as well as ultraviolet radiation, are reflected (US DOE Office of Energy Saver 2021c; Efficient Windows Collaborative 2021c; Efficient Windows Collaborative 2011).

There are two main categories of low-E coatings: sputtered low-E and pyrolytic low-E. Sputtered low-E coatings, also often referred to as soft coats, are designed to be placed inside a double-glazed insulated glass unit because they cannot be exposed and must be protected from contact and humidity. Pyrolytic low-E coatings are hard and durable enough to be exposed internally and are therefore also often referred to as hard coats or room-side coats. They can either be applied to single-pane glass or used in a double-glazed insulated glass unit. Comparatively, sputtered low-E coatings generally perform better than pyrolytic with respect to solar control and the reduction of heat transfer (Glassworks 2018; Efficient Windows Collaborative 2021a).

This spectral selectivity can also be achieved with high-performance tints. Tints are generally the result of colorants added to the glass during production. Some tints are also produced by adhering colored films to the glass following production (Whole-Building Design Guide 2016). Tinted glass is used to reduce glare from the outdoors and solar heat transmission through the glass. Tinted glass retains its transparency from the inside, although the brightness of the outward view is reduced, and the color is changed. As traditional bronze and gray tinted glass reduces the amount of daylight entering the room, a solution to applications where daylighting is desirable is to use clear glass with spectrally selective low-E coatings (Figures 31–33). Alternatively, high-performance tints with a relatively high VLT can be used in conjunction with low-E coatings (Efficient Windows Collaborative 2011).

In addition to the insulating capabilities of double glazing and low-E coatings, gas fills and additional glazing layers can further improve the insulating value of a glass unit. Triple-glazed and quadruple-glazed windows are designed to have their inner layers consisting either of glass or of suspended plastic films. Triple-glazed windows may have two glazing layers and one suspended plastic films, whereas quadruple-glazed ones may have two glazing layers and two suspended plastic film (Efficient Windows Collaborative 2021e). These inner layers bring down the U-value of the unit by dividing the inner air space into two or three chambers, which can be filled with insulating gas, such as 0.5-inch argon fill or 0.25-inch krypton fill. In addition to the U-value, solar heat gain can be effectively reduced. However, as a side effect of having these additional layers, there is some reduction in visible light transmission, as shown in Figure 31.

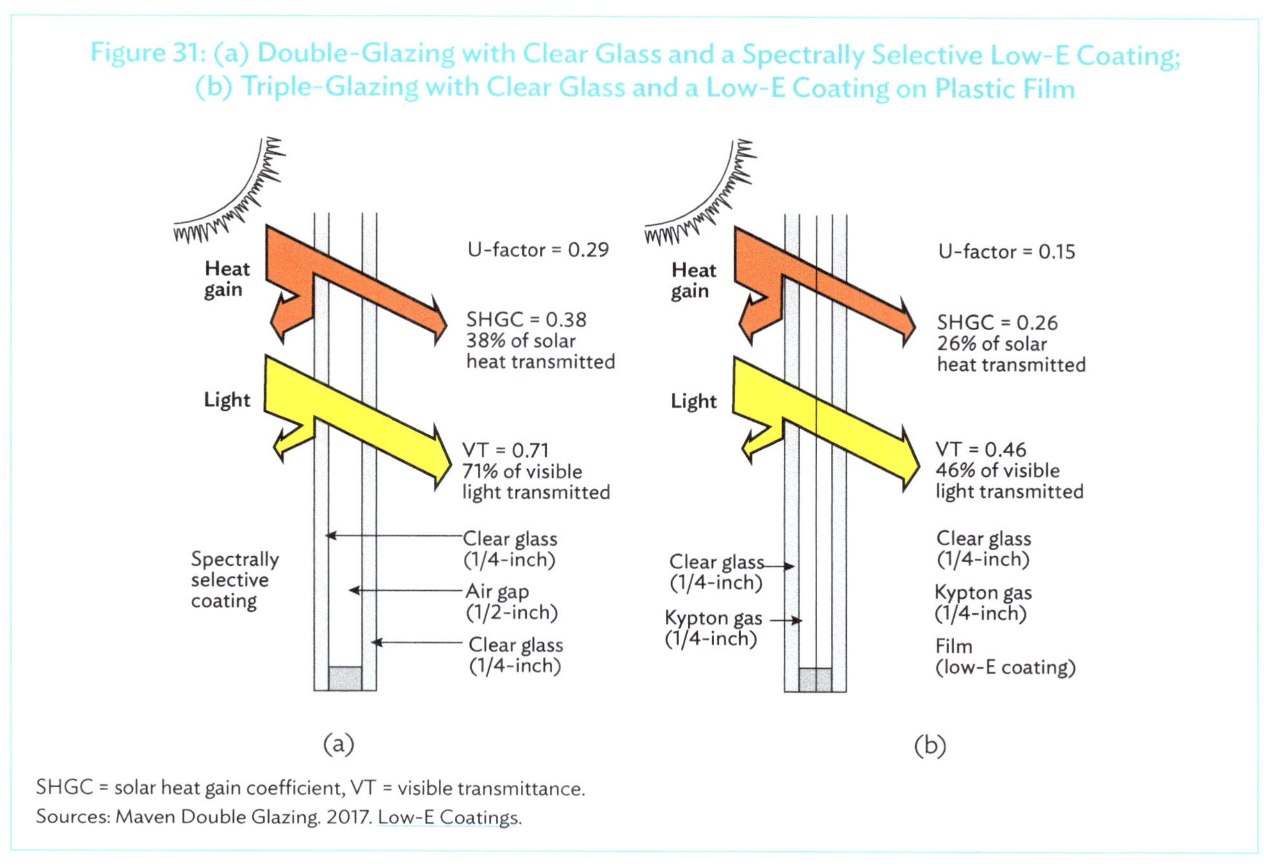

SHGC = solar heat gain coefficient, VT = visible transmittance.
Sources: Maven Double Glazing. 2017. Low-E Coatings.

In general, from the perspective of reducing solar heat gain, single-glazing with clear or tinted glass is not recommended due to its high SHGC. For cooling-dominated places such as ADB members in tropical and subtropical climates, the ideal type of glazing is double-glazed low-solar-gain low-E glass with argon gas fill (with or without a room-side low-E coating). For climates with both cooling and heating loads, such as ADB members in East Asia and Central West Asia, the ideal types of glazing include double-glazed moderate-solar-gain low-E glass with argon gas fill (with or without a room-side low-E coating), and triple-glazed low-solar-gain low-E glass. All these types provide reasonably high VLT values, which are good for daylight harvesting. Table 5 summarizes the performance characteristics of these types. The values are indicative of the glazing alone (center-of-glass). The choice of frame can make a substantial difference in performance.

Table 5: Ideal Glazing Types for Buildings in Climates That Mainly Require Cooling or Climates with Both Cooling and Heating Loads

Climate Zone	Type of Glazing	U-Value (Btu/hr·ft²·°F)	SHGC	VLT
Cooling-dominated climates	Double-glazed low-solar-gain low-E glass	0.24	26%	64%
	Double-glazed low-solar-gain low-E glass with a room side (fourth surface) low-E coating	0.20	27%	63%
Climates with both cooling and heating loads	Double-glazed medium-solar-gain low-E glass	0.25	42%	72%
	Double-glazed medium-solar-gain low-E glass with a room side (fourth surface) low-E coating	0.20	41%	70%
	Triple-glazed low-solar-gain low-E glass	0.15	24%	51%

Btu = British thermal unit, hr = hour, ft = foot, F = Fahrenheit, SHGC = solar heat gain coefficient, VT = visible transmittance.

Sources: Efficient Windows Collaborative. 2021c. Window Technologies: Glazing Types - Double Low-E Glazing; Efficient Windows Collaborative. 2021d. Window Technologies: Glazing Types - Double Low-E Glazing with Roomside (4th surface) Low-E; Efficient Windows Collaborative. 2021e. Window Technologies: Glazing Types - Triple Low-E Glazing.

In addition to these properties, heat gain or loss through a window caused by air leakage (infiltration and exfiltration) around the frame and sash and through gaps in movable window parts can have a considerable impact on building energy use. This effect is measured by the amount of air that passes through a unit area of window under given pressure conditions. Infiltration varies slightly with wind-driven and temperature-driven pressure changes. In tropical and subtropical climates, air leakage may lead to an increased cooling load. When outdoor humidity is high, air leakage may have an adverse impact on moisture mitigation and indoor air quality, thus causing greater needs for ventilation and dehumidification. All these will translate into more energy consumption for the air-conditioning system. Limiting air leakage is key to improving energy efficiency. Making windows and other components of the building envelope more airtight can save 5%–40% of cooling or heating energy (IEA 2013).

At the whole-building level, an important variable to be considered at the design stage is the window-to-wall ratio, which is also known as the glazing ratio. This ratio plays a key role in determining facade thermal properties and building energy performance because heat gain and loss through windows can be as much as 20 to 30 times that through walls (Fairconditioning 2021b). For cooling-dominated climates, a higher ratio implies more heat gain and, consequently, a higher cooling load and energy consumption of the air-conditioning system. Meanwhile, the high ratio may bring the benefits of better natural ventilation and daylight harvesting, which collectively contribute to lowering building energy consumption. If, however, a lower ratio is applied, the building is exposed to less solar radiation and heat gain. This is in favor of reducing cooling loads but goes against daylight harvesting and natural ventilation which may be desired. A lower ratio may also reduce the quality of the view. Either way, there are trade-offs, and therefore some compromises must be made. Moreover, the architecturally aesthetic value of a building often carries a substantial weight in deciding a building's facade design, especially in the case of commercial buildings. Therefore, the impact of the window-to-wall ratio needs to be carefully examined during the design stage to achieve a reasonable balance across these performance aspects. As a general recommendation, buildings in climates that are either cooling-dominated or have significant cooling demand during summer should avoid having a very high window-to-wall ratio. Usually, residential buildings have much lower window-to-wall ratios than nonresidential buildings, such as office buildings, commercial buildings, etc.

For example, in India, design guidelines for energy-efficient multistory residential buildings recommend that the window-to-wall ratio should be 10%–15% in bedrooms and 30% in living rooms for residential buildings in composite and hot-dry climates (Bureau of Energy Efficiency 2014), and 10% in bedrooms and 20% in living rooms for residential buildings in warm-humid climates (Bureau of Energy Efficiency 2016). In the PRC, the design standard for energy efficiency of residential buildings in hot-summer-and-warm-winter zones requires the window-to-wall ratio of a single facade to be no more than 0.4 for south–north orientation and no more than 0.3 for east–west orientation (Ministry of Housing and Urban–Rural Development 2013). For nonresidential buildings such as office buildings and commercial buildings in the PRC, the applicable design standard imposes specific requirements on the window-to-wall ratio and energy-properties of glazing (Ministry of Housing and Urban–Rural Development 2015), as summarized in Table 6. In the Philippine Green Building Code, the window-to-wall ratio ranges from 10% to 90%, with each ratio value having a maximum SHGC and a minimum VLT of glazing. The higher the ratio, the lower the required SHGC (Department of Public Works and Highways 2015). IFC's greening building certification system, EDGE, sets a baseline window-to-wall ratio of 30% for residential buildings and 55% for nonresidential buildings, based on a study of the facades of building typologies across various regions (IFC 2019b).

Table 6: Window-to-Wall Ratio, U-value, and Solar Heat Gain Coefficient Requirements for Public Building in the People's Republic of China

Climate zone	Window-to-wall ratio (WWR)	U-value (W/m²·K)	SHGC
Cold (typical cities: Beijing, Tianjin)	0.40 < WWR ≤ 0.70	≤2.7	–
	WWR > 0.70	≤2.4	
Hot summer and cold winter (typical cities: Shanghai, Nanjing)	0.40 < WWR ≤ 0.70	≤3.0	≤0.44
	WWR > 0.70	≤2.6	
Hot summer and warm winter (typical cities: Guangzhou, Shenzhen)	0.40 < WWR ≤ 0.70	≤4.0	≤0.44
	WWR > 0.70	≤3.0	

SHGC = solar heat gain coefficient, Watt/square meter (W)/m².°K (European metric).
Source: Ministry of Housing and Urban-Rural Development. 2015. *Design Standard for Energy Efficiency of Public Buildings (GB 50189-2015)*.

Cool Roof

The roof of a building may receive up to four times more solar radiation per m² than the walls during summer (Bureau of Energy Efficiency 2014). The exposed roof surface can become very hot, with dark surfaces reaching temperatures as high as 80°C (Global Cool Cities Alliance 2012; Bureau of Energy Efficiency 2017). The excessive solar heat gains through the roof result in a high indoor air temperature, create thermal discomfort in the occupied space, and increase cooling demand to be met by the air-conditioning system or equipment.

A cool roof is one that strongly reflects solar energy and also cools itself by efficiently emitting some of the absorbed heat back into the atmosphere (Cool Roof Rating Council 2021). Under the same conditions, a cool roof could stay cooler by as much as 28°C to 36°C than a conventional dark roof (Global Cool Cities Alliance 2012; US DOE Office of Energy Saver 2021a). Accordingly, the amount of heat conducted into the building underneath the roof can be reduced substantially. As a result, the indoor environment can be kept cooler and more comfortable than otherwise. Meanwhile, for an air-conditioned building, a cooler indoor environment helps reduce the need for air-conditioning, save energy, and potentially provide an opportunity to downsize a new or replacement air-conditioning system or equipment. Beyond individual buildings, cool roofs contribute to alleviating the heat island effect in dense urban areas and also reducing peak electricity demand, which can help prevent power outages (Global Cool Cities Alliance 2012; NZEB Alliance 2021b; US DOE Office of Energy Saver 2021a; Cool Roof Rating Council 2021; Green Building Alliance 2021).

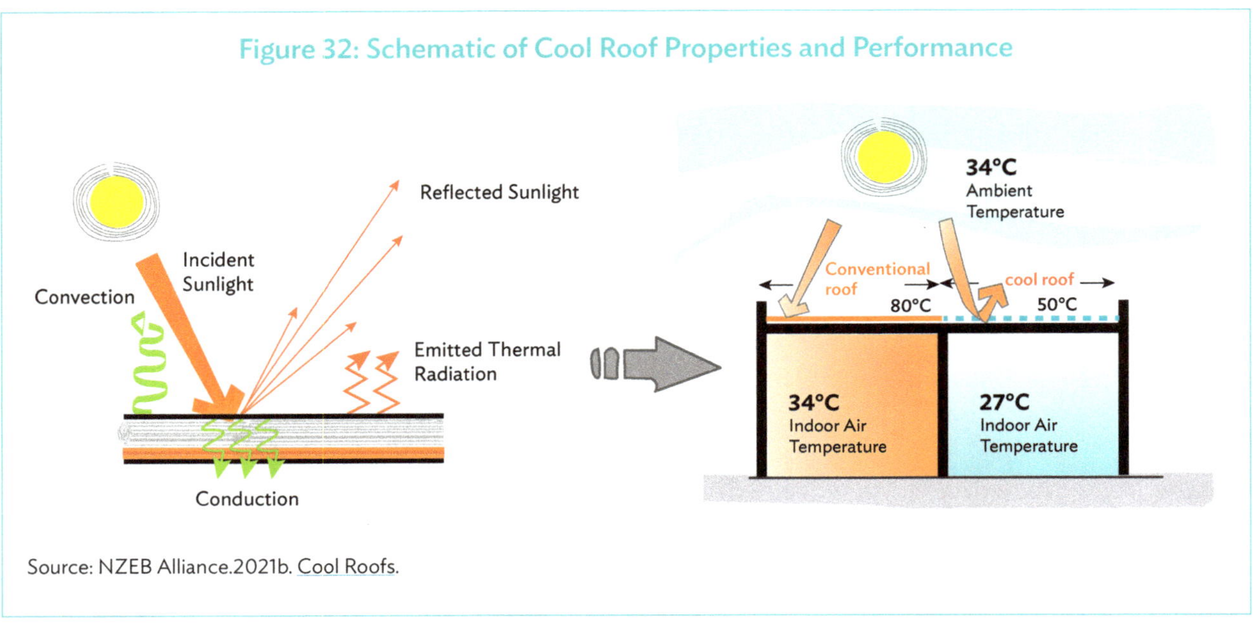

Source: NZEB Alliance.2021b. Cool Roofs.

The term "cool roof" refers to the outer layer or exterior surface of a roof, which acts as the key reflective surface. The two primary thermal properties characterizing a roof surface are solar reflectance and thermal emittance. Solar reflectance, also known as reflectivity, is the fraction of solar energy that is reflected by a roof's surface back to the sky, expressed with a number between 0 and 1. White surfaces have the highest solar reflectivity, while black surfaces have the lowest. The term "albedo" is often used interchangeably with solar reflectance. Surfaces with a high albedo or reflectance stay much cooler than those with a low albedo or reflectance. Thermal emittance, also known as infrared emittance or emissivity, is the relative ability of a roof surface material to radiate absorbed heat, expressed with a number between 0 and 1. High thermal emittance helps a surface cool by radiating heat to its surroundings. Nearly all nonmetallic surfaces have high thermal emittance, usually between 0.80 and 0.95. Uncoated metal has low thermal emittance. Radiant energy incident on a roof surface that is not reflected is absorbed by it. This absorbed heat is eventually radiated by the roof or transferred through conduction. Compared with traditional roofs, a cool roof can reduce the solar heat gain of a building by first reflecting a substantial amount of incoming radiation and then quickly reemitting the absorbed portion. As a result, the cool roof stays cooler.

Another important indicator to evaluate a roof's ability to shield the building beneath it from solar radiation, namely, the "coolness" of the roof, is the solar reflectance index (SRI), which incorporates both solar reflectance and thermal emittance in a single value. Developed by the Lawrence Berkeley National Laboratory (LBNL), SRI quantifies how hot a surface would get relative to standard black and standard white surfaces. It is defined such that a standard black (reflectance 0.05, emittance 0.90) is 0 and a standard white (reflectance 0.80, emittance 0.90) is 100. By internalizing the trade-off between reflectance and emittance into a single value, SRI helps simplify the comparison of roof products.

Cool roof options and materials available to a building depend mainly on the type of building and its roof structure. Broadly, roofs can be categorized into low-sloped and steep-sloped. The term "low-sloped roofs" typically refers to roofs with a profile of no more than 5 centimeters (cm) vertically over 30 cm horizontally, or a pitch no more than 2:12. Most commercial and high-rise residential buildings have low-sloped or even flat roofs with little visibility from the ground. This makes it relative easier and less costly to apply cool roofs, either at new construction or during routine retrofits or scheduled replacement of roofing covering. In contrast, roofs with a pitch higher than 2:12 are considered "steep-sloped," as in the case of many low residential buildings. As steep-sloped roofs can be more easily seen from the ground, often it is the aesthetic preference or concern, rather than the surface solar reflectance

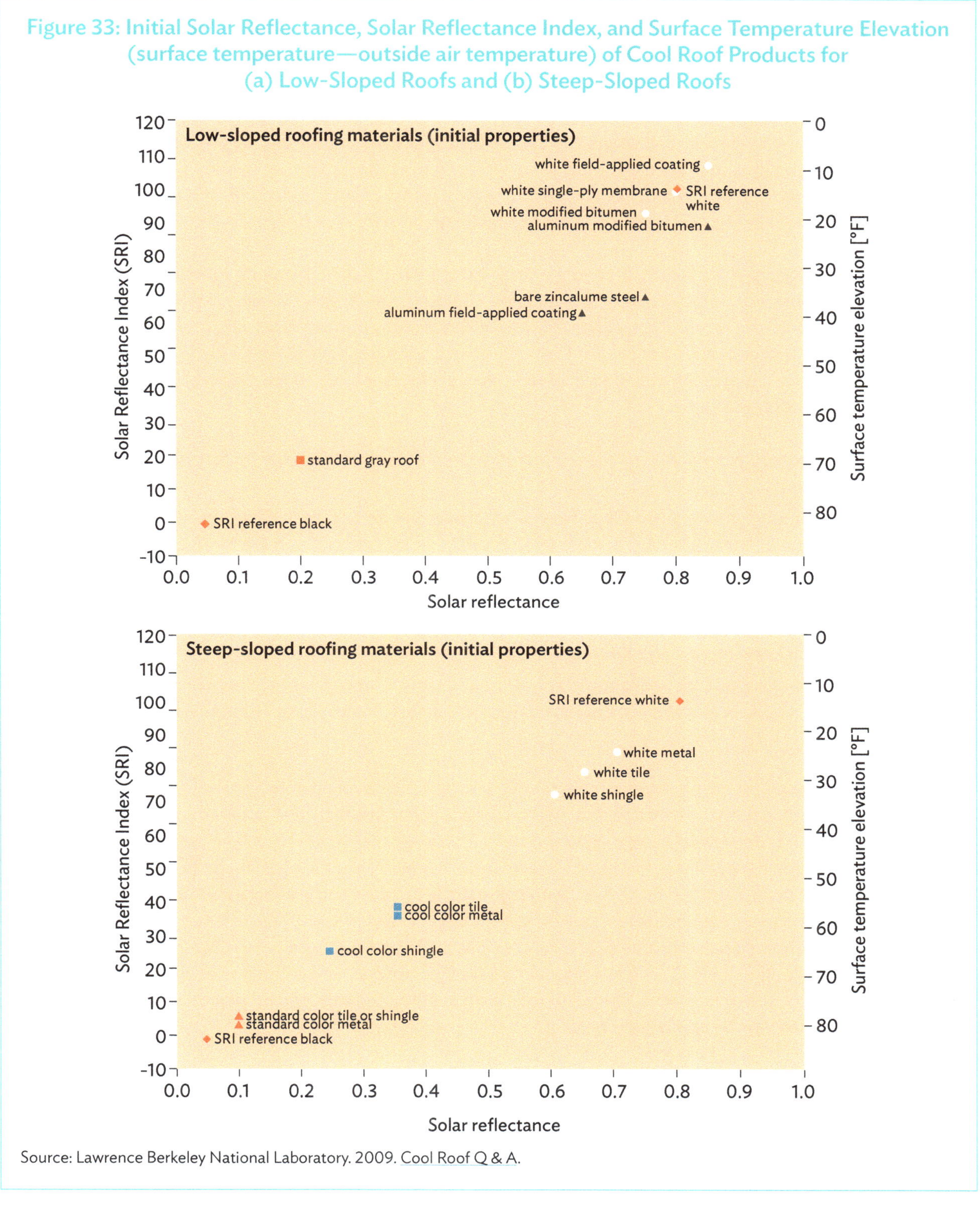

Figure 33: Initial Solar Reflectance, Solar Reflectance Index, and Surface Temperature Elevation (surface temperature—outside air temperature) of Cool Roof Products for (a) Low-Sloped Roofs and (b) Steep-Sloped Roofs

Source: Lawrence Berkeley National Laboratory. 2009. Cool Roof Q & A.

of a roofing product that plays the decisive role in selecting the cool roof option. This explains why white roofing products do not gain popularity for residential roofs. There are, however, a wide range of roof materials using alternative cool colors, such as green, red, and gray, that can strongly reflect the invisible near-infrared solar radiation and make the roof cooler than conventional roofs but are not as cool as white roofing products.

Figures 34–36 show the initial solar reflectance, SRI, and surface temperature elevation (surface temperature minus outside air temperature on a summer afternoon) of some cool roof products for low-sloped roofs and steep-sloped roofs. This is followed by a more detailed introduction of various types of cool roofs (APEC Energy Working Group 2012; US DOE Office of Energy Saver 2021a; Designing Buildings 2020; Green Building Alliance 2021).

Coated roofs. A coated roof is coated with a paint-like finish to help enhance the roof's adhesion, durability, and longevity. Roof coatings can be divided into two categories: field-applied and factory-applied. Field-applied coatings are applied directly onto the roof surface, either on a new roof assembly or over an existing roof surface. Factory-applied coatings are applied at the factory prior to distribution. Cool coatings are best for low-sloped roofs and can be added to a variety of surfaces, including asphalt cap sheet, gravel, metal, and other single-ply materials. Typical cool roof reflective coatings include:

- White roof coatings, which are opaque and reflective, consist of polymeric materials and some types of white pigment.
- Pigmented coatings, which are made from similar materials and come in darker shades such as red, green, and blue. These coatings are less efficient than white coatings and are more commonly used for aesthetic purposes in residential buildings.
- Aluminum roof coatings, which are made from resin containing flakes of aluminum leaf. These coatings provide the visual effect of an aluminum sheet, masking the material layer underneath while providing effective solar reflectance and thermal emittance.
- Elastomeric roof coatings, which are acrylic-based materials, help protect roofs from extreme weather conditions. These coatings can be applied directly to an existing roof, depending on the condition of its surface.

Single-ply membranes. Single-ply membranes are prefabricated sheets rolled onto a low-sloped or steep-sloped roof. They can be firmly set on the roof and attached with mechanical fasteners and adhered with chemical adhesives, or they can be loosely laid and weighted down with ballast or pavers. These membranes serve to protect the structure from extreme weather and can provide insulation as well. They are often made from felt, fiberglass, polyester, or other materials and can be covered with pigments to increase solar reflectance. Generally, there are four main types of single-ply membrane for cool roofs, including (i) ethylene propylene diene monomer (EPDM) rubber; (ii) chlorosulfonated polyethylene (CSPE), also known as Hypalon; (iii) polyvinyl chloride (PVC) polymers; and (iv) thermoplastic olefin (TPO). The first and second are thermosets that cannot be hot air welded and must be sealed using tape or contact cement. The third and fourth are thermoplastics consisting of compounded plastic polymers, to which heat can be applied for melding.

Shingles and tiles. Roof shingles and tiles are roofing products based on the same concept, which is to put together individual elements to form a roof. These elements are normally flat rectangular shapes laid in rows without the side edges overlapping. With available products made of various materials, shingles and tiles are commonly used for residential buildings and increasingly for commercial buildings. A typical example is asphalt shingle that is surfaced with light-colored or cool-colored granules, if a darker color is preferred. It is also possible to field-coat previously installed asphalt shingle roofs to make them cool. Other shingle-based products include polymer shingles that are factory-colored with light-colored or cool-colored pigments and wood singles and shakes, which are naturally cool-colored materials. As for using tiles as cool roofs, concrete tiles can be integrally colored with pigments, surface-colored with a slurry coating, which mixes pigment with white cement, or with a pigmented polymer coating.

Unglazed clay tiles in a natural cool color or clay tiles with a factory-fired light- or cool-colored glaze can provide the desired coolness. Alternatively, glazed clay tiles can also be retrofitted with a field-applied light- or cool-colored polymer coating. An advantage of roofing products made of shingles and tiles is their heavy thermal mass owing to their dense and earthen composition. This may yield energy savings on top of what can be achieved by their solar reflectance and thermal emittance properties.

Metal roofs. Cool metal roofing is a family of energy-efficient roofing products comprising unpainted metal, factory-painted metal, and metal that is factory-surfaced with mineral granules. It is available in a wide variety of finishes, colors, textures, and profiles to suit different built forms and structures. Mill-finish metal roofing has high solar reflectance, but poor thermal emittance. Metal roofing with oven-cured, pre-painted organic coatings with special cool pigmentation can offer high solar reflectance and high infrared emittance even with darker colors. Emissivity as high as 90% can be achieved for painted and granular-coated roofing. Metal roofing can be broadly categorized into structural and nonstructural. Structural metal roofing is attached directly to purlins or lathe boards without requiring underneath support. It can be further categorized into low-slope and steep-slope products. Low-slope structural metal roofing consists of interlocking panels or overlapping panels, which can have a painted, mill-finish, or clear acrylic finish. Steep-slope structural metal roofing is available in both vertical and horizontal profiles. Nonstructural metal roof panels require a solid substrate beneath them, typically plywood, oriented strand board, or a metal roof deck. It has a variety of styles, including vertical standing seams and horizontal panels, which are often shaped to resemble the look of asphalt shingles, wood shakes, slate, and tiles, or to fit unique curvatures.

Built-up roofs. A built-up roofing system is made from multiple layers that commonly include a base sheet, fabric reinforcement layers, and a protective surface layer (Figure 34). Cooling options for built-up roofs can vary by building type. One method is to embed thermally reflective materials into asphalt or coal tar. Another method is to cap the built-up roof with mineral-surfaced sheets consisting of reflective mineral granules or applied reflective coatings. Built-up roofs mostly appear light gray or tan and have a solar reflectance of 0.15 to 0.25.

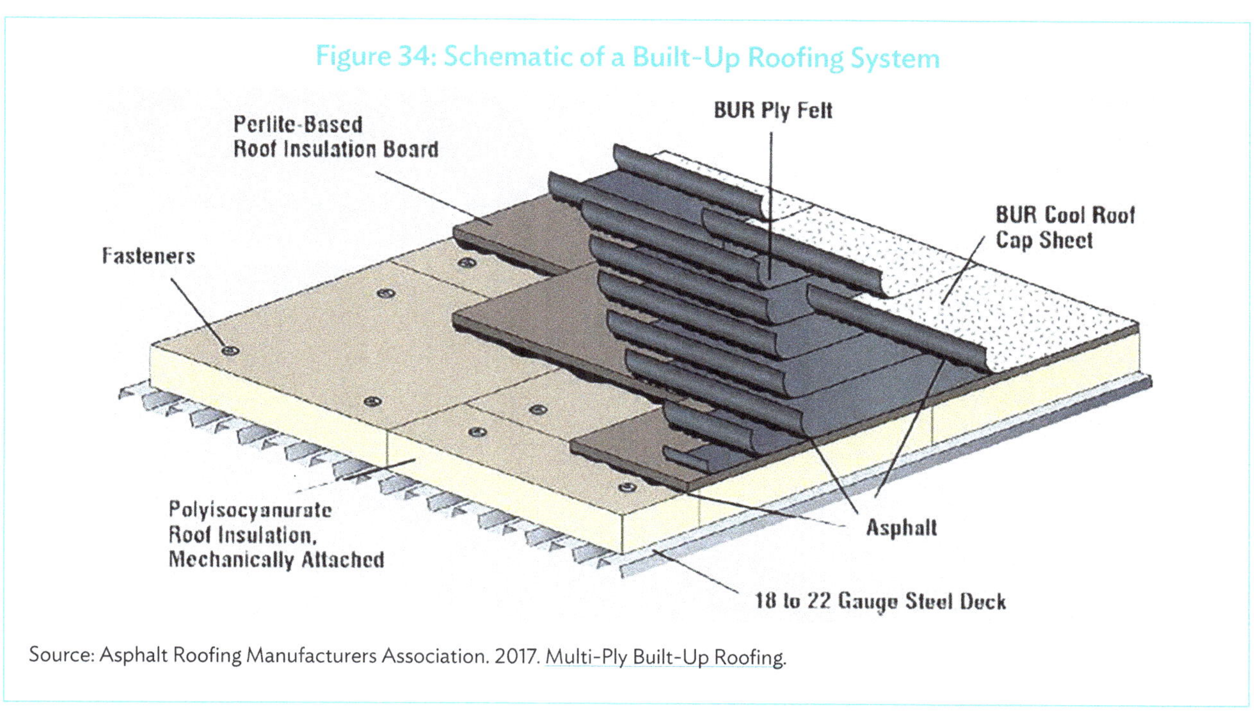

Figure 34: Schematic of a Built-Up Roofing System

Source: Asphalt Roofing Manufacturers Association. 2017. Multi-Ply Built-Up Roofing.

Modified bitumen. Modified bitumen is bitumen (asphalt or tar) modified with plastic, layered with reinforcing fabrics, and then topped with a surfacing material. The radiative properties of modified bitumen are determined largely by the surfacing material. So, the cooling performance of modified bitumen roofing is realized using a cap sheet with a factory-applied reflective mineral surface or reflective coating. Modified bitumen is similar to built-up roofing but with a higher degree of elasticity. It is suitable for both warm and cold temperatures. Modified bitumen roofing can be installed in four main ways, i.e., torch-applied, hot-mopped, cold-applied, or with self-adhesives. The tough and tenacious properties of the modified bitumen blend, coupled with high-tensile polyester and/or fiberglass reinforcements, provide strong resistance to foot traffic, punctures, dropped tools, hailstorms, and other potential damages. This makes these products ideal for buildings with high-traffic roofs, such as those with mechanical systems installed on the rooftop. Modified bitumen sheet membranes can also be used to surface a built-up roof to produce a hybrid roof.

Foam roofs. A foam roof, topped with a foam-like substance for insulation purposes, is an affordable cool roof technique with a long history of applications. The foam is generally made from two liquid chemicals that mix to react and expand to form a solid, flexible, and lightweight piece that adheres to the roof. Foam roofs can be field-applied or factory-applied. Field-applied foam roofs, similar to field-applied coatings, are sprayed on in liquid form and harden as they set on top of the roof. Factory-applied foam roofs are formed into rigid panels and coated with a reflective coating.

In general, cool roofs are cost-competitive with conventional roofing options, having a short payback period based on the benefit of energy savings (Global Cool Cities Alliance 2012; Natural Resources Defense Council 2020). Usually, in the case of new buildings or existing buildings for which the roof surfaces need to be replaced anyway, the costs of cool roof options would be similar to or only slightly more than similar traditional non-cool alternatives. This means the cost premiums for cool roofs are limited or, at times, none (Bureau of Energy Efficiency 2017; Global Cool Cities Alliance 2012). For a traditional roof that is in good condition and not due to be retrofitted or replaced, converting it into a cool roof would incur a higher incremental cost than converting a roof scheduled to be re-roofed anyway (US DOE Office of Energy Saver 2021a). The costs of conversion depend on the existing roof's surface and condition. During a cool roof's service lifetime, the recurring maintenance cost is usually similar to that of traditional non-cool roofs. It is, however, worth noting that cool roofs tend to have an extended lifetime because they maintain a lower average temperature, which slows heat-related degradation of roofing materials (Akbari 2005; Global Cool Cities Alliance 2012). For example, a coated cool metal roof could have higher durability and work in good condition for a longer period of time as compared to a similar coated dark metal roof (Global Cool Cities Alliance 2012). This benefit contributes to enhancing the cost-competitiveness of cool roofs. Table 7 presents the new installation cost premiums for cool roof options as alternatives to some common traditional roof types and technologies.

Table 7: Solar Reflectance and Cost Premiums of Cool Roof Options for Some Common Traditional Roof Types

Traditional Roof Types	Traditional Roof Options	Non-cool Roof Solar Reflectance	Applicable Cool Roof Options	Cool Roof Solar Reflectance	Cost Premium ($ per square foot)
Asphalt shingle	Black or dark brown with conventional pigments	0.05–0.15	"White" (actually light gray) or cool color shingle	0.25	0.00–0.75
Built-up roof	Dark gravel	0.10–0.15	White gravel	0.30–0.50	0.00
	aluminum coating	0.25–0.60	White smooth coating	0.75–0.85	0.50–1.50
Clay tile	Dark color with conventional pigments	0.20	Unglazed red tile (terracotta)	0.40	0.00
			Colored with cool pigments	0.40–0.60	0.00
			White tile	0.70	0.00
Concrete tile	Dark color with conventional pigments	0.05–0.35	Colored with cool pigments	0.30–0.50	0.00
			White tile	0.70	0.00
Liquid applied coating	Smooth black	0.05	Smooth white	0.70–0.85	0.00
Metal roof	Unpainted corrugated	0.30–0.50	White painted	0.55–0.70	0.00–1.00
	Dark-painted corrugated	0.05–0.10	Colored with cool pigments	0.40–0.70	0.00
Modified bitumen	With mineral surface cap sheet	0.10–0.20	White coating over a mineral surface	0.60–0.75	0.50
Single-ply membrane	Black (PVC or EPDM)	0.05	White (PVC or EPDM)	0.70–0.80	0.10–0.15
			Colored with cool pigments	0.40–0.60	0.00
Wood shake	Dark color with conventional pigments	0.35–0.50	Left bare	0.40–0.55	0.00

Source: Global Cool Cities Alliance. 2012. A Practical Guide to Cool Roofs and Cool Pavements; Hewitt, V., E. Mackres, and K. Shickman. 2014. Cool Policies for Cool Cities.

Box 5: Case Study of Cool Roof Application for Office Buildings in Hyderabad, India

A demonstration project to quantify and record the benefits of cool roofs was carried out at two office buildings in Hyderabad, India. The study team included the International Institute of Information Technology and Lawrence Berkeley National Laboratory (LBNL), supported by the United States Agency for International Development (USAID) and SPM Thermoshield.

The complex for the application included two office buildings that were nearly identical. This setup facilitated the investigation by ensuring identical parametric values for floor area, number of floors, roofing material and system, occupancy and schedules, and cooling systems. Each building had two stories, with a roof area of approximately 700 square meters (m²). For one building, its roof was painted black. For the other building, a white reflective cool roof coating was applied to its roof.

Weather towers, temperature sensors, current transducers, and data loggers were deployed to continuously monitor the weather condition (outdoor temperature, relative humidity), building temperatures (heat flux through the roof, surface temperature, roof underside temperature, indoor air temperature), and energy use (whole-building electricity use, cooling energy use). Results showed that the application of a cool roof reduced the average daily roof surface temperature by 20°C (Box Figure 5.1) and the cooling energy use by 15% to 20% during hot summer days (Box Figure 5.2).

Box Figure 5.1: Surface and Under-Surface Temperatures of Roof Assembly Before and After Cool Roof Coating Application

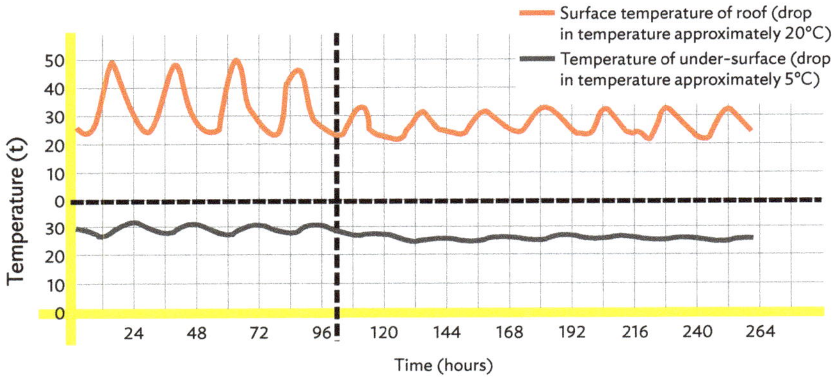

Box Figure 5.2: Heat Flux and Cooling Energy Use Before and After Cool Roof Coating Application

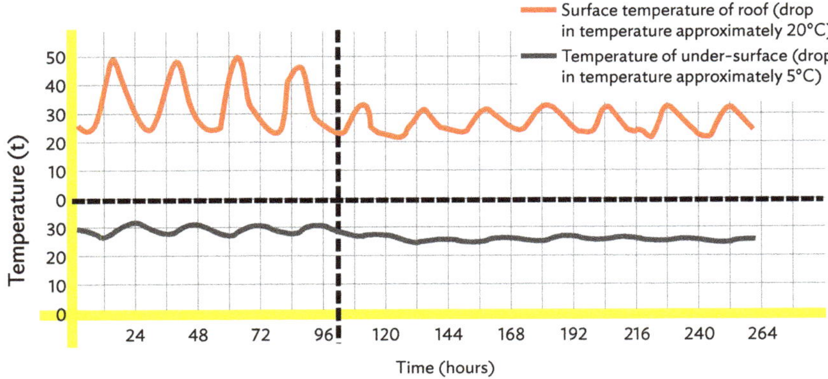

Source: Bureau of Energy Efficiency. 2017. Cool Roofs for Cool Delhi.

Ventilative Cooling

Natural Ventilation

Natural ventilation can serve a range of purposes in buildings, from delivering fresh air to maintain indoor air quality to ventilative cooling to limit temperatures in the summer. In naturally ventilated buildings, the driving force for air movement can be generated by a range of techniques that harness the potential of the stack effect, using air passages at different heights, and wind effects. As it involves no energy for driving airflow, natural ventilation is also known as passive ventilation. The complexity of building designs employing natural ventilation can vary substantially, depending upon specific application contexts, from simple designs of manually opened windows to sophisticated automated mixed-mode systems to those involving induced airflow through technologies such as passive downdraught evaporative cooling.

A major reason to take advantage of natural ventilation and ventilative cooling is the potential for lower energy consumption, CO_2 emissions, and operational costs compared with mechanical ventilation. Studies have found that buildings using mixed-mode ventilation and air-conditioning are major CO_2 emitters. In these buildings, CO_2 emissions resulting from fans and pumps required to transport conditioned air accounted for up to 50% of the emissions associated with space cooling and heating. Also, as the built form of air-conditioned buildings tends to be a deep plan, artificial lighting is more commonly used than buildings of other forms, thereby causing more energy consumption and CO_2 emissions.

In addition to energy savings and emission reduction, there are a range of other motives for choosing natural ventilation over other options. Naturally ventilated buildings tend to create an indoor environment more comfortable and desirable to occupants than mechanically ventilated and/or air-conditioned buildings. Usually, a lower prevalence of sick building syndrome (SBS) is reported by occupants of naturally ventilated buildings (Figure 35). Besides, naturally ventilated buildings, if carefully designed, can cost less to construct and maintain than buildings that are more heavily mechanically serviced. Maintenance is generally infrequent and of a relatively straightforward and unspecialized nature.

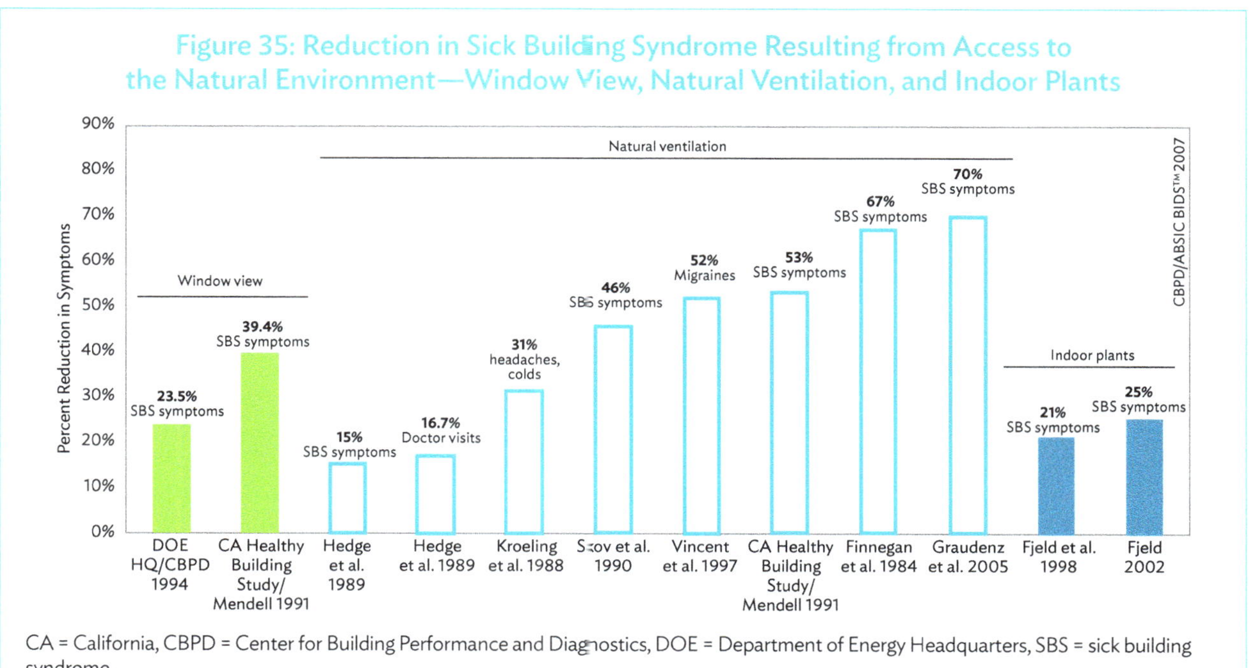

Figure 35: Reduction in Sick Building Syndrome Resulting from Access to the Natural Environment—Window View, Natural Ventilation, and Indoor Plants

CA = California, CBPD = Center for Building Performance and Diagnostics, DOE = Department of Energy Headquarters, SBS = sick building syndrome.

Source: Kellert, S.R., J. Heerwagen, and M. Mador. 2008. Biophilic Design: The Theory, Science and Practice of Bringing Buildings to Life. Wiley.

Heat gains, especially in nondomestic buildings located in urban heat islands, are a critical factor in the success of design strategies for natural ventilation. It is also important to identify, at an early stage, and minimize, to the extent possible, other external environmental conditions that may constrain the use of natural ventilation, such as high levels of ambient noise and air pollution. Their presence may require windows to be closed and the operation of a mechanical ventilation system, resulting in higher energy consumption and CO_2 emissions. Further, controls of natural ventilation need to be ergonomically efficient and rapidly responsive to occupants' needs.

The most important aspects of natural ventilation are the form and layout of the buildings, which can enhance airflow. The airflow path defines different ventilation modes, such as background ventilation, single-sided ventilation, cross-ventilation, stack ventilation, and sub-slab distribution. While sometimes mechanical equipment and devices are added to enhance ventilation system performance and control, these systems are invariably conceived as variants of the basic modes of ventilation strategies. Figure 36 provides an illustration.

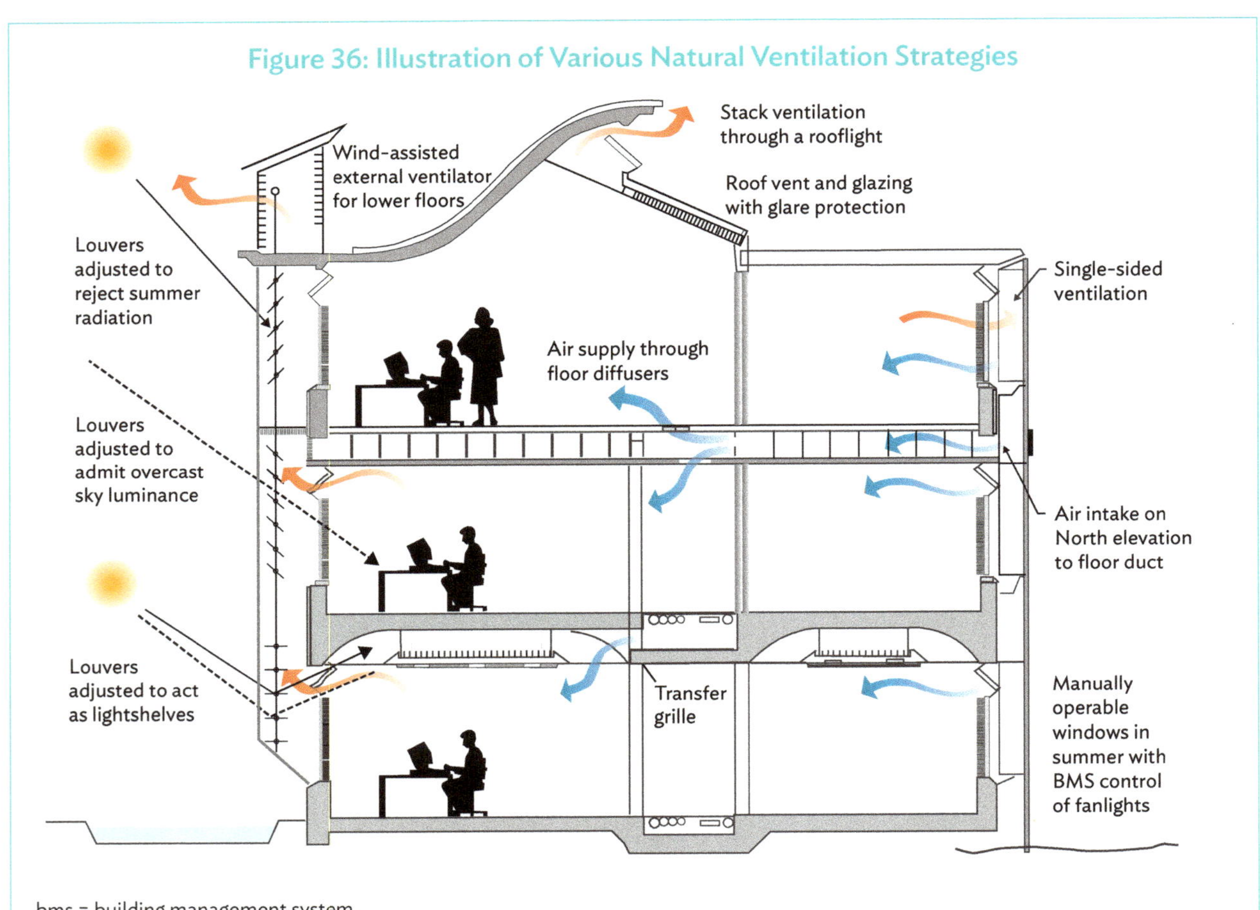

bms = building management system.
Source: Pennycook, K. 2009. *The Illustrated Guide to Ventilation: BG 2/2009*, Bracknell, Berkshirem, UK: Building Services Research and Information Association (BSRIA).

Background Ventilation

Background ventilation, also known as trickle ventilation, serves to provide the necessary minimum fresh air rate to satisfy the minimum ventilation requirements during room occupation, particularly in winter, but without increasing heat loss (Figure 37). At higher airspeeds, trickle ventilation can also provide external airflow for ventilative cooling outside of the occupation period.

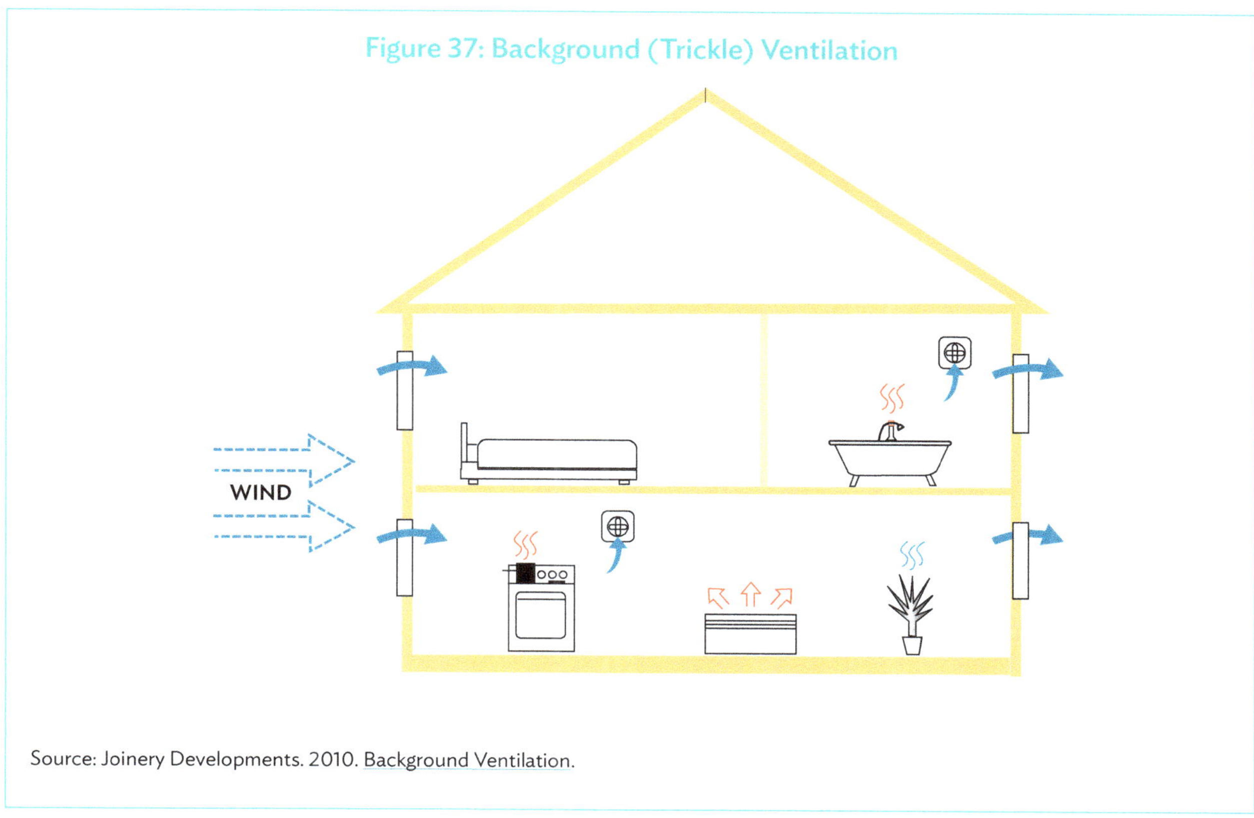

Figure 37: Background (Trickle) Ventilation

Source: Joinery Developments. 2010. Background Ventilation.

To minimize cold draughts, trickle vents are usually controllable and should be placed at a high level, typically 1.75 meters above floor level, and designed to promote rapid mixing within the room. Installation locations of trickle vents vary—they may be mounted between window lintel and frame, be surface-mounted at the window frame or sash, or be glazed in. Trickle vents may be passive, with the airflow driven by pressure drop, or be equipped with a fan serving as as an exhaust vent. Technical challenges that should be considered in designing and operating trickle vents include the prevention of excessive airflow in cases of high wind pressure, reasonable U-values to address heat loss, noise protection, and fire resistance. Figure 38 provides an example of a commercially available trickle vent.

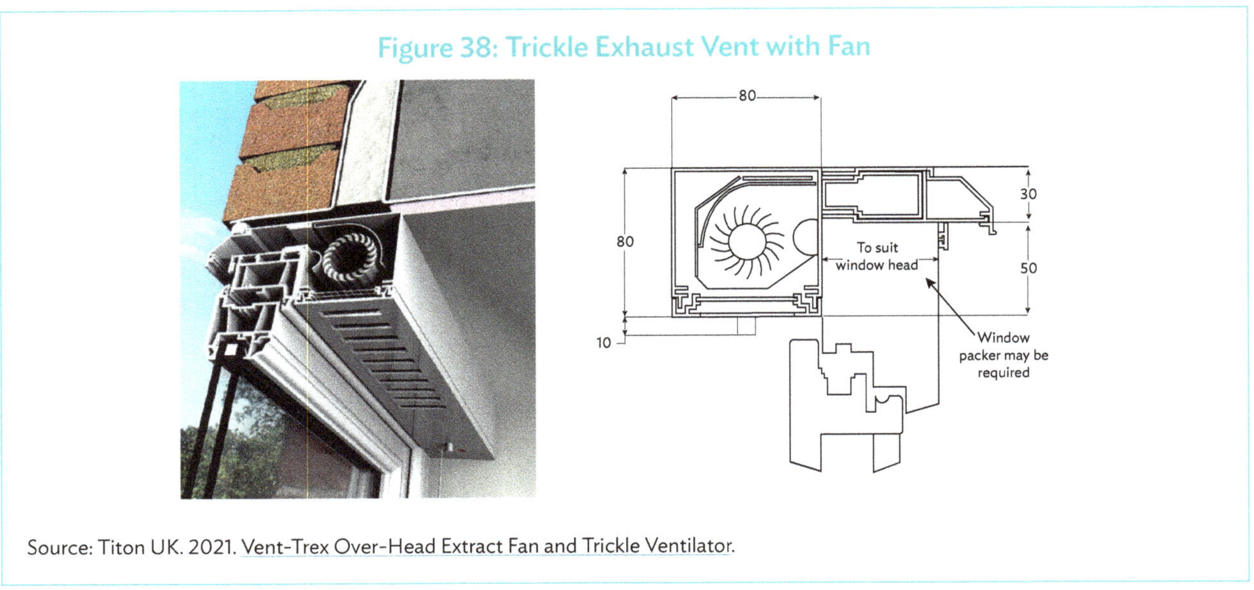

Figure 38: Trickle Exhaust Vent with Fan

Source: Titon UK. 2021. Vent-Trex Over-Head Extract Fan and Trickle Ventilator.

Single-Sided Ventilation

Single-sided ventilation typically serves single rooms and uses vents on only one side of the enclosure (Figure 39). It is often found in many multi-room buildings with opening windows on one side of the room and a closed internal door on the other side.

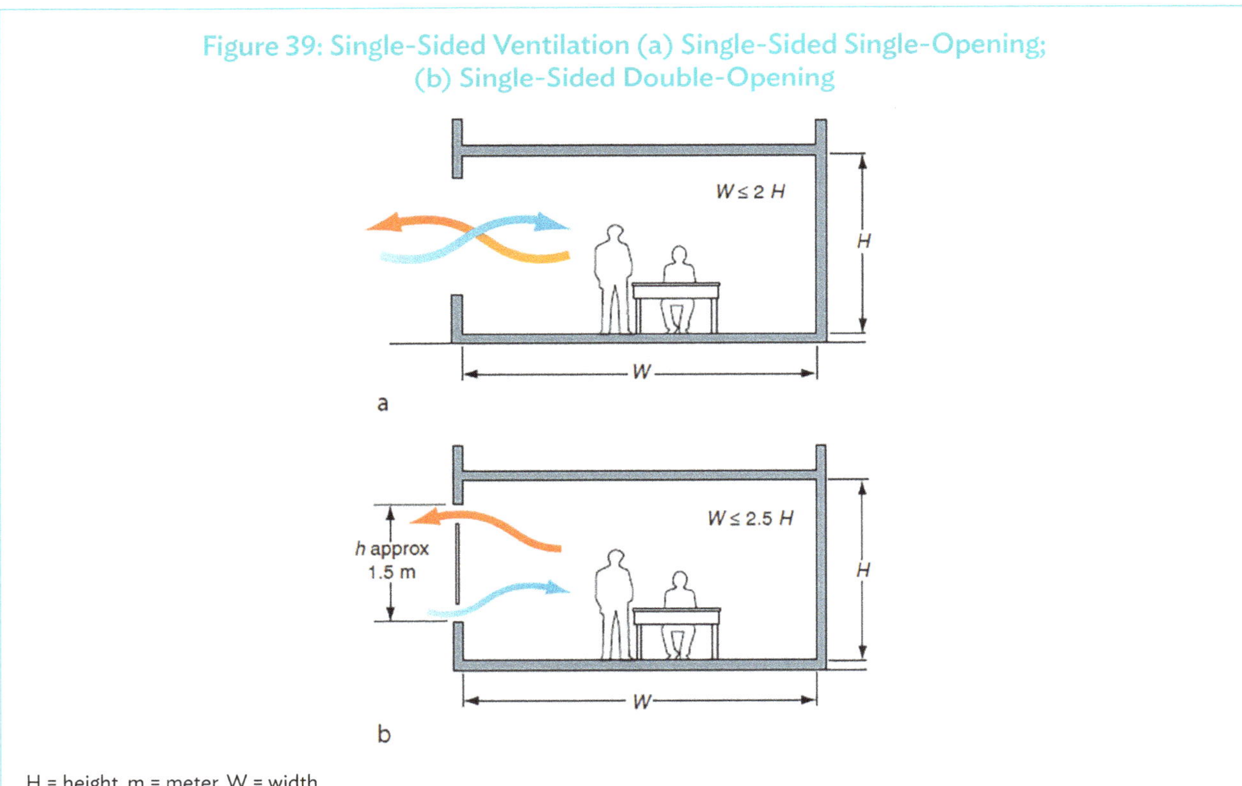

Figure 39: Single-Sided Ventilation (a) Single-Sided Single-Opening; (b) Single-Sided Double-Opening

H = height, m = meter, W = width.
Source: Yang, T. and D.J. Clements-Croome. 2013. Natural Ventilation in Built Environment. In V. Loftness and D. Haase, eds. *Sustainable Built Environments*. Springer.

Compared to other natural ventilation strategies, a single-sided ventilation with a single opening is less attractive due to lower ventilation rates and smaller distances of ventilating air penetrating into the room. It is mainly driven by wind turbulence, which generates reversing flows in and out of the room. When the single opening is tall, the exchange of air can also be driven by stack effects generated by temperature differences.

In the case of double openings, the driving force of air exchange can be enhanced by room-scale stack effects. The stack-induced airflow is a function of the vertical separation of the openings and the inside and outside temperature differences. Compared to a single opening, there is an increased probability of pressure differences occurring between two openings, thus improving ventilation effectiveness to some extent. In winter, to minimize cold drafts for occupants at the perimeter, care should be exercised to position the inlets within the room.

Single-sided ventilation should only be used when the building form or location limits ventilation to one facade. As a rule of thumb, the limiting depth for effective ventilation is about two times the floor-to-ceiling height for single openings, and 2.5 times the floor-to-ceiling height for double openings, respectively. When heat gain is low, single-sided ventilation could be effective over a depth of up to 10 meters and may be incorporated into open-plan offices.

Cross-Ventilation

Cross-ventilation is usually driven by wind-generated pressure differences, inducing positive pressures on windward surfaces and negative pressures on leeward surfaces (Figure 40). Airflow characteristics are determined by the combined effect of wind and temperature differences. The floor depth in the direction of ventilation must be limited to effectively prevent heat and pollutant buildup in the space during ventilation. As a rule of thumb, the limiting depth for effective ventilation is about five times the floor-to-ceiling height, implying a relatively narrow plan depth for the building.

An effective measure to improve wind-driven natural ventilation, such as cross-ventilation, is to use windcatchers, which are vertical building-integrated structures, often roof-mounted, that induce outdoor fresh air into indoor space by harnessing the pressure difference over the building and across the device openings. Major components of a windcatcher include a louvered terminal, a base, and damper assemblies that enable automatic or manual adjustment of the ventilation. With compartmentalized vertical vents, the pressure difference across the segments facing the wind drives outdoor air into the room, and the suction created by the negative pressure on the leeward segments draws indoor air out of the room (Figure 41).

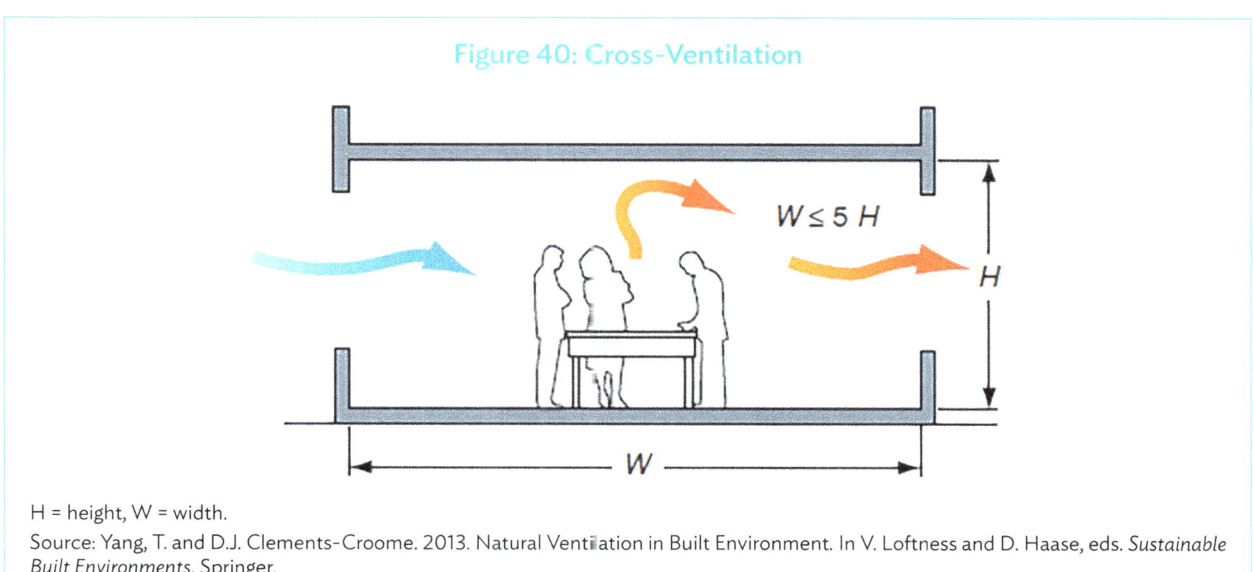

H = height, W = width.
Source: Yang, T. and D.J. Clements-Croome. 2013. Natural Ventilation in Built Environment. In V. Loftness and D. Haase, eds. *Sustainable Built Environments*. Springer.

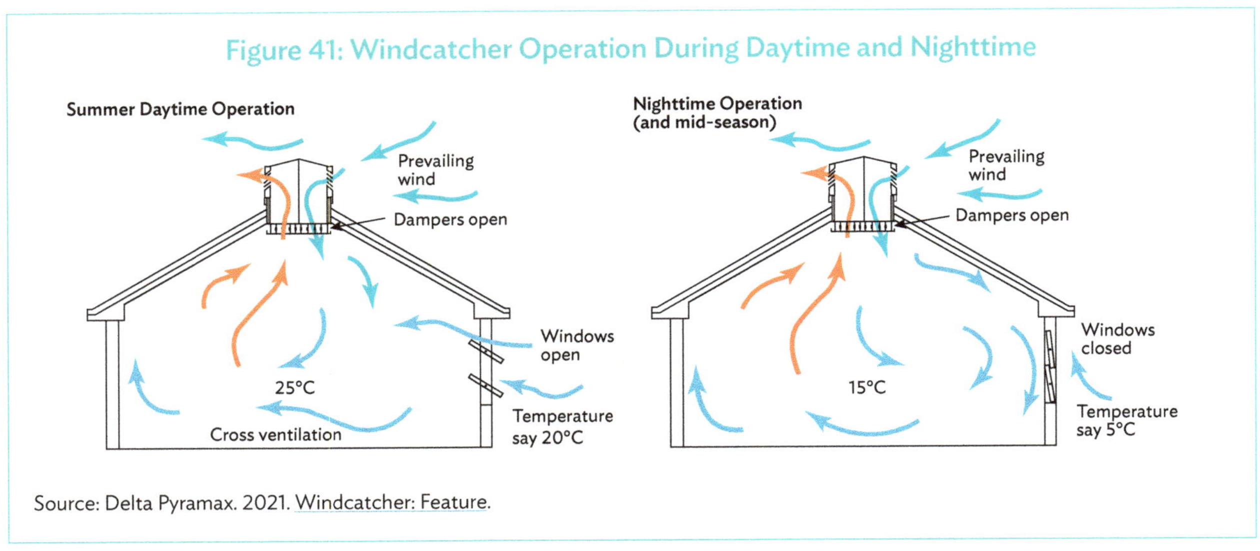

Source: Delta Pyramax. 2021. Windcatcher: Feature.

Roof-mounted windcatchers are particularly suitable for schools, libraries, and health and community centers—all of which require low operational and maintenance costs. The performance of windcatchers depends upon the positioning and alignment of windcatchers, the shape of louvers, the size of louvered area, the height of louvers relative to roof level, the resistance of dampers, and the airflow driving forces (pressure difference and temperature difference) in the context of prevailing wind conditions. Other factors such as acoustic liners, which might be required in applications in areas of heavy local traffic, and the airflow network within the ventilated room, which is affected by room layout and heat gains, should also be considered.

Figure 42 shows the Airscoop terminal manufactured by Passivent, UK, which is a roof-mounted windcatcher particularly designed for large or deep-plan spaces where direct perimeter ventilation through the building facade is limited or not possible. Each unit of the Airscoop terminal is divided into four separate internal ducts so that prevailing wind from any direction can be captured and channeled down through the windward chambers. The cooler and denser outdoor fresh air flows down into the building, while the warmer and lighter stale air from inside is displaced upward and out via the leeward chambers. The complete separation of chambers provides improved distribution and mixing of air and prevents short-circuiting of inward and outward airflows around the unit. To cater to the needs of buildings with intermediate ceilings, such as offices and classrooms, the inlets and outlets of an Airscoop unit can be connected via flexible ductwork to four diffusers mounted to the ceiling of the ventilated space. This flexibility enables the ceiling diffusers to be positioned for maximum ventilation and air distribution benefits in relation to space utilization.

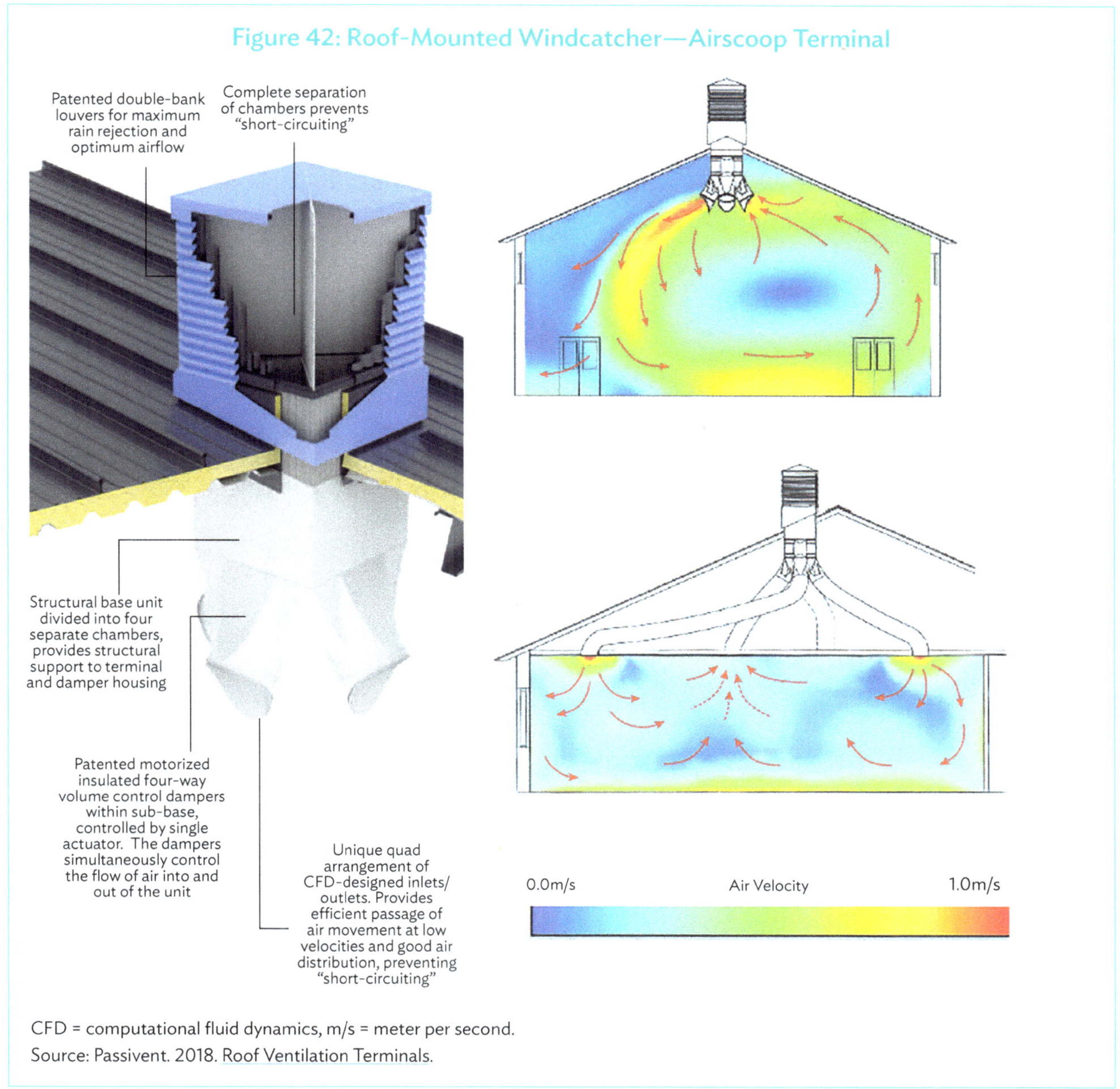

Figure 42: Roof-Mounted Windcatcher—Airscoop Terminal

CFD = computational fluid dynamics, m/s = meter per second.

Source: Passivent. 2018. Roof Ventilation Terminals.

Box 6: Effects of Arrangement and Spacing of Windcatchers on Ventilation Performance

The layout and spacing of windcatchers have substantial impacts on ventilation performance in terms of indoor airflow distribution, ventilation rate, and carbon dioxide (CO_2) concentration. Calautit, O'Connor, and Hughes (2014) conducted a computational fluid dynamics (CFD) analysis to numerically investigate the effects of spacing and arrangement of multiple windcatchers located on the same building (Box Figure 6.1). The CFD model of an isolated windcatcher as the benchmark was validated using experimental wind tunnel measurements. A total of six different cases, consisting of two different arrangements with three different spacing lengths between windcatchers, were simulated.

Box Figure 6.1: Parallel and Staggered Arrangements of Roof-Mounted Windcatchers

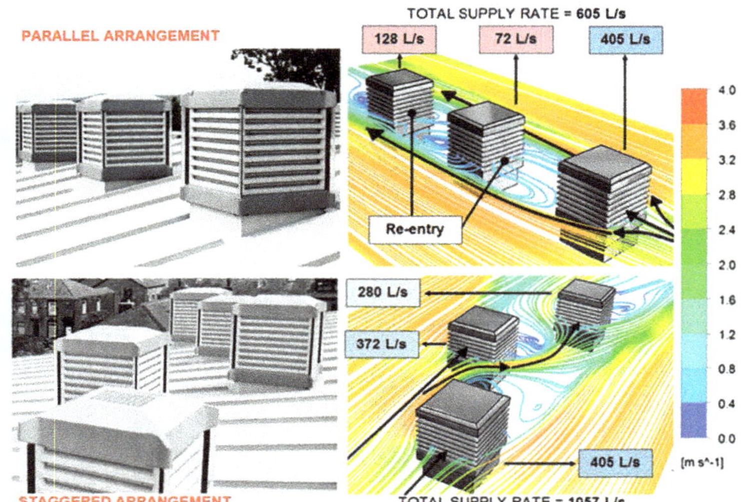

Source: Calautit, J.K., D. O'Connor, and B.R. Hughes. 2014. Determining the Optimum Spacing and Arrangement for Commercial Wind Towers for Ventilation Performance. Building and Environment. 82. pp. 274–287. .

It was found that a parallel arrangement of windcatchers was not effective for ventilating an occupied volume, regardless of the spacing between two windcatchers, when the incident wind direction was parallel to the arrangement. The maximum supply rate for the leeward windcatcher in parallel arrangement at a spacing of 5 meters was just over 50% of the British regulation rate of 10 liters per second per occupant (L/s/occupant) and 40% of the supply rate of an isolated windcatcher as the benchmark. Decreasing the spacing between the parallel windcatchers to 3 meters further reduced the supply rate to 2.4 L/s/occupant, and the device was observed to be operating in reverse—airflow entering from a leeward opening. As the angle of wind increased, an improvement in air supply rates was seen. For a staggered arrangement, the leeward windcatcher was able to supply the recommended ventilation rates at all tested spacing lengths.

The average indoor CO_2 concentration of the ventilated space with the leeward windcatcher was substantially higher in the parallel arrangement than the staggered arrangement at a zero-degree wind angle. As shown in Box Figure 6.2, with the parallel arrangement, the average CO_2 concentration was 28–50 parts per million (ppm) higher than the outdoor concentration. With the staggered arrangement, which effectively minimized the reentry of pollutants, the indoor CO_2 concentration was brought down to just 1–3 ppm above the outdoor level.

continued on next page

Box 6 *continued*

Box Figure 6.2: Contours of Carbon Dioxide Concentration (ppm) Inside the Occupied
and Unoccupied Room Ventilated with Parallel (above) and Staggered (below)
Arrangements with Spacing of 5 Meters

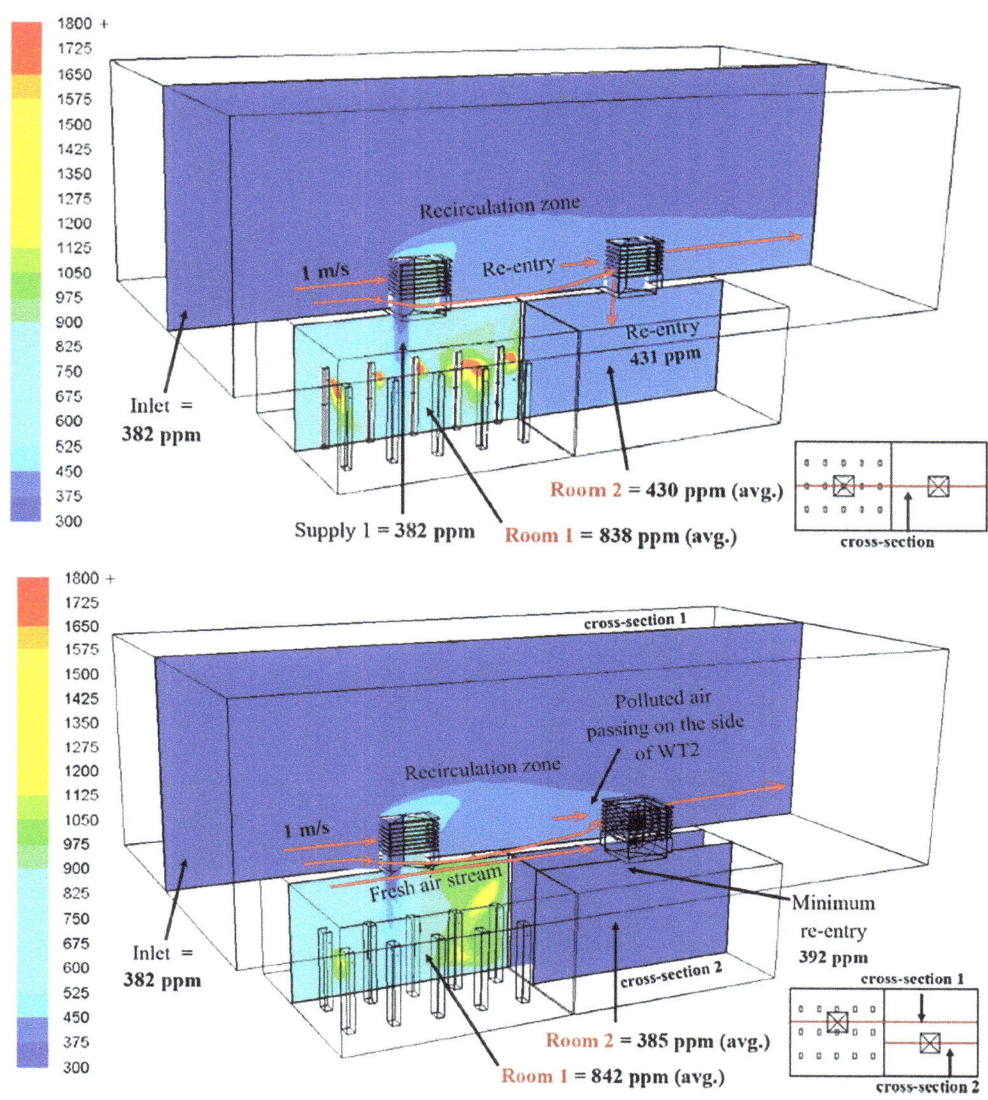

ppm = parts per million.
Source: Calautit, J.K., D. O'Connor, and B.R. Hughes. 2014. Determining the Optimum Spacing and Arrangement for Commercial Wind Towers for Ventilation Performance. Building and Environment. 82. pp. 274–287.

Source: Compiled by authors.

Box 7: Case Study of Cross-Ventilation—Bayalpata Hospital, Achham, Nepal

Bayalpata Hospital is in the mountains of northern Nepal. Through three phases of expansion, the previously obsolete and overrun medical facility was transformed into a model of sustainable rural health care. Now, with a built area of 4,225 square meters (m^2) spreading across a 7.5-acre hilltop in Achham, Nepal, this medical complex comprises five medical buildings that provide the functions of outpatient, inpatient, and antenatal units, along with emergency facilities for 70 beds, operation theaters, a pharmacy, a radiology lab, and other laboratory spaces. It also has an administration block with offices and a 60-seat canteen. Fourteen single-family houses and an eight-bedroom dormitory provide accommodation to the hospital's staff and their families (Box Figure 7.1).

Box Figure 7.1: View of Bayalpata Hospital, Achham, Nepal

Source: Sharon Davis Design. 2019. Bayalpata Hospital.

The design and construction of the hospital responded closely to the local site and climatic conditions. Its architecture maintains a vernacular and modest scale through setbacks, gabled roofs, and low-cost heat-storing materials. The low-rise, one- and two-story structures of the complex are organized around landscaped courtyards. Built from on-site materials and with low-tech construction methods, the complex managed to minimize the cost-prohibitive transportation of building materials in this mountainous region, thus contributing to low-carbon design from a life cycle perspective.

Box Figure 7.2: Courtyards of Bayalpata Hospital

Source: Transsolar Energietechnik. 2021. Bayalpata Regional Hospital, Achham, Nepal.

continued on next page

Box 7 *continued*

Except for the operating theater and laboratories, all the other spaces of the complex are cooled and heated passively. The structures of the complex consist of massive, rammed earth walls with insulated roofs. Materials with thermal mass retain daytime heat gain in winter, while keeping the interiors cool by preventing overheating during summer. Cross-breezes through courtyards, assisted by clerestory ventilation and ceiling fans, facilitate natural ventilation, and improve indoor thermal comfort conditions with minimal energy consumption (Box Figure 7.3). For electricity, a grid-connected, 100-kilowatt photovoltaic system installed across all south-facing roofs is the major source of electricity supply to the hospital. A solar-powered water heating system installed can heat up to 3,300 liters of water per day for both medical and domestic use. Electricity for artificial lighting is reduced by harvesting daylighting throughout all clinical areas, which is realized through tall, narrow windows, south-facing glazed clerestories, and skylights in the buildings.

Box Figure 7.3: Ventilation Strategy of Bayalpata Hospital

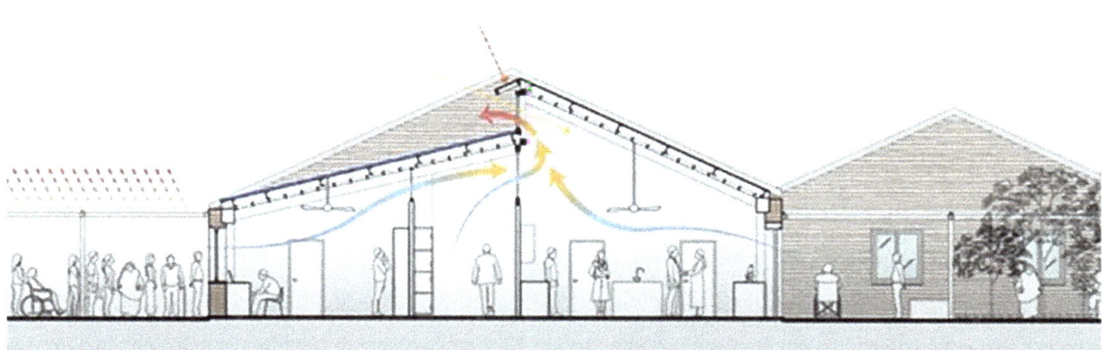

Source: NZEB Alliance. 2021a. Bayalpata Hospital Achham, Nepal.

Source: NZEB Alliance.2021a. Bayalpata Hospital Achham, Nepal; Sharon Davis Design. 2019. Bayalpata Hospital; Transsolar Energietechnik. 2021. Bayalpata Regional Hospital, Achham, Nepal.

Stack Ventilation

For buildings with deeper floor plans, stack ventilation provides a solution. This is achieved via purposefully designed and built vertical ducts, shafts, solar chimneys, internal atria, etc. These various types of vertical spatial continuity within the building provide passage for air at higher temperatures, thereby generating pressure differences that give rise to the stack effect. Warmer and less dense indoor air rises and is exhausted via high-level vents. The upward air movement produces a negative indoor pressure at the bottom and a positive indoor pressure at the top. Thus, it is important to ensure that the roof vents are always positioned in a negative pressure zone with regard to wind-induced pressure. This can be achieved through designing the roof profile so that the vents are in a negative pressure zone for all wind directions (the Venturi effect), installing a single vent system, that turns to face away from the wind, or installing an automatically controlled multi-vent system that opens and closes outlets depending upon wind directions so that the opened ones are always in a negative pressure zone.

Figure 43 and Figure 44 illustrate how stack ventilation works. In Figure 43, ventilation shafts, stack towers, and an atrium are used, which are simple architectural features to enhance ventilation. Figure 44 is based on solar-induced ventilation. The double facade system not only promotes access to daylight, but also causes a higher pressure difference between the lower inlets and higher outlets of the air, creating a special form of solar chimney. The heat gain from solar radiation through south-facing tilted skylights strengthens the stack effect in the atrium.

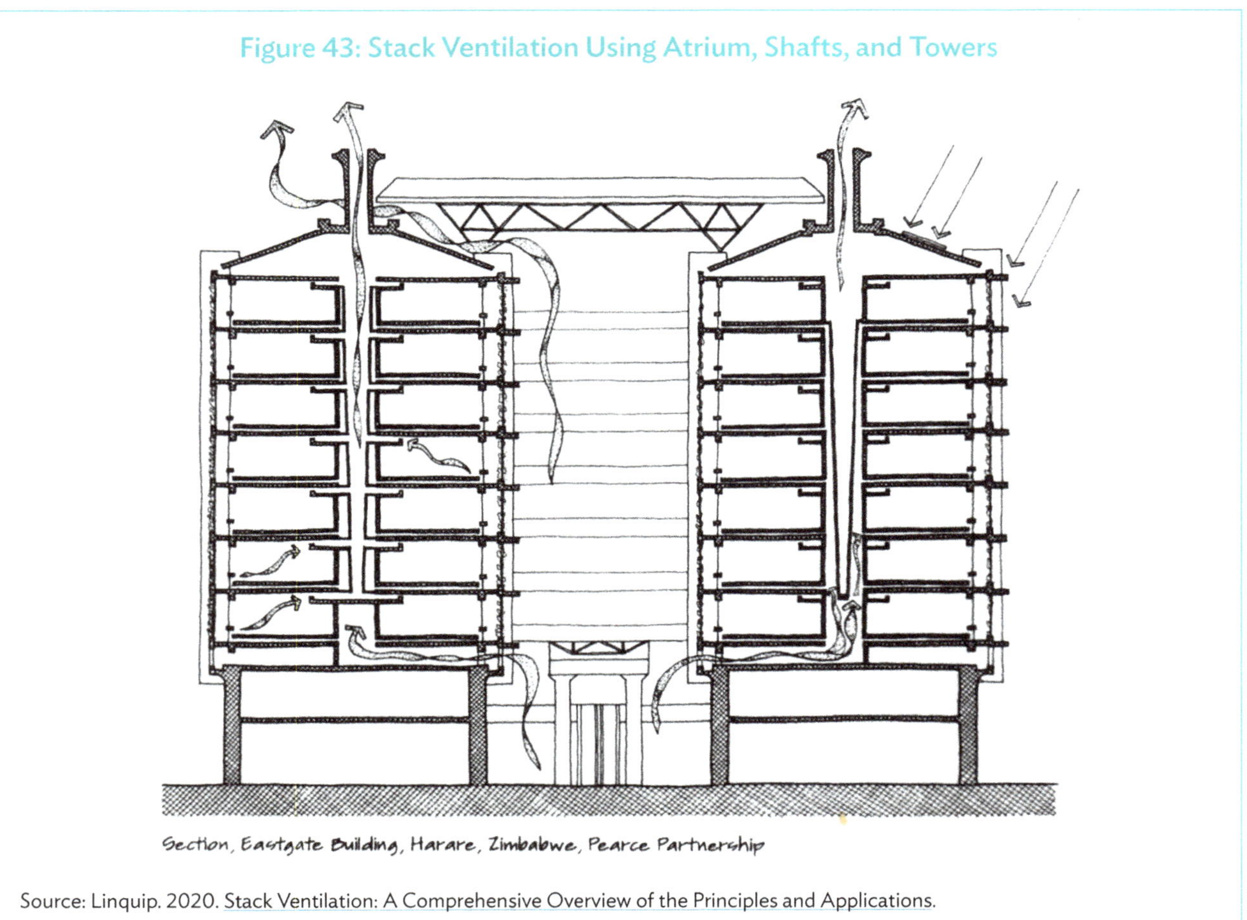

Figure 43: Stack Ventilation Using Atrium, Shafts, and Towers

Source: Linquip. 2020. Stack Ventilation: A Comprehensive Overview of the Principles and Applications.

Figure 44: Solar-Induced Ventilation

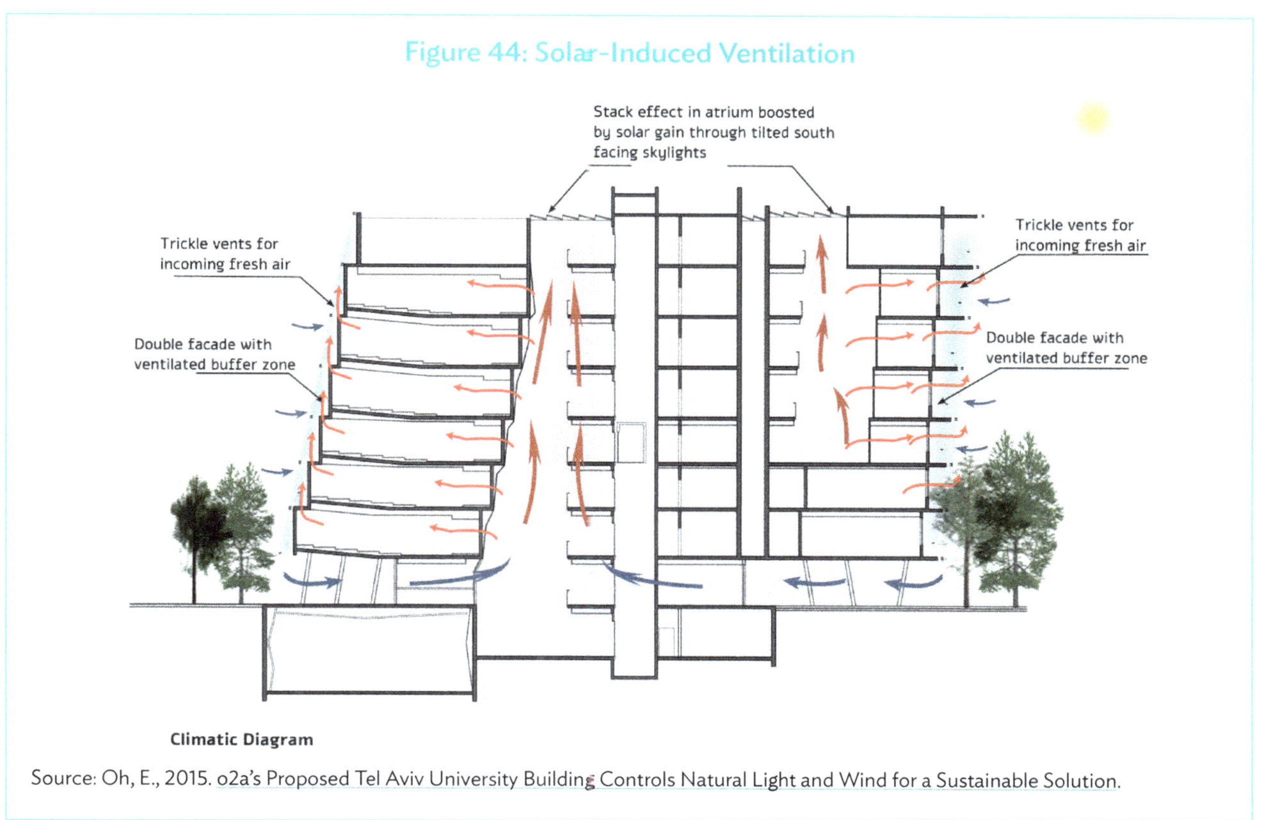

Source: Oh, E., 2015. o2a's Proposed Tel Aviv University Building Controls Natural Light and Wind for a Sustainable Solution.

Figure 45: Different Forms of Stack Ventilation

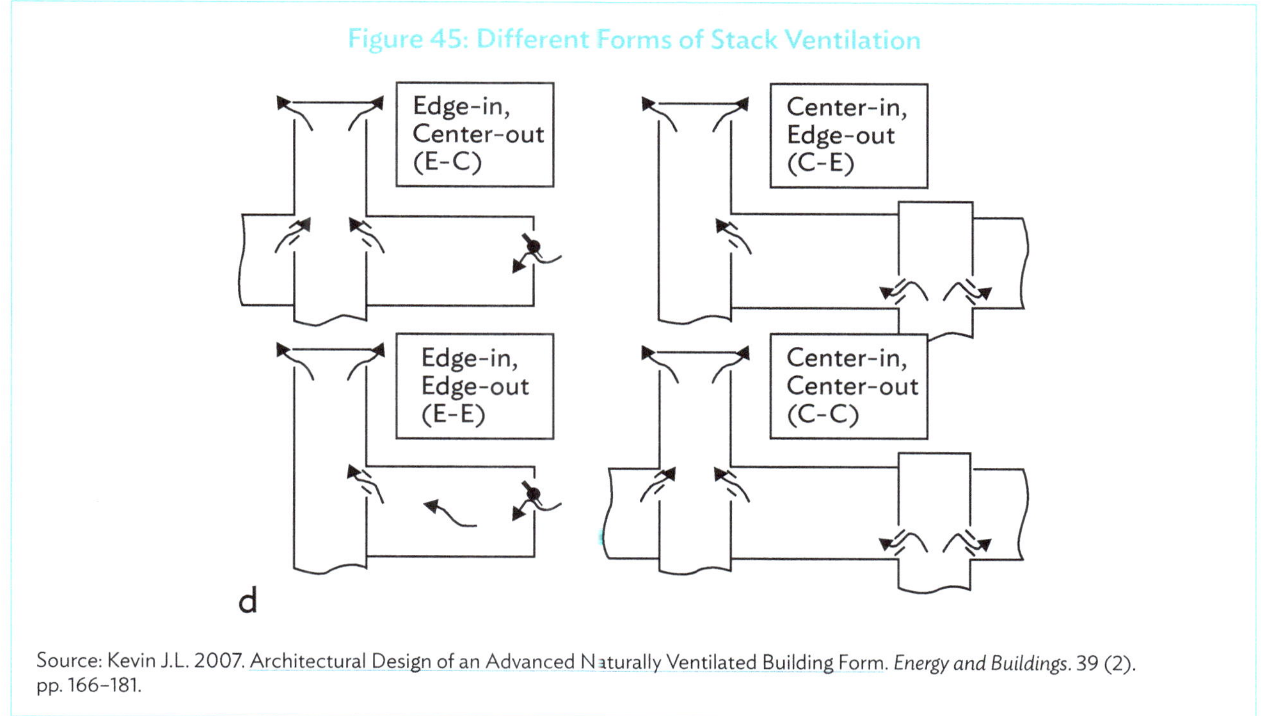

Source: Kevin J.L. 2007. Architectural Design of an Advanced Naturally Ventilated Building Form. *Energy and Buildings.* 39 (2). pp. 166–181.

Depending upon the locations where air flows in to and out of a building, stack ventilation can be categorized into four forms: (i) edge-in, center-out; (ii) edge-in, edge-out; (iii) center-in, edge-out; and (iv) center-in, center-out (Figure 46). The edge-in forms are susceptible to local air pollution, noise, and security concerns associated with natural ventilation design for buildings in urban areas. These issues can be mitigated by the center-in forms, which enable the external facade of the buildings to be sealed. With edge-in forms, the use of operable windows can enhance the airflow throughout the building during summer and can also allow outside air to be preheated by perimeter heating elements in the building. The edge-in, center-out form allows for natural ventilation design for deep-plan buildings. With center-in forms, the central stack, while being the air supply route, can also introduce daylight into a deep-plan building. The center-in, edge-out location locates the exhaust stacks at the perimeter of the building and therefore allows more flexible internal space planning.

Box 8: Case Studies of Stack Ventilation

CASE 1: BSKYB BROADCAST CENTER, LONDON, UNITED KINGDOM

The BSkyB broadcast center located in West London, houses the world's first naturally ventilated television studios (Box Figure 8.1). The building has 13 giant ventilation chimneys, including nine lining the building's eastern elevation and another four mounted to the west facade. The construction of concrete boxes within boxes provides a solution to eliminate external noise as well as naturally ventilate the studios to remove excessive heat generated by studio lights. Fresh air is supplied through acoustically lined tunnels built in between the underside of the studio's concrete floor and the floor of the surrounding box from street level (Box Figure 8.2). This construction form allows large air paths to minimize resistance to air movement as well as eliminate all influxes of noise.

To prevent a common stack ventilation problem—air cooling in the flue and dropping back into a room—the flues are lined and insulated on the inside. In an intermediate mode, the ventilation system will run on extract only to pull the air up the chimney and warm it. When the right flue surface temperature is reached, the air's natural buoyancy will take over, and the system then automatically switches to natural ventilation mode.

At the south end of the building, a glazed atrium houses a series of meeting rooms, a café, and breakout spaces. The office areas (8 meters in depth) on the west elevation are ventilated using single-side natural ventilation. The offices (15 meters in depth) on the eastern side utilize three atrium-line chimneys in the center of the building to help draw air across the floor plates. Access to daylight through these atria is an additional benefit of the design.

Box Figure 8.1: BSkyB Broadcast Center

Source: ARUP Associates. 2013. BskyB Sky Studios.

Box Figure 8.2: Natural Ventilation Flow within BSkyB Television Studios

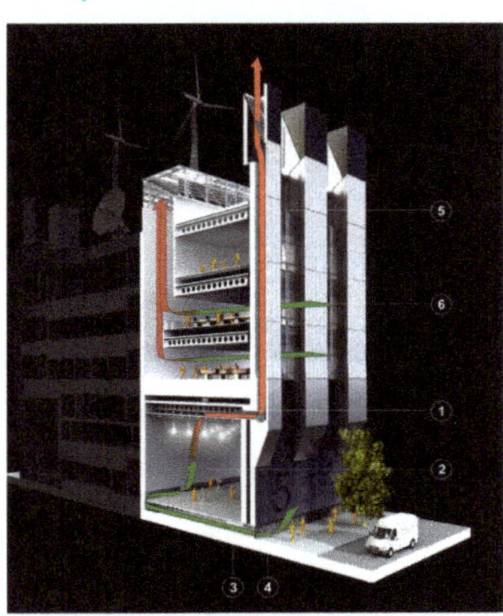

1. Waste heat from the studio lights rises through the studio ventilation chimneys.
2. As waste heat rises, a small negative pressure is set up in the studios.
3. This pressure drop overcomes the resistance of the sound attenuators, drawing in fresh cool air from the exterior.
4. Exterior intake grilles
5. When external conditions are inappropriate for natural ventilation, mechanical ventilation and cooling of the studio spaces can be implemented using the same chimneys.
6. Office natural ventilation chimney follows similar principles

Source: ARUP Associates. 2013. BskyB Sky Studios.

Box 8 *continued*

CASE 2: COMMERZBANK, FRANKFURT, GERMANY

The Commerzbank in Frankfurt, Germany, is a successful example of a sustainable skyscraper integrating natural ventilation, daylight, and exterior views (Box Figure 8.3). The building's triangular-shaped plan provides a rigid structural support with high-rise functional cores at each corner of this triangle. The layout of the floor levels creates a helix with 4-story-high cutouts, creating space for landscaped terraces, called sky gardens. The center atrium of the buildings provides light from the glass roof at the atrium's top, as well as from the sky garden facades to the office areas. Sky gardens rotate around the facade and allow for ventilation through the atrium, which is divided into sections. Daylight is brought directly into the center of the building. Offices facing the center are provided with daylight and outdoor views through these green, natural spaces. The outer shell of the facade has slotted openings through which the outside air flows into the cavities between the layers and then enters rooms through open windows. This way, the wind pressure is brought down to appropriate levels to provide a natural ventilation of airflow. During winter, exterior vents are closed to allow heat to build up in the cavity spaces and improve the thermal insulation properties of the windows (Box Figure 8.4).

Box Figure 8.3: Commerzbank, Frankfurt, Germany

Source: Urban Systems Design. 2021. Commerzbank HQ Frankfurt, Germany.

Box Figure 8.4: Ventilation System of Commerzbank, Frankfurt (a) Double Skin facade (vented cavity); (b) Passive Heating, Cooling, and Ventilation System; (c) Active Heating, Cooling, and Ventilation System

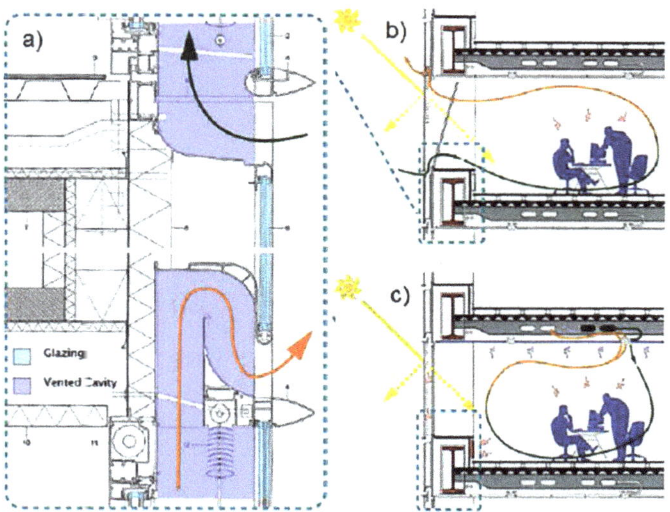

Source: Penić, M., N. Vatin, and V. Murgul. 2014. Double Skin Facades in Energy Efficient Design. *Applied Mechanics and Materials.* 680. pp. 534–538.

continued on next page

Box 8 continued

CASE 3: QUEEN'S BUILDING, DE MONTFORT UNIVERSITY, UNITED KINGDOM

The Queen's Building at De Montfort University was built in 1993 (Box Figure 8.5). It was awarded Green Building of the Year 1995 by *The Independent*, and since then it has been serving as a demonstration project to showcase innovative technologies and demonstrate ways of achieving significant carbon reductions in modern buildings through refurbishment. The building features large ventilation chimneys, heavy thermal mass, a shallow floor plan, operable windows, and large ceiling heights to facilitate natural ventilation and daylighting. This traditional brick building has wide insulation-filled cavity walls and concrete slabs in the ceilings, which serve as a buffer for the indoor environment when outdoor temperatures vary drastically. Glazed ventilators also help to provide as much natural lighting as possible. In the auditoria, fresh air enters through louvers in the north facade by means of plenums below the raked wooden floor and wall inlets, which are controlled by the building energy management system (Box Figure 8.6).

Box Figure 8.5: The Queen's Building, De Montfort University

Source: Royal Academy of Engineering 2010. *Engineering a Low Carbon Built Environment: The Discipline of Building Engineering Physics.*

Box Figure 8.6: Natural Ventilation Strategies of the Queen's Building

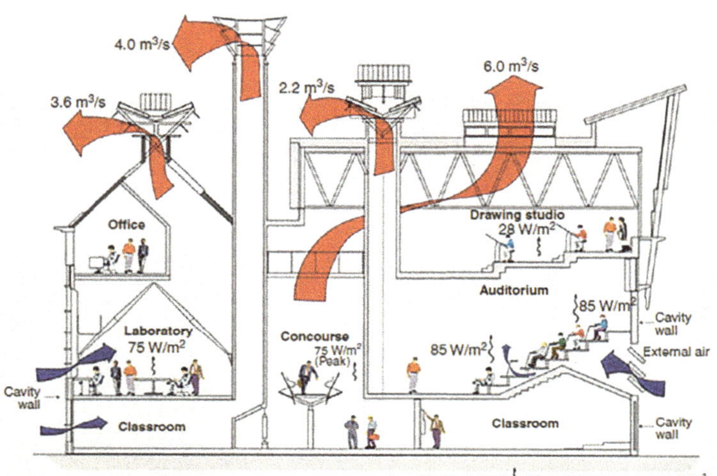

Source: Yang, T. and D.J. Clements-Croome. 2013. Natural Ventilation in Built Environment. In V. Loftness and D. Haase, eds. *Sustainable Built Environments*. Springer.

CASE 4: THE SMART GHAR-III PROJECT (GREEN HOMES AT AFFORDABLE RATE), RAJKOT, INDIA

The Pradhan Mantri Awas Yojana (PMAY)—Housing for All (Urban)—was the Government of India's flagship program to address the housing shortage and fulfill housing demand for the urban poor. This mission target was 12 million affordable, eco-friendly houses between 2015 and 2022, supported by central financial assistance and subsidies through the states and union territories with different implementing mechanisms. The targeted construction was estimated to account for almost 11% of the predicted urban residential built-up area, housing 20%–30% of the urban population in 2022 (Chetia et al. 2020).

Rajkot, a mid-sized city in the Indian State of Gujarat, is the site of one project under the PMAY Untenable Slum Redevelopment program. The city falls in the composite climate zone with peak summer daytime temperatures reaching 41°C–43°C. Indoor temperatures on a typical summer day can reach 38°C and remain above 30°C for more than 6,200 hours per year. The project, named Smart GHAR III for "Green Homes at an Affordable Rate," comprises 11 residential towers and 1,176 dwelling units, each of which has a built-up area of 33.6 m². Each tower has seven stories with stilt parking.

continued on next page

Box 8 *continued*

Box Figure 8.7: The Smart GHAR-III Project in Rajkot, India

Source: Swiss Agency for Development and Cooperation. 2018. *Thermally Comfortable and Climate-Friendly Affordable Housing in India: The Smart GHAR III Project*.

One of the key design strategies to achieve indoor thermal comfort was to maximize the potential of natural ventilation for cooling. Although the wind speed in Rajkot is good, the design and layout of the residential towers and dwelling units are such that wind flow might not reach all flats as desired, and there would exist situations where very low or even no natural ventilation is available. This issue is addressed by using the existing service shafts between flats and exhaust fans to improve the natural ventilation. Adequate negative pressure is created in the shaft to improve air circulation and increase the air change rate in the flats. In addition to the assisted ventilation, the design strategies also include the use of a highly reflective china mosaic finish for roofing, tall and partially glazed casement windows, and lightweight concrete block walls with a low U-value. The window-to-wall ratio is low. Collectively, these measures effectively reduce heat gain through the building envelope.

Box Figure 8.8: Concept Sketch of Assisted Ventilation Through Common Shaft Between Flats

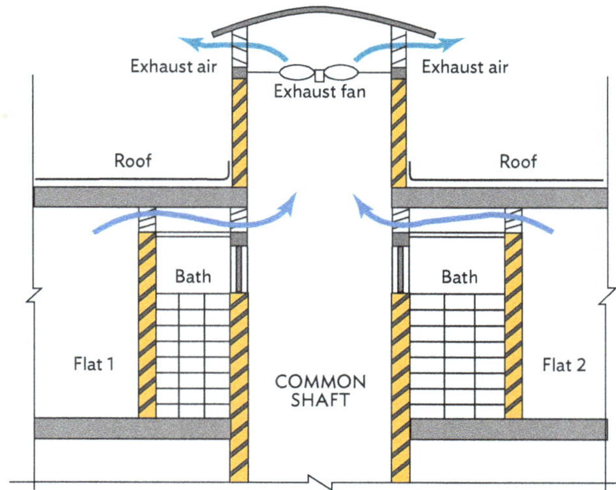

Source: WWF. 2021. Rajkot – Efficient Cooling is Key.

By adopting the above design strategies, the peak summer indoor temperature is reduced by more than 5°C compared to conventional housing in the same climate. The number of thermal discomfort hours per year drops from 6,200 to 2,500 (Gruner and Zinecker 2020). This improvement also helps to avoid or significantly reduce the need for air-conditioning units installed by the tenants, which leads to electricity savings, CO_2 emission reductions, and reduced energy bills. This project provides a good demonstration that careful design and appropriate building materials can achieve energy efficiency and thermal comfort in low-cost housing with a reduced climate impact (Swiss Agency for Development and Cooperation 2018).

Source: Compiled by authors.

Night Ventilation

The motive for night ventilation is to avail of free night cooling benefits made possible by removing excess heat stored in the building structure through natural or mechanical ventilation during the nighttime (Figure 46). A key factor in night ventilation is to ensure good thermal contact between the ventilating air and the thermally massive building structure, e.g., the underside of floor slabs. During the day, the exposed floor slabs provide radiant cooling by absorbing the internal heat gains. During the night, the absorbed heat gains are dissipated as the exposed slabs are cooled by ventilating air. On the following day, the slabs can absorb more heat than they would otherwise, thereby significantly reducing or even avoiding the need for mechanical cooling and improving the indoor airflow and thermal conditions of the building. In this process, the thermal capacity of the slabs (and other thermal mass as part of the building structure) serves as a reservoir for incidental heat gains.

When mechanical ventilation must be used, the design of night ventilation needs to take account of the energy consumed by fans to transport the air through the building structure. It is important to use low-energy mechanical ventilation systems to avoid fan energy consumption exceeding the energy savings brought by the night cooling. It is also essential that the controls of night ventilation ensure that the summer night purge is undertaken during the coolest part of the night. In general, a well-insulated building with sufficient thermal mass exposed to night ventilation under good control strategies can reduce peak daytime temperatures by 2℃ to 3℃. However, it should be noted that there is a risk of overcooling and subsequent reheating or thermal discomfort of the space on the following day. This needs to be controlled carefully. In addition, urban heat island effects in large cities present challenges to the effectiveness of night ventilation, as the nighttime air temperature is not low enough to cool down the exposed building thermal mass.

Figure 46: Night Ventilation Concept (left) and Exposed Slab (right)

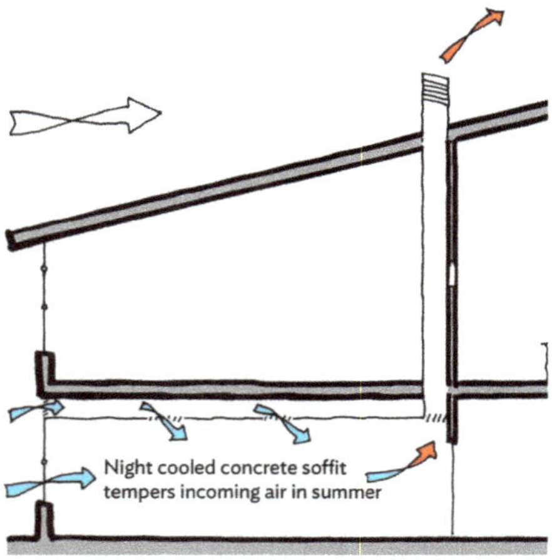

Sources: Passivent. 2021. Night cooling; European Concrete Platform. 2009. *General Guidelines for Using Thermal Mass in Concrete Buildings*.

Mixed-Mode Ventilation

As natural ventilation is passive, it is likely that a purely natural ventilation system may not perform as expected, due to unavailable or suboptimal natural driving forces, such as wind for wind-driven systems, or pressure and temperature differences for stack ventilation systems. Therefore, it is necessary to use mechanical ventilation to complement natural ventilation and improve ventilation performance. This leads to mixed-mode ventilation. There are three distinct approaches to mixed-mode ventilation, including contingency design, concurrent mixed-mode design, and zoned design.

Contingency design. This approach provides for the future addition or removal of mechanical ventilation or cooling systems. For example, a naturally ventilated building may be planned for the ease of adding mechanical ventilation and/or air-conditioning. Equally, an air-conditioned building may be planned so that natural ventilation, or a combination of natural and mechanical systems, could easily be used when such a need arises. This approach is particularly useful in situations where the occupancy and/or activities in the building may change.

Concurrent mixed-mode. This approach has both natural and mechanical ventilation systems present. If well-designed and operated, the combination of the two systems can achieve desirable performance at a cost less than what otherwise would have been required by a full air-conditioning system.

Zoned design. This approach uses different systems, or combinations of systems, in parts of the building where the requirements for ventilation and cooling vary because of planning, location, occupancy, and/or usage. Zoning is particularly efficient where problems can be grouped. For example, parts of a building may be similarly poorly located for passive cooling but have heat gains from solar radiation, extensive equipment use, and high occupancy. In this situation, local air-conditioning may provide a solution.

It is important to ensure that natural and mechanical systems are properly designed and operated to avoid clashes, wasteful, and inefficient operation. In general, the strategies to achieve this include the following three ways.

Concurrent operation. An intrinsically efficient mechanical ventilation system, with or without cooling, operates in parallel with the natural ventilation system. With some care, this can be an effective and energy-efficient solution. For example, in a building designed for high thermal stability, the use of background mechanical ventilation, coupled with efficient fans, night ventilation, and heat recovery, may be able to maintain a steady indoor thermal environment in a cost-effective manner. While the use of mechanical ventilation reduces the need to open windows, openable windows can provide natural ventilation when required and can enhance indoor air quality and the thermal comfort of occupants. Moreover, the availability of natural ventilation helps to avoid over-sizing of the mechanical system and save costs.

Changeover operation. In a building, natural and mechanical ventilation systems are both available and can operate together in different ways depending upon weather, time, occupancy, activities, etc. The underlying principle is that the use of natural ventilation is maximized to the extent reasonably possible, with the mechanical system used only as and when necessary. That is, natural and mechanical ventilation systems do not necessarily operate at the same time. For example, night ventilation in a building may be assisted by the mechanical system to enhance the cooling effect of the building structure. The night cooling effect may be further enhanced by mechanical refrigeration and cooling. Another example is local changeover, e.g., when windows are opened, mechanical cooling such as fan coils or chilled panels are switched off automatically.

Alternate operation. This is similar to the changeover operation but based on a longer-term operation and management choice. For example, when mechanical cooling is operated during the summer season, occupants are asked not to open windows when the cooling is running, e.g., during the daytime. Another example is that the system is operated to cater to the specific needs of occupancy and use, e.g., full air-conditioning with closed windows. When there is a change of occupancy and use at the end of tenancy, the system operation may be changed to full natural ventilation or a hybrid operation.

Figure 47 presents a flowchart for selecting an appropriate mixed-mode strategy in various application contexts.

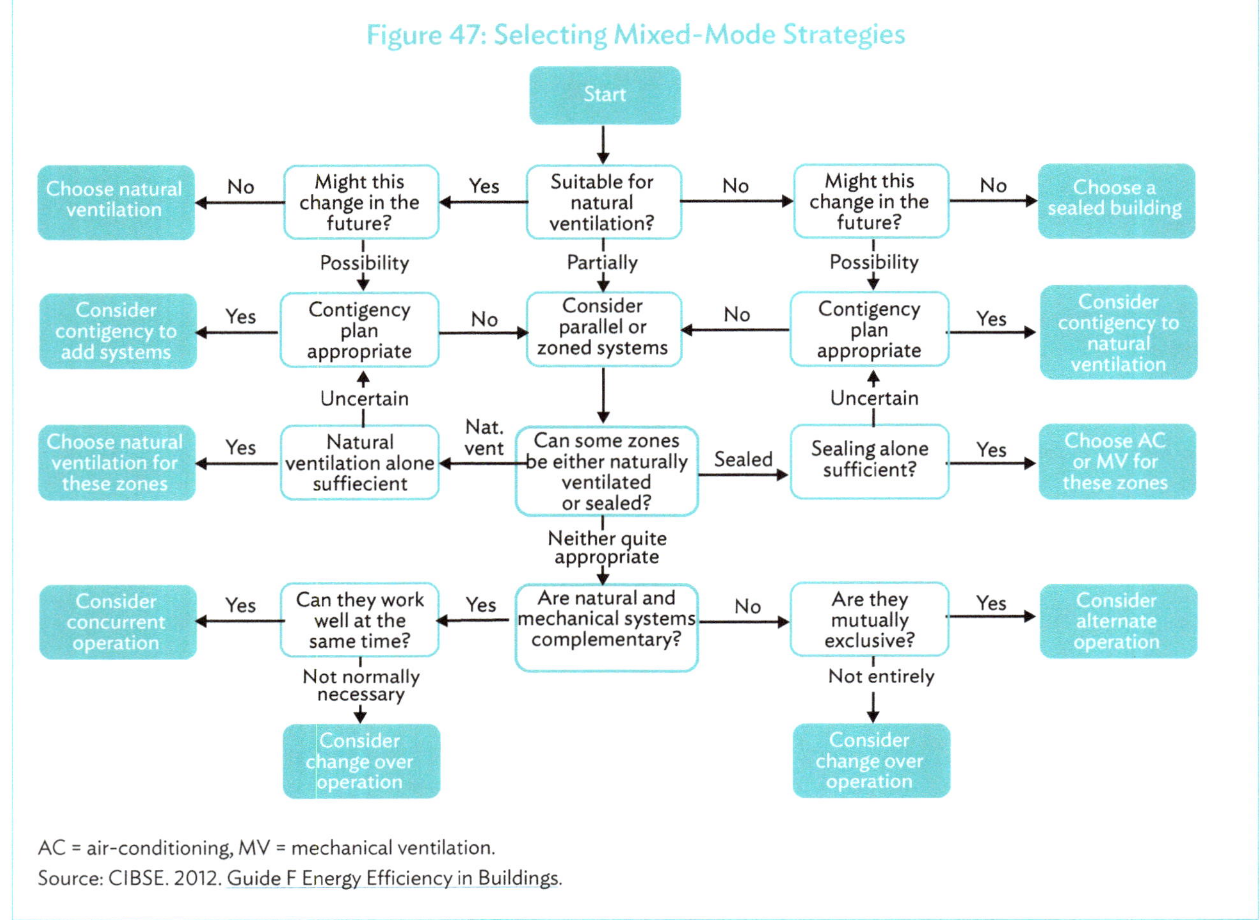

Figure 47: Selecting Mixed-Mode Strategies

AC = air-conditioning, MV = mechanical ventilation.
Source: CIBSE. 2012. Guide F Energy Efficiency in Buildings.

Box 9: High-Performance Air Systems

A high-performance air system (HPAS) is a variable-air-volume (VAV) system that optimizes energy efficiency, comfort, and indoor air quality, incorporating ventilation, cooling, and heating in a single ducted delivery system (Box Figure 9.1). HPASs perform significantly better than minimally code-compliant VAV systems by integrating the strategies and best practices of rightsizing, optimal zoning, free cooling, and coil cleaning using ultraviolet germicidal lamps. Also, HPASs minimize static-pressure drop, system leakage, and system effects.

Key strategies and technologies of HPASs include VAV, small temperature control zones, high-efficiency fans, economizers, air-to-air energy recovery, low leakage dampers and ducts, spiral or oval static regain ducting, low-pressure-drop components, variable flow compressors, and diagnostic sensors for system performance monitoring. All these contribute to energy savings. Advanced control technologies further increase energy savings through building-automation strategies such as demand-controlled ventilation, supply-air-temperature reset, and static-pressure reset.

An HPAS's high energy efficiency leads to low life cycle costs. Cooling energy cost savings are significant, as free cooling requires less cooling. Fan energy savings are also significant due to a lower system design static-pressure and optimal fan sizing. Additional energy savings come from on-and-off control via scheduling, the use of high-efficiency motors and variable-frequency drives, and demand-controlled ventilation. Continuous automated monitoring systems can lower life cycle costs by identifying operational events that could affect building performance. In addition to saving energy, the benefits also include higher worker productivity and an improved ability to lease the space. These benefits contribute to lowering life cycle costs.

Compared to ductless systems, other benefits of an HPAS include a low risk of refrigerant leakage (centralization of refrigerant in an equipment room or rooftop unit), improved comfort, flexibility, and efficiency (small temperature control zones), improved indoor air quality (use of minimum efficiency reporting value-13 or better filters), reduced noise (fans in a central location away from occupied spaces), higher adaptability (moving diffusers is easier than moving fan coil units), and safe and easy maintenance (no scheduled maintenance in occupied spaces).

Box Figure 9.1: Schematic of a High-Performance Air System

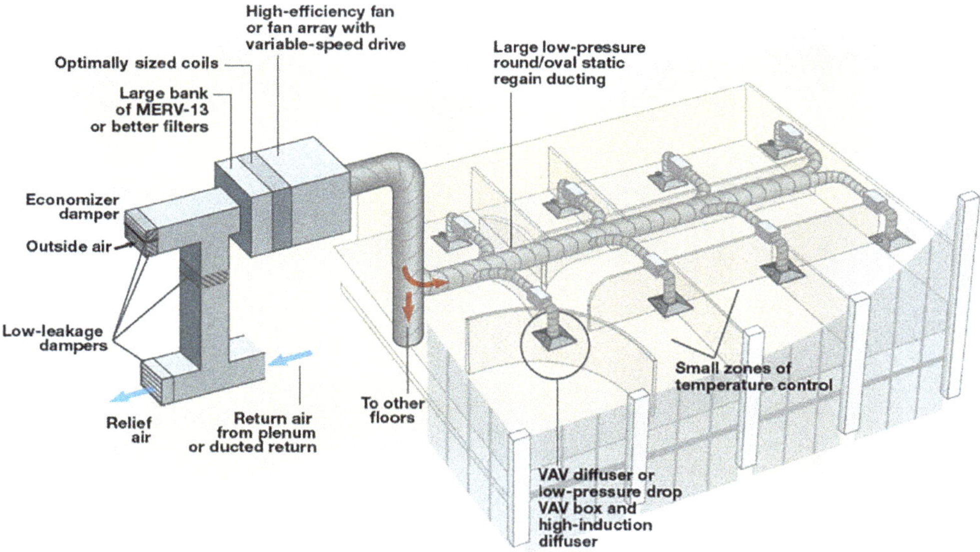

MERV = minimum efficiency reporting value, VAV = variable air volume.
Source: Air Movement and Control Association. 2017. *Introducing High Performance Air Systems*; Cowie, J. HPAC Engineering. 2017. *The Case for High-Performance Variable-Air-Volume (VAV) Systems*; Graham, C.I. 2016. *High-Performance HVAC*.

Box 10: Case Study of Mixed Ventilation and Microclimatic Envelope

Parkview Green FangCaoDi is one of the largest sustainable architecture projects in the People's Republic of China (PRC). This LEED Platinum and China 3-Star certified project, with a gross floor area of 200,000 square meters, comprises four buildings that include a luxury hotel, a retail mall, grade-A commercial office space, and a basement carpark. All buildings are designed with atria spaces, sky gardens, terraces, and link bridges to fit within a pyramidal glazed envelope (Box Figure 10.1). The four multistory buildings are separated by 24 meters to take advantage of natural light in the atrium. Offices are planned with a maximum 15-meter depth from facade to core and a clear floor-to-ceiling height of 2.9 meters for optimal daylight to reduce energy for lighting.

Box Figure 10.1: Parkview Green FangCaoDi, Beijing

Integrated Design Associates. 2021. Parkview Green (FangCaoDi).

Building.com.hk. 2011. *Parkview Green Beijing: Microclimatic Envelope*.

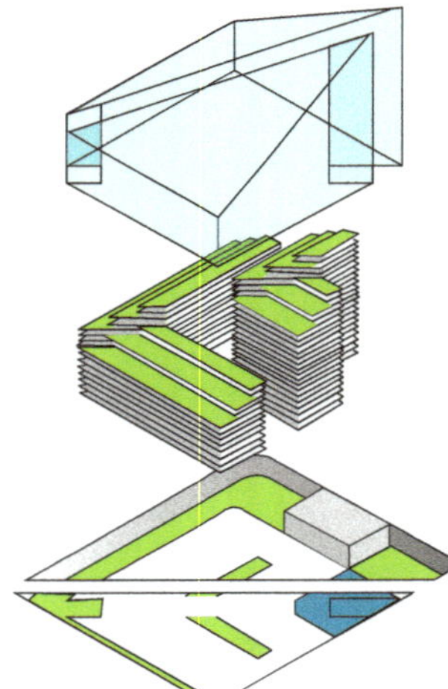

Building.com.hk. 2011. *Parkview Green Beijing: Microclimatic Envelope*.

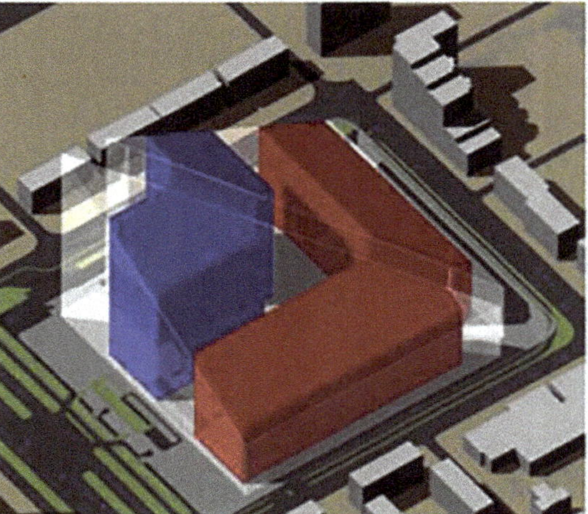

Building.com.hk. 2011. *Parkview Green Beijing: Microclimatic Envelope*.

The project aimed to set the standard for a new paradigm of sustainable building in the PRC through a series of specific objectives, including sustainable and environmentally friendly occupied areas, low natural resource consumption and energy efficiency, minimal social and environmental impact on surrounding building occupants, cost-effective green technologies, maximized use of hybrid ventilation, low environmental impact, and good indoor environmental quality, among others.

continued on next page

Box 10 *continued*

At the heart of the set of design strategies to achieve these objectives is the pyramidal envelope, which is made up of a layer of single-glazing for walls and air-filled ethylene tetrafluoroethylene (ETFE) cushions for the roof, all supported by a steel structure that wraps the entire development. Within the envelope, the buildings are encased and shielded from the outer environment. This created an air buffer zone—the microclimate within the zone is relatively uniform and easily controllable. As a weather protection layer, the buffer zone controls the microclimate by way of thermal insulation formed in the airspace between the outer skin and the buildings. Also, as a passive breathing apparatus, the buffer zone regulates the enclosed environment in response to the extreme seasonal temperature variations in Beijing. It limits the need for air-conditioning in hot summers and reduces heat loss during freezing winters.

Box Figure 10.2: Section View, Ventilation Intake, and Air Exhaust of Parkview Green FangCaoDi

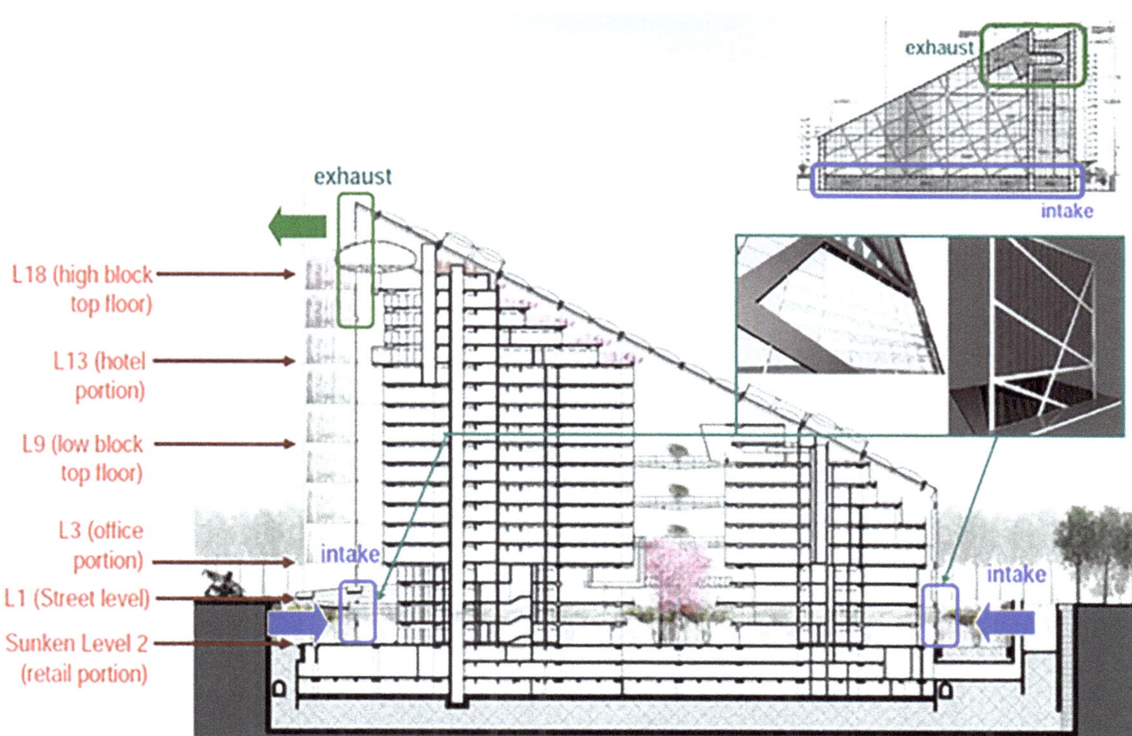

ARUP. 2008. Integrated Building Environmental Design—Beijing Parkview Green.

The pyramidal form of the envelope lends itself to natural air movement through the heat stack effect (Box Figure 10.2). Fresh air is drawn in at the base of the building as heated air rises through the atrium and into the roof void. The ETFE roof is set at a constant distance of 3 meters from the inner buildings to maintain air passage. Cool fresh air or warm air heated by solar radiation is being fed into the internal areas of the four buildings as required to regulate the temperature inside the office space. Operable windows and ETFE cushions are built into the microclimatic envelope for natural ventilation. These operable vents are computer-controlled and designed to work under different prevailing wind and environmental conditions. With the microclimatic envelope maintaining the atrium several degrees warmer in winter and cooler in summer, space cooling and heating loads and therefore the energy consumption of the HVAC system of the project are reduced substantially. The design and operation of the microclimatic envelope design are illustrated in box

continued on next page

Box 10 *continued*

figure 10.3.

Box Figure 10.3: Microclimatic Envelope Design (Summer Season) of Parkview Green FangCaoDi

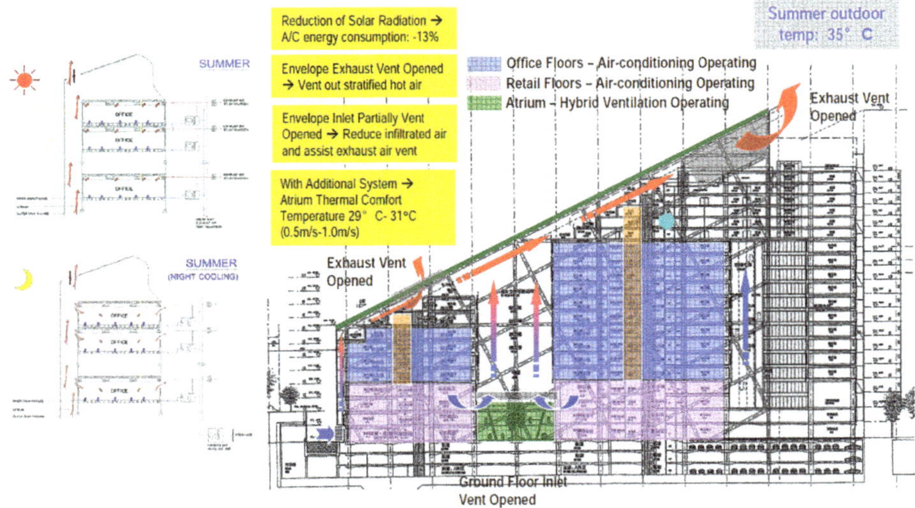

ARUP. 2008. Integrated Building Environmental Design—Beijing Parkview Green.

Box Figure 10.4: Microclimatic Envelope Design (Winter Season) of Parkview Green FangCaoDi

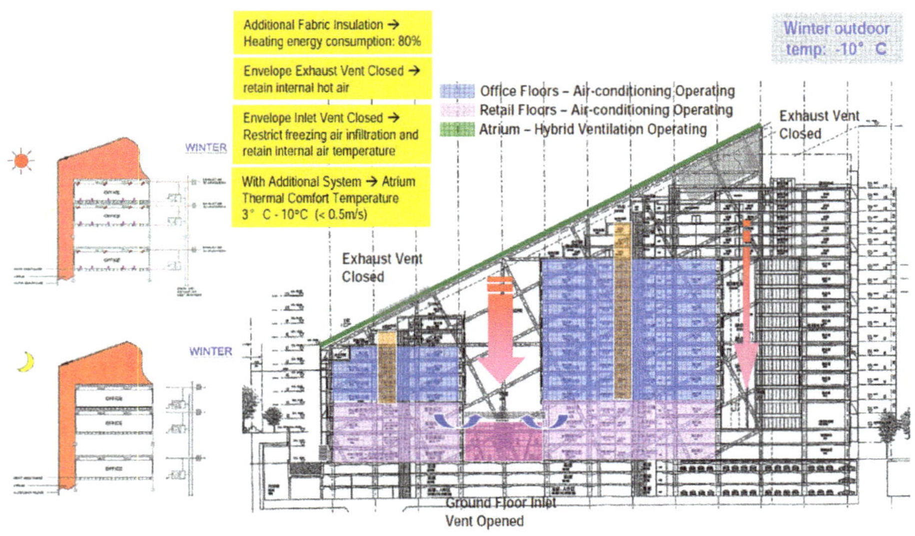

Source: ARUP. 2008. Integrated Building Environmental Design—Beijing Parkview Green; Integrated Design Associates 2021;
The Skyscraper Center. 2016. Parkview Green FangCaoDi.

Low-Energy Cooling Technologies

Heat Recovery Ventilation

The principle of heat recovery is simple. Heat energy that would otherwise be released into the environment and therefore wasted is recovered for useful purposes. The recovered heat can take the form of sensible heat (temperature), latent heat (moisture), or a combination of both. Air-to-air heat recovery in ventilation and air-conditioning systems is the process of recovering heat and/or moisture between supply and return airstreams at different temperatures and humidities. The enthalpy of the supply air is decreased by the heat recovery device during warm or hot weather and increased during cold weather. This process helps to ensure the desired level of thermal comfort and indoor air quality while maintaining low operational energy costs and improving overall system efficiency. In some applications, heat recovery measures can reduce the system capacity and, thus, the overall capital cost of the building services. It is advisable to integrate heat recovery into the system at the design stage, as retrofitting the system at a later stage will be less cost-effective.

The key to the technical and economic feasibility of heat recovery is that the otherwise-wasted, recoverable heat should be of sufficient amount to justify the added complications of system design and operation. Additional power consumption by fans to overcome the additional air resistance of heat exchangers is a key factor to consider. As long as the ventilation and air-conditioning systems are in operation, additional fan power consumption is needed. However, the amount of recoverable heat may vary substantially throughout a year and sometimes need to be bypassed. Therefore, it is important to ensure the total energy savings from heat recovery exceeds the additional power consumption. Sometimes, under certain weather conditions, temperature differences between supply and return air streams may be reversed. In such a situation, reversed energy transfer should be avoided. In some applications, such as high moisture content in return air or full fresh air supply without recirculation, possible cross-contamination between air streams is an important design consideration.

Ideally, an air-to-air heat exchanger should function well to allow temperature-driven heat transfer and partial pressure-driven moisture transfer between supply and return air streams. It should provide optimal energy recovery performance to minimize pressure drop, at reasonable dimensions, weight, and cost. The ideal heat exchanger should also be able to minimize cross-stream transfer of air, particulates, and contaminants. In real-world applications, the main types of heat recovery devices include plate heat exchangers, run-around coils, heat pipes, and heat wheels.

Plate Heat Exchangers

A plate heat exchanger contains a box with a series of parallel plates constructed from aluminum, steel alloys, or polymers, which allow the return air from the conditioned space to bypass the supply air for heat transfer. The plates are formed with spacers or separators constructed into the plates or with external separators. Plate spacing usually ranges from 0.1 to 0.5 inches, depending on the design and applications. The air streams are divided by the plates and have no contact with each other. Heat is transferred directly from the warm air stream through the separating plates into the cool air stream. Normally, only sensible heat is transferred. When water vapor permeable materials are used, such as microporous polymeric membranes, moisture may also be transferred, therefore enabling both sensible and latent heat (enthalpy) exchange.

Plate heat exchangers have only a primary heat transfer surface area separating the air streams and are not subject to the additional secondary resistance inherent in some other exchanger types, e.g., pumping liquid, in run-around systems or transporting a heat transfer medium. Almost any airflow direction and pattern of the supply and exhaust air streams can be handled by plate exchangers. In real applications, crossflow exchangers are the most commonly used configuration, due to design, construction, and cost factors. Sometimes additional counterflow patterns can be applied to increase heat transfer effectiveness. As simple static devices, plate heat exchangers require no motive power and are therefore easy to commission and maintain. There is no or little leakage between air streams. Thus, the risk of cross-contamination is low. Normally, the efficiency of plate heat exchangers is in the range of 30% to 70%.

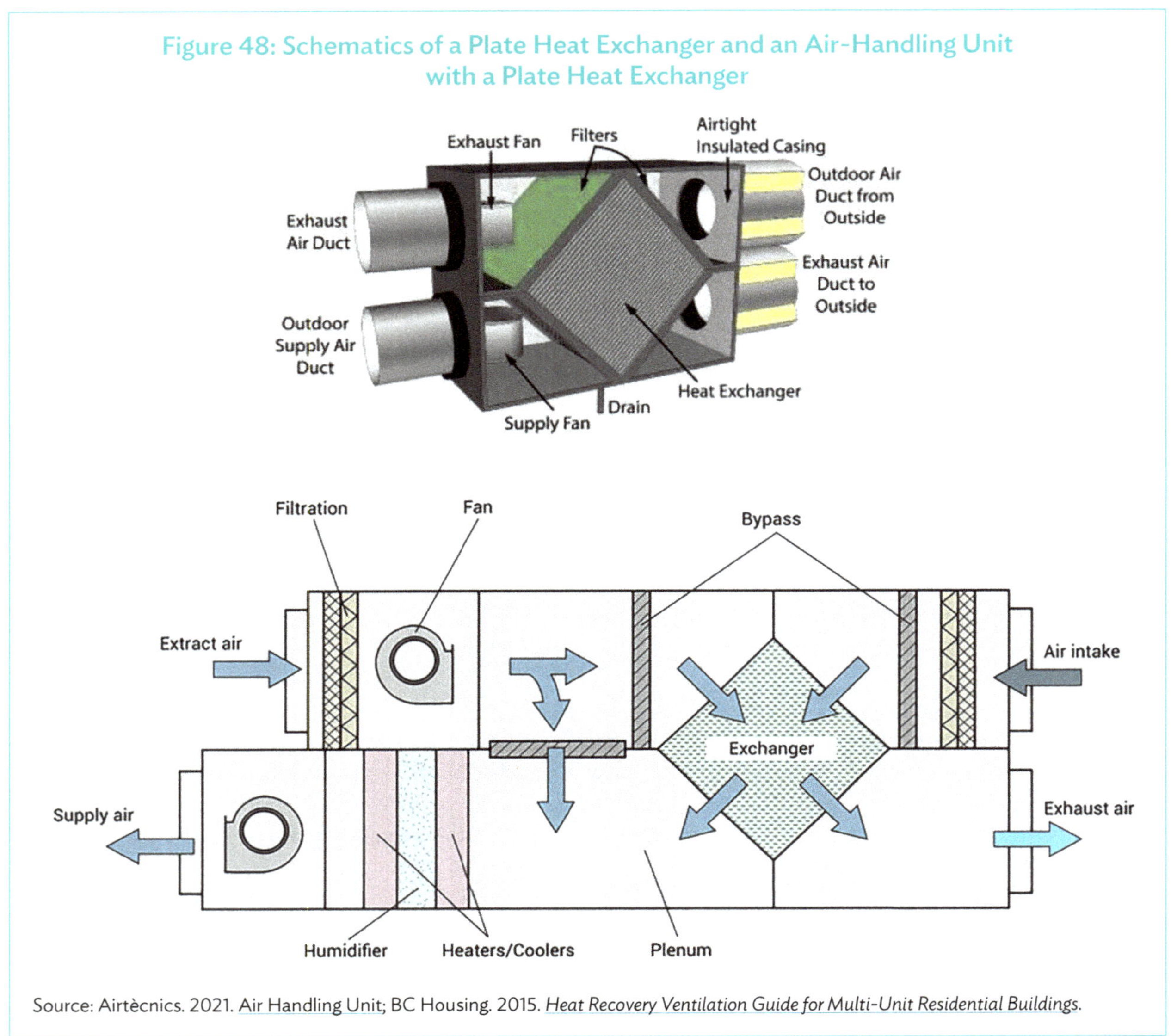

Figure 48: Schematics of a Plate Heat Exchanger and an Air-Handling Unit with a Plate Heat Exchanger

Source: Airtècnics. 2021. *Air Handling Unit*; BC Housing. 2015. *Heat Recovery Ventilation Guide for Multi-Unit Residential Buildings*.

Run-Around Coils

Run-around coils, also called coil energy recovery loops, are composed of extended surface, finned-tube copper coils located in the supply and return air streams of a building. The coils are connected in a closed loop by pipework, through which an intermediate heat transfer fluid is pumped and circulated. Typically, the fluid is water or an antifreeze solution. Sensible heat transfer takes place between air streams indirectly, via the fluid. Latent heat, i.e., moisture, cannot be transferred between air streams.

Run-around coils are particularly well-suited to applications where the supply and return air units are set up separately from one another due to layout constraints or strict hygienic requirements. The complete separation of supply and return air streams avoids the mass transfer of bacteria, contaminants, odors, etc. The high flexibility of the loop configuration also allows simultaneous heat transfer involving multiple sources and uses in a single loop. Another advantage of run-around coils is their seasonal reversibility, which makes it possible to precool in the summer and preheat in the winter.

When assessing the technical and economic feasibility of run-around coils, additional air resistance, pumping energy consumption, temperature conditions, and frost protection are important factors to be considered. The typical efficiency of run-around coils ranges from 45% to 65%, depending on the number and spacing of coil rows and the prevailing temperatures.

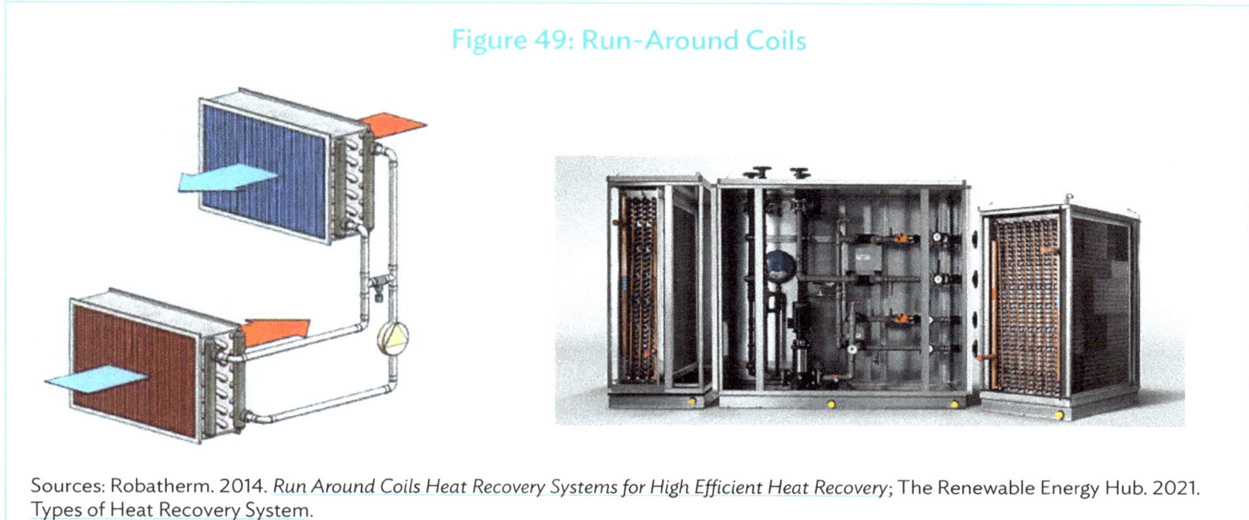

Figure 49: Run-Around Coils

Sources: Robatherm. 2014. *Run Around Coils Heat Recovery Systems for High Efficient Heat Recovery*; The Renewable Energy Hub. 2021. Types of Heat Recovery System.

Heat Pipes

Heat pipes use a closed fluid cycle within a sealed tube to provide a reversible heat recovery system. A heat pipe has a condensation at one end and an evaporation section on the other end. Hot air flowing over the evaporation section transfers heat to the working fluid, which then vaporizes. A vapor pressure gradient drives the vapor to the condensation section, where the vapor condenses and releases the latent heat of vaporization. The condensed working fluid flows back to the evaporation section for re-vaporization. This way, the working fluid keeps operating in a closed-loop evaporation and condensation cycle, as long as there is a temperature difference as the driving force.

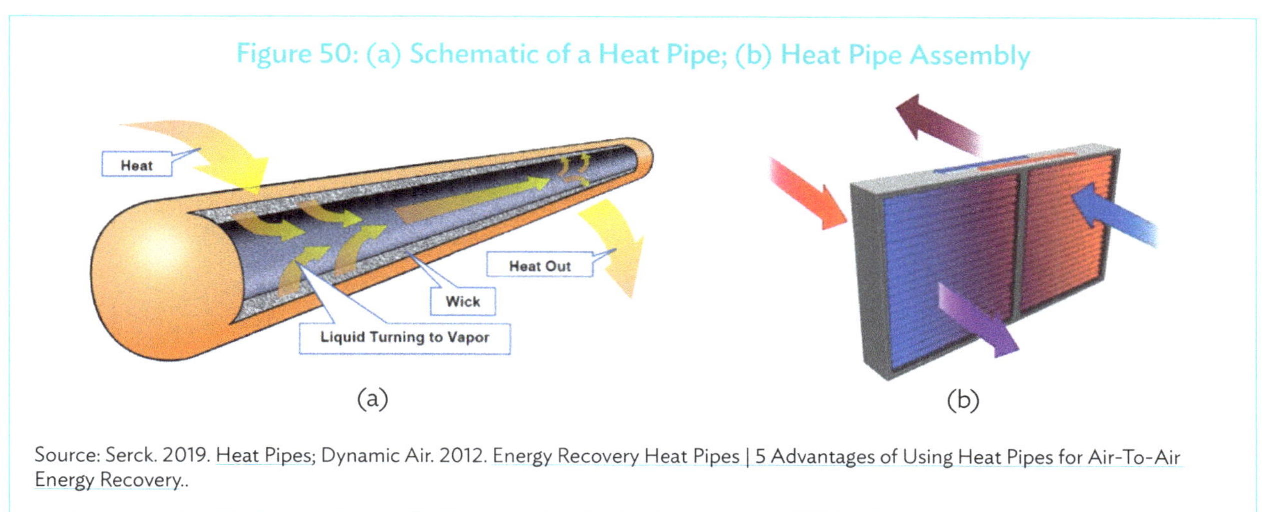

Figure 50: (a) Schematic of a Heat Pipe; (b) Heat Pipe Assembly

Source: Serck. 2019. Heat Pipes; Dynamic Air. 2012. Energy Recovery Heat Pipes | 5 Advantages of Using Heat Pipes for Air-To-Air Energy Recovery..

For ventilation and air-conditioning applications, heat pipe heat exchangers can be used to retrofit an existing system for improved energy recovery and dehumidification. During summer, the hot supply air stream is precooled by a heat pipe heat exchanger before it enters the cooling coil. This allows more dehumidification along with cooling by the cooling coil. Therefore, the peak demand and total energy consumption of the system can be reduced. For new installations, using heat pipes can have the benefit of reducing the air-conditioning system. During winter, the process is reversed. The warm return or exhaust air stream from the conditioned space transfers heat to the cold supply air stream indirectly via the heat pipe heat exchangers. This process preheats the supply air before it enters the heating coil.

Heat pipes can also be designed specifically for dehumidification of the supply air. As shown in Figure 51, air passing through the evaporator coil is cooled to a low temperature, resulting in significant moisture removal. The "over-cooled" air is then reheated to a desirable temperature by the reheat heat pipe section, using the heat transferred from the precool heat pipe. This entire process of precooling and reheating is accomplished passively, without mechanical active parts or the input of additional energy. This offers the ability to remove considerably more moisture than regular systems.

The working fluid is critical to the operational performance of a heat pipe. Its key characteristics should include high latent heat of vaporization, high surface tension, low liquid viscosity, and thermal stability under operating conditions. For ventilation and air-conditioning applications, normally this fluid is commercial refrigerant. Copper or aluminum are typical materials for heat pipe tubes and fins. Supply and exhaust air ductwork need to be adjacent to install the heat pipes and allow heat transfer through them. Typically, the efficiency of heat pipes is in the range of 50% to 65%.

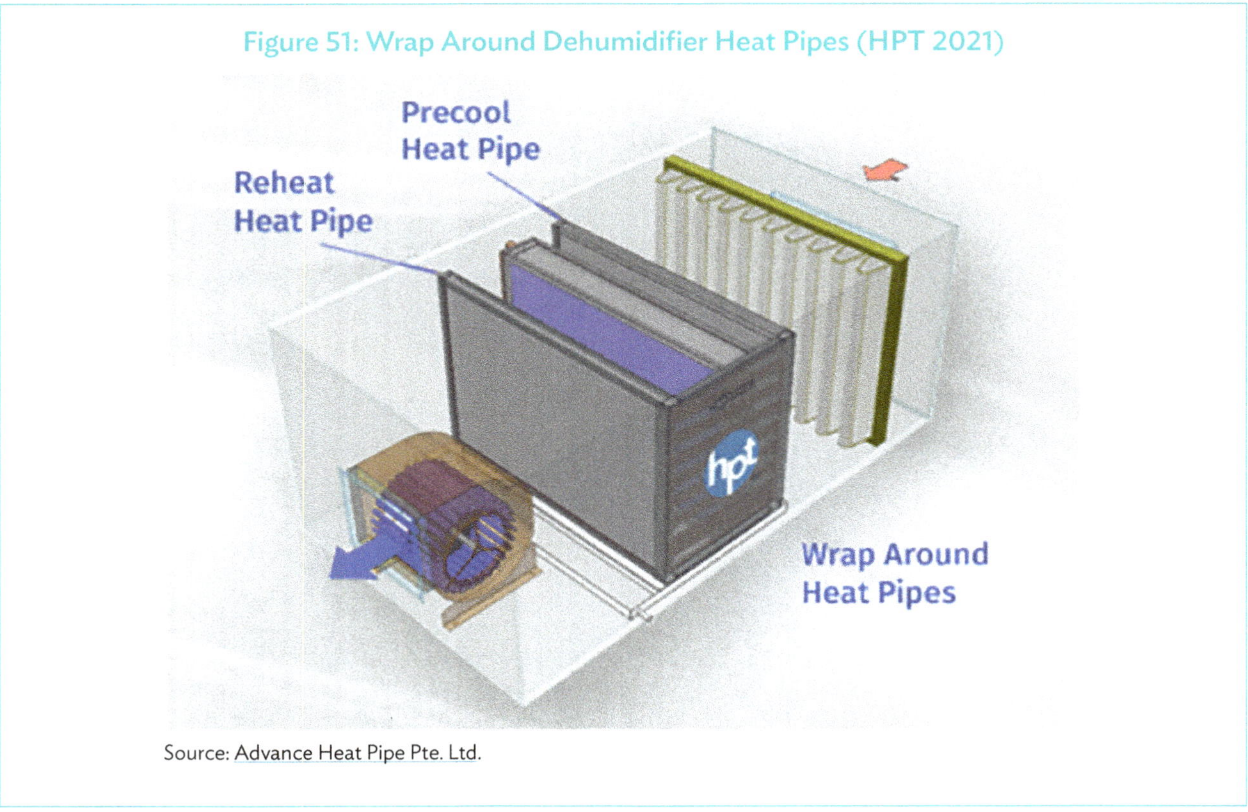

Figure 51: Wrap Around Dehumidifier Heat Pipes (HPT 2021)

Source: Advance Heat Pipe Pte. Ltd.

Thermal Wheel

Compared to these heat recovery devices, a thermal wheel, also known as a rotary air-to-air heat exchanger, usually has a higher heat recovery efficiency, which can reach 60% to 85%, depending on the medium construction of the wheel.

A thermal wheel has a revolving cylinder filled with an air-permeable and heat-absorbing medium that has a large internal surface area. The wheel rotates slowly while adjacent supply and exhaust air streams from an air-handling unit flow through half of the wheel in a counterflow arrangement. Sensible heat is transferred from the hot air stream to the medium for temporary storage and subsequent release to the cold air stream (O'Connor, Calautit, and Hughes 2016). During summer, the hot supply air is precooled by this setup. During winter, the cold supply air is preheated. Figure 52 shows the schematic of a thermal wheel operating for precooling outdoor air during summer.

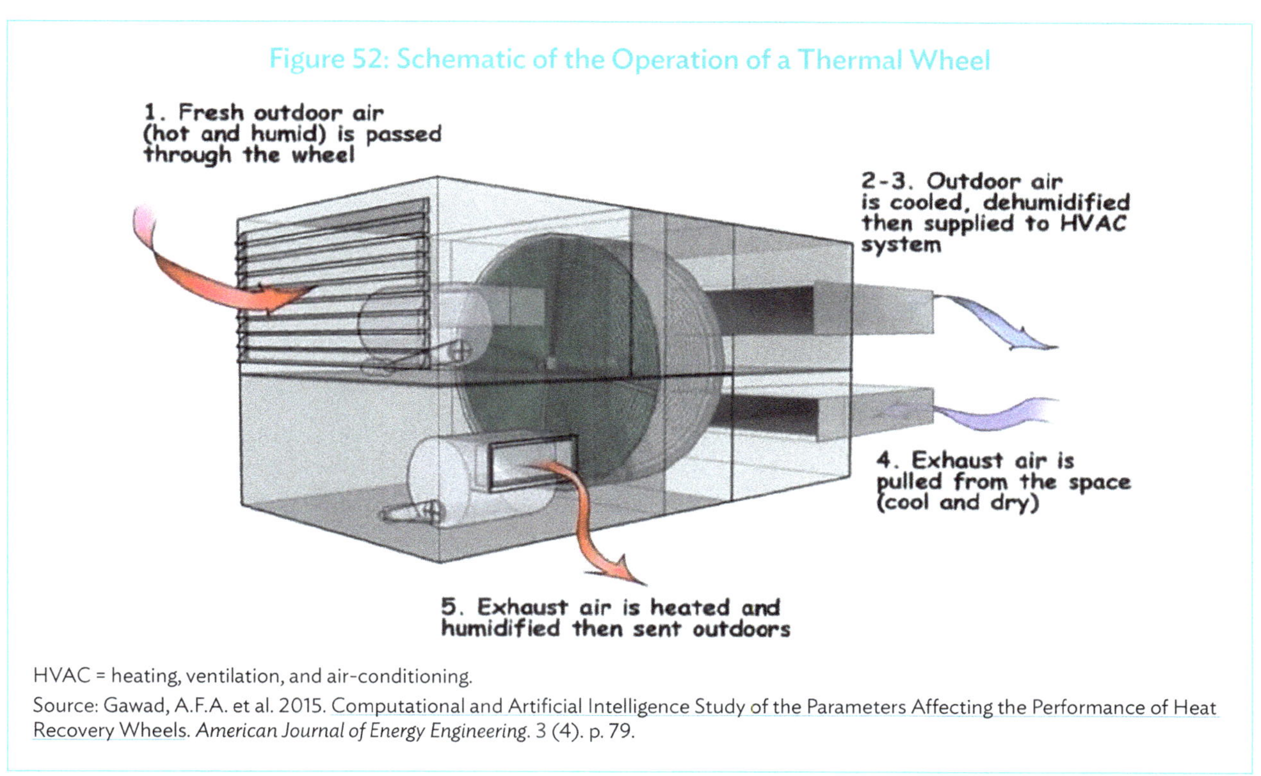

Figure 52: Schematic of the Operation of a Thermal Wheel

HVAC = heating, ventilation, and air-conditioning.
Source: Gawad, A.F.A. et al. 2015. Computational and Artificial Intelligence Study of the Parameters Affecting the Performance of Heat Recovery Wheels. *American Journal of Energy Engineering*. 3 (4). p. 79.

Depending on the heat transfer medium, a thermal wheel can also recover both sensible and latent heat, i.e., total heat. When a hygroscopic medium is used, latent heat is transferred as the medium adsorbs water vapor from the higher-humidity air stream and desorbs moisture into the lower-humidity air stream. This process is driven by the vapor pressure difference between the air stream and medium. Simultaneously, sensible heat transfer also occurs.

Air temperature, dew point, and contaminants are key factors that influence the choice of materials for the casing, rotor structure, and medium of a thermal wheel. Common choices include metal, mineral, or synthetic materials. Aluminum, steel, and polymers are the usual structural, casing, and rotor materials for normal ventilation and air-conditioning systems. The operation of a thermal wheel can be regulated by controlling the amount of air bypassing the wheel and varying the rotational speed of the wheel. There is the possibility of cross-contamination from the exhaust air to the supply air, which can normally be mitigated using a purge section.

It should be noted that thermal wheels are different from desiccant wheels, which are for the specific purpose of dehumidifying the supply air stream. Often, desiccant wheels and thermal wheels are used in series within a system for heat recovery and dehumidification. Section 4.3.4 of this handbook discusses applications of desiccant wheels.

Evaporative Cooling

Climatic conditions may often provide a source of "renewable" energy to provide a cooling effect. In dry and high-temperature climates, air systems using evaporative cooling can be an energy-efficient, cost-effective, simple, and environmentally friendly solution as a good supplement to conventional mechanical vapor-compression cooling systems (Heidarinejad et al. 2009; ASHRAE 2017).

Evaporative cooling works under the basic principle that water evaporated in non-saturated air will produce a drop in dry-bulb temperature and an associated rise in moisture content. When the warm ambient inlet air comes into contact with water added to the inlet air, it transfers some of its heat to the liquid water and evaporates some of the water. The process of heat transfer from inlet air to water cools the inlet air. The increase in water vapor content means a higher relative humidity in the air. An additional benefit of this process is that the water can clean the inlet air stream of airborne particulates and soluble gases (ASHRAE 2016). Theoretically, the inlet air can be cooled up to 98% of the difference between its ambient dry-bulb and wet-bulb temperatures. Thus, the most cooling can be achieved in hot and/or dry weather. The primary disadvantage of evaporative cooling is the large quantity of water consumption, which is a major component of system operating costs. There are two types of evaporative cooling: direct evaporation and indirect evaporation. Both direct and indirect systems have significant potential to save energy by eliminating the need for mechanical cooling, or by reducing the operating hours of chillers.

For direct evaporative cooling, moisture is added directly to the supply air stream to allow the latent heat of evaporation to cool the air. The dry-bulb temperature of the air is lowered, whereas the wet-bulb temperature remains unchanged. The relative humidity of the air has increased. This process is shown in the psychrometric chart in Figure 53.

In real applications, instead of directly adding water to the air, a wetted element is used as the cooling medium. A typical method is to expose the supply air to a thin film of water on the extended surface of honeycomb-like wetted media made of cellulose paper, which is also referred to as an evaporative cooling panel or pad. Water is

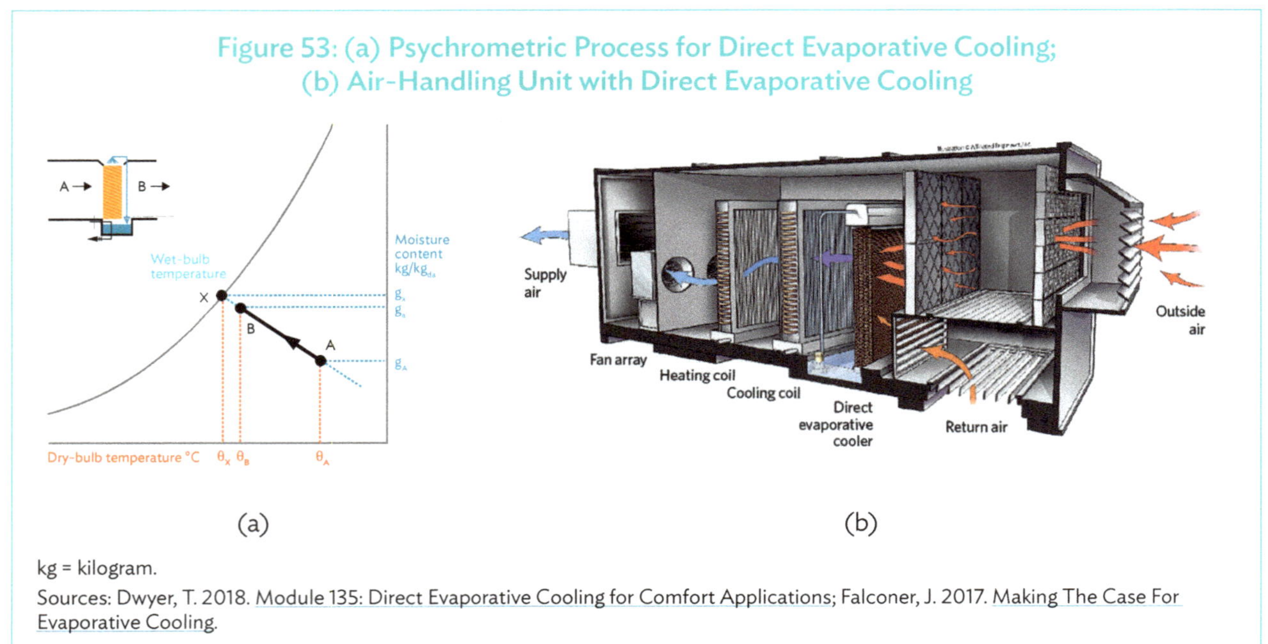

Figure 53: (a) Psychrometric Process for Direct Evaporative Cooling; (b) Air-Handling Unit with Direct Evaporative Cooling

kg = kilogram.
Sources: Dwyer, T. 2018. Module 135: Direct Evaporative Cooling for Comfort Applications; Falconer, J. 2017. Making The Case For Evaporative Cooling.

supplied through a manifold at the top of the pad and allowed to trickle down as air flows through the pad. Water used for this purpose may or may not require treatment, depending on its quality and the pad's specifications. An alternative method is termed fogging, which is to add moisture to the supply air by spraying fine droplets of water (Figure 54). High-pressure fogging systems can be designed to produce droplets of variable sizes, depending on the desired evaporation time and ambient conditions. Typically, the water used by fogging systems is of higher quality than that used by wetted media systems.

With direct evaporative cooling, the humidity of the supply air and that of the conditioned space will increase. From the perspective of thermal comfort and indoor air quality, a high relative humidity may still be acceptable so long as the dry-bulb temperature is within a reasonable range. Usually, the relative humidity should not exceed 70%. In many applications, a high relative humidity may not be desired and therefore needs to be mitigated.

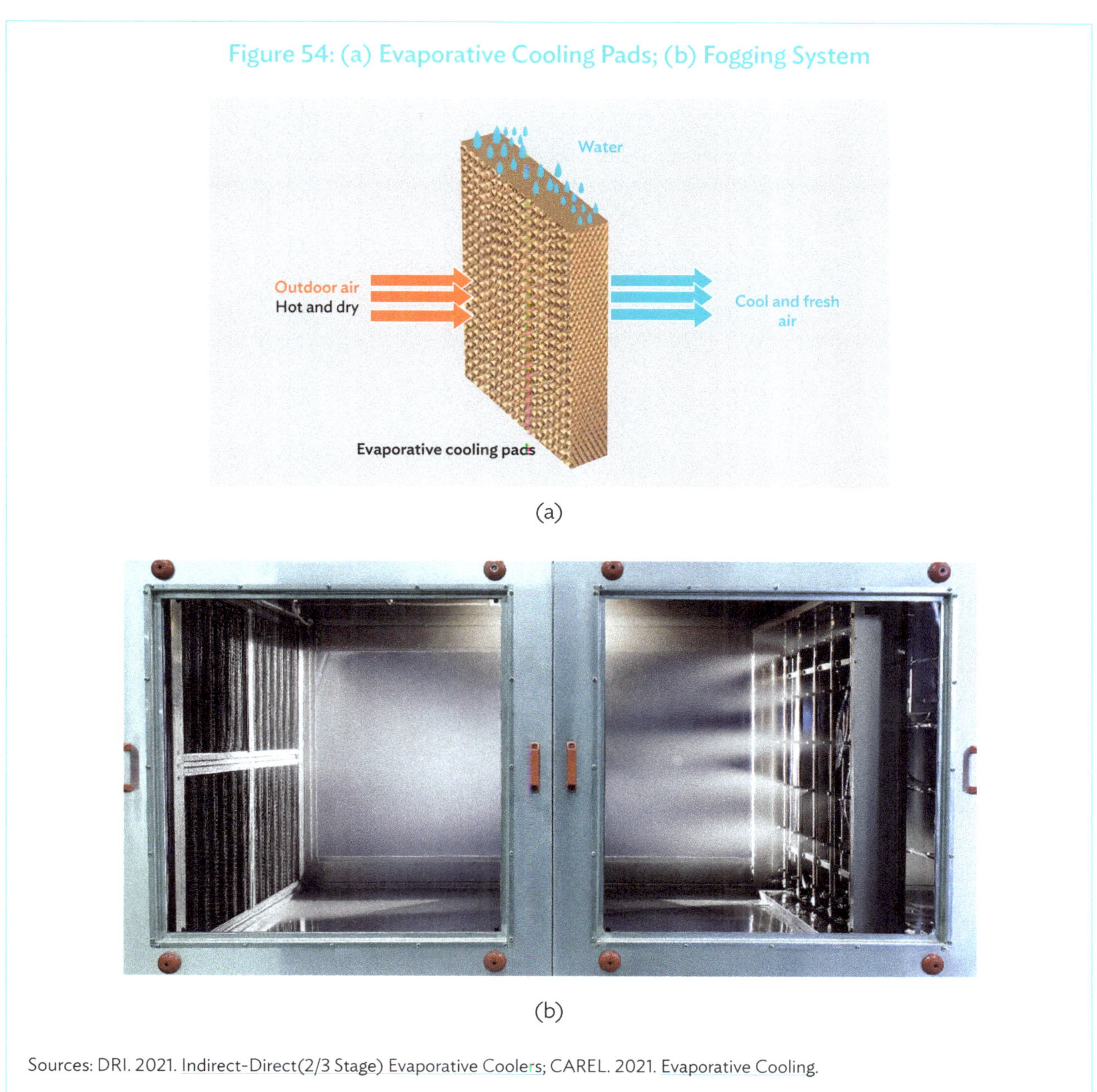

Figure 54: (a) Evaporative Cooling Pads; (b) Fogging System

(a)

(b)

Sources: DRI. 2021. Indirect-Direct(2/3 Stage) Evaporative Coolers; CAREL. 2021. Evaporative Cooling.

One way is to mix the cooled and humidified supply air with outdoor air that has bypassed the evaporative cooler. Another method, which is more popular, is to use indirect evaporative cooling. A secondary air stream, which can be either return air from the conditioned space or outdoor air, is first evaporatively cooled and then used to precool, through a heat exchanger, the incoming outdoor air to reduce its dry-bulb temperature without directly affecting its moisture content. The wet-bulb temperature of the supply air is reduced. This offers the benefit of allowing the supply air (with a reduced wet-bulb temperature) to be further cooled, through direct evaporation, to a lower dry-bulb temperature than what is possible by only direct evaporative cooling of the ambient air, without first using indirect cooling (ASHRAE 2016). There are, however, disadvantages to indirect evaporative cooling. When humidity is high, indirect evaporative cooling cannot reduce the dry-bulb temperature as low as what is possible with conventional chillers. Compared with direct cooling, the additional capital cost of an indirect heat exchanger is required for indirect evaporative cooling. Also, indirect evaporative cooling has a higher total parasitic load than chillers and direct evaporative cooling, due to electricity consumption for pushing air through the direct evaporative cooling sections and the heat exchanger (ASHRAE 2016). Indirect evaporative cooling has an efficiency of 60%–70%, whereas direct evaporative cooling can reach an efficiency of 85% (Sarbu and Sebarchievici 2017b). Figure 55 shows the psychrometric path, a schematic of a simple heat exchanger, and an air-handling unit with indirect evaporative cooling.

Evaporative cooling is often used in conjunction with a conventional mechanical cooling system. It can also be combined with desiccant dehumidification to create a hybrid system to cope with both sensible and latent cooling loads in the conditioned space.

Figure 55: (a) Psychrometric Process for Indirect Evaporative Cooling; (b) Schematic of Indirect Heat Exchanger; (c) Air-Handling Unit with Direct Evaporative Cooling

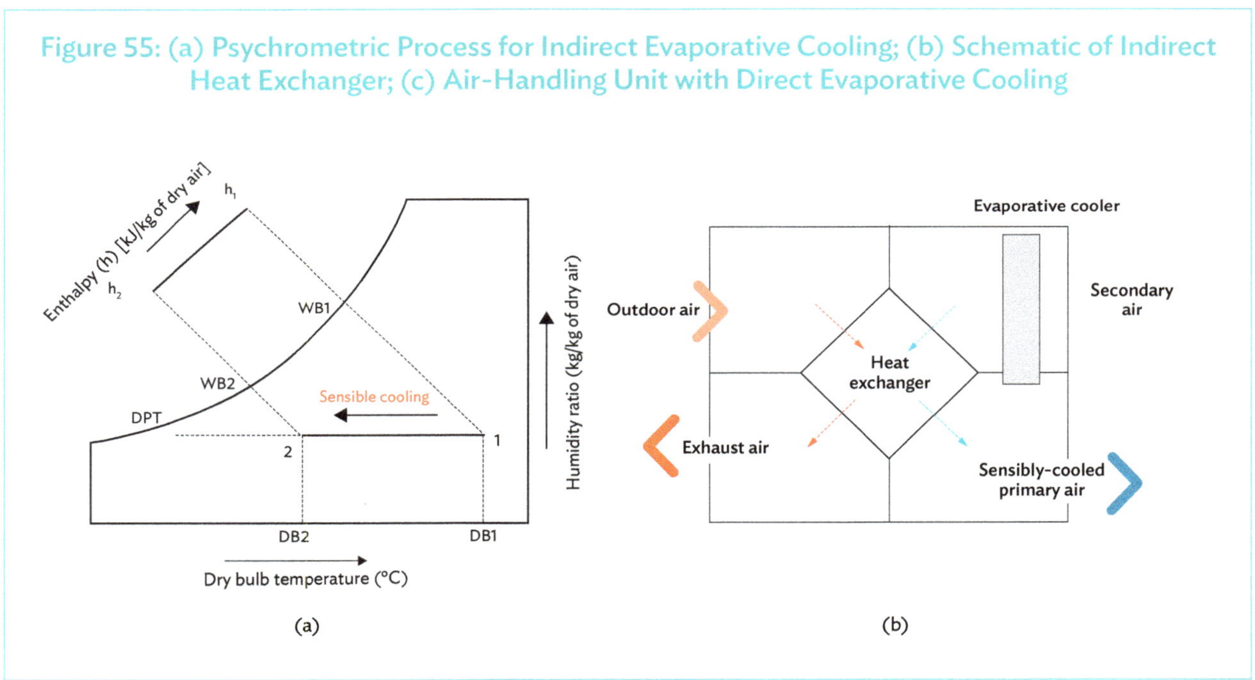

continued on next page

Figure 55 *continued*

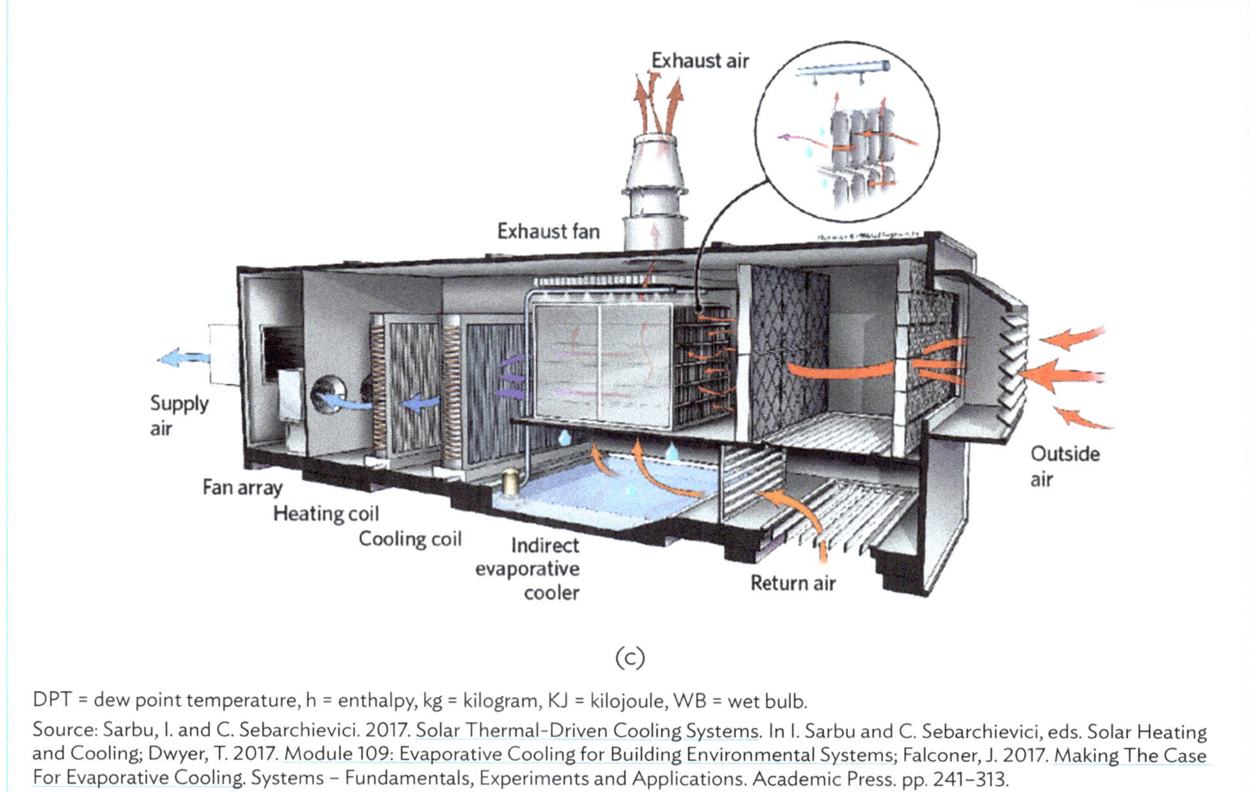

DPT = dew point temperature, h = enthalpy, kg = kilogram, KJ = kilojoule, WB = wet bulb.

Source: Sarbu, I. and C. Sebarchievici. 2017. Solar Thermal-Driven Cooling Systems. In I. Sarbu and C. Sebarchievici, eds. Solar Heating and Cooling; Dwyer, T. 2017. Module 109: Evaporative Cooling for Building Environmental Systems; Falconer, J. 2017. Making The Case For Evaporative Cooling. Systems – Fundamentals, Experiments and Applications. Academic Press. pp. 241–313.

Sorption Cooling

Sorption technologies are typically used in thermal-driven cooling systems. There are two types of sorption systems, i.e., open ones, and closed ones. Open sorption systems, also called desiccant cooling systems, are used for air dehumidification. Open systems are introduced in a separate section. This section focuses on closed sorption systems for refrigeration and cooling purposes, including absorption systems and adsorption systems. Closed systems use chemical or physical attraction between a pair of substances, called the sorption working pair, to produce the refrigeration effect. The working pair is isolated from the ambient, hence the term "closed." For the absorption cooling system, the common working pairs are water-lithium bromide (LiBr) and ammonia-water. For the adsorption cooling system, the common working pairs are water-zeolite, water-silica gel, ammonia-calcium chloride ($CaCl_2$), etc.

Absorption Cooling

As a thermal-driven cooling system, absorption cooling has the advantage of being able to utilize low-grade heat, such as solar thermal heat and industry waste heat. Like conventional vapor compression-based cooling systems, the cooling effect of absorption cooling systems also comes from the evaporation of a liquid refrigerant. The difference between the two systems is the way in which the refrigerant vapor is pressurized. In conventional vapor-compression cooling systems, the refrigerant vapor is pressurized by mechanical compression. In absorption cooling systems, the refrigerant vapor is pressurized by the solution circulating under two pressure conditions. This is realized through operating the absorption cooling system with a binary working fluid, i.e., a working pair that contains the refrigerant and the absorbent. The refrigerant has a lower boiling temperature than the absorbent. The equilibrium concentration of the working pair varies with temperature and pressure.

Figure 56 presents a schematic diagram of a typical absorption cooling system. The system consists of four components, including an absorber, where sorption takes place; a generator, where desorption takes place; a condenser; and an evaporator. The system operation involves two circulation loops, i.e., a solution loop and a refrigerant loop. In the solution loop, the solution flows through the generation (desorption), throttling (expansion valve), absorption, and pumping processes. In the refrigerant loop, the refrigerant flows through the generation (desorption), condensation, throttling (expansion valve), evaporation, and absorption processes. The two loops are not separate but connected by the absorption and generation (desorption) processes.

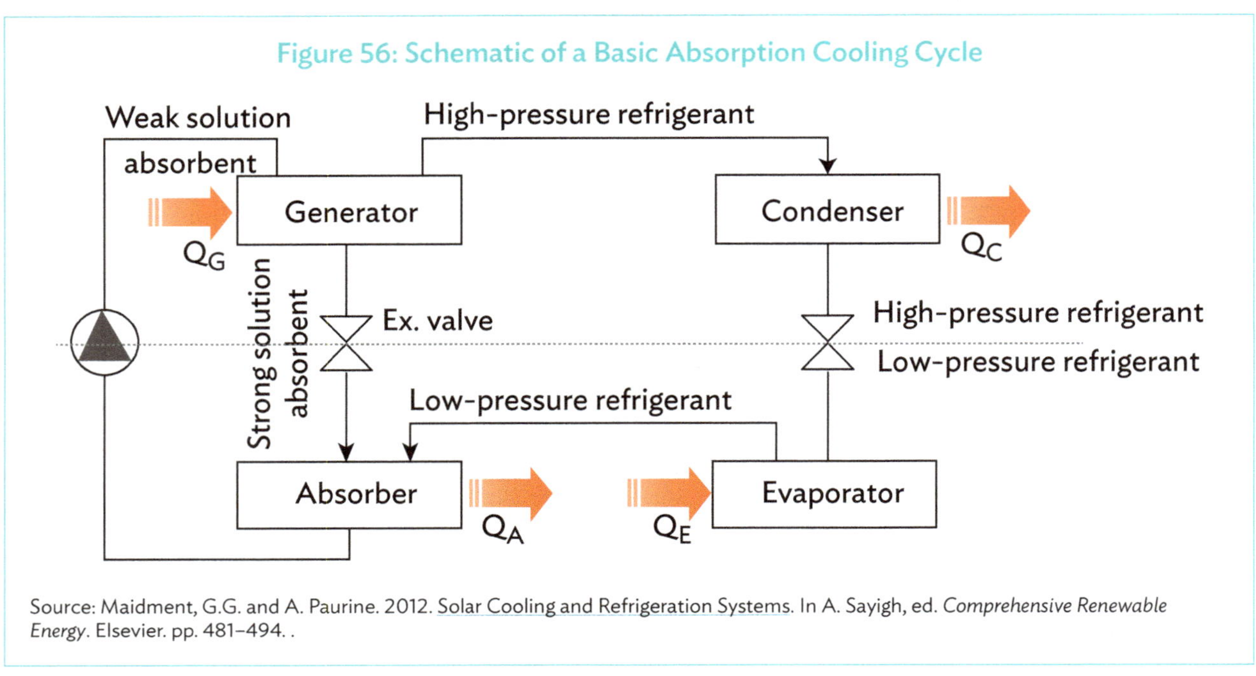

Figure 56: Schematic of a Basic Absorption Cooling Cycle

Source: Maidment, G.G. and A. Paurine. 2012. Solar Cooling and Refrigeration Systems. In A. Sayigh, ed. *Comprehensive Renewable Energy*. Elsevier. pp. 481–494. .

High-pressure liquid refrigerant leaving the condenser flows through an expansion valve for throttling, which reduces its pressure, and then flows into the evaporator. In the evaporator, the low-pressure refrigerant evaporates and removes heat from the cooling load (air or water). The resultant low-pressure refrigerant vapor is then sent to the absorber, where it is absorbed into a strong solution absorbent, and the sorption heat is released to the ambient. The evaporation and absorption processes have the same pressure. The weak solution with a high percentage of refrigerant leaving the absorber is recirculated to the generator, where it is boiled into a strong solution absorbent. At the same time, the refrigerant vapor is boiled off. This process is under high pressure and requires heat, which can be supplied via solar thermal energy or industrial waste heat. After this step, the strong solution is passed back to the absorber via an expansion valve, and the high-pressure refrigerant vapor is sent to the condenser for condensation. The generation and condensation processes have the same pressure. Essentially, the solution loop acts as a thermal-driven compressor to regenerate the high-pressure refrigerant vapor, which is analogous to the electric compressor used in a conventional system. Therefore, from an energy-saving perspective, the attractiveness of an absorption cooling system is the very low electricity consumption required for the solution loop and the ability to use low-grade heat sources for the desorption process in the generator, such as solar heat and waste heat. The condensation and evaporation processes work under the same principle as a conventional system.

The performance of absorption cooling is substantially dependent on the properties of its working pair. Typical working pairs for absorption cooling systems include water-LiBr and ammonia-water. In a water-LiBr solution, water is the refrigerant and the water-LiBr solution is the absorbent. In an ammonia-water solution, ammonia is the refrigerant, and the ammonia-water solution is the absorbent. Systems using water-LiBr solutions are often used for air-conditioning, whereas systems using ammonia-water solutions are more suitable for refrigeration in industrial applications. The operation of water-LiBr-based absorption systems is subject to limits on the evaporating temperature and the absorber temperature due to the freezing of the water and the solidification of the LiBr-rich solution. While this is not an issue for ammonia-water systems, the toxicity of ammonia is a concern, and ammonia is often used for systems with large capacities. Table 8 compares the advantages and disadvantages of the two working pairs.

Table 8: Comparison Between Water-LiBr and Ammonia-Water as Working Pairs for Absorption Systems

Working pair	Advantages	Disadvantages
Water-LiBr	• High coefficient of performance • Low operation pressures • Environmentally friendly and innoxious	• Risk of congelation • Need for a device for anticrystallization • Relatively expensive (LiBr)
Ammonia-water	• Evaporative at temperatures below 0°C	• Toxic and harmful to health (ammonia) • Need for rectifier for refrigerant purification • Operation at high pressures

Source: Sarbu, I. and C. Sebarchievici. 2017b. Solar Thermal-Driven Cooling Systems. In I. Sarbu and C. Sebarchievici, eds. Solar Heating and Cooling Systems – Fundamentals, Experiments and Applications. Academic Press. pp. 241–313.

There are several variants of absorption chillers, including half-effect, single-effect, double-effect, and triple-effect systems, with a working pair of water-LiBr or ammonia-water. The main practical difference between these systems is the driving temperature required for regenerating the solution and their relative coefficients of performance (COPs) as well as capital cost and complexity. A common configuration is the single-effect water-LiBr system that requires heat at around 80°C and typically achieves a COP of around 0.7 for a chilled water application.

As the desorption process requires low-grade heat, solar thermal heat provides an ideal heat source to drive an absorption chiller and produce a cooling effect. Options for solar collectors include a flat plate, an evacuated tube, a parabolic trough collector, a parabolic concentrator, and so on. Different solar collectors produce heat at different temperatures, while different configurations of absorption chillers require different driving temperatures. For example, a flat-plate solar collector may work well for a single-effect system, but not for a double-effect system which requires a much higher driving temperature. Therefore, temperature is key to the proper combination of solar collector and absorption chiller. Sometimes, an auxiliary heat source is used to complement the solar collector to enhance the system's ability to operate continuously.

A single-effect absorption chiller has the simplest configuration and the highest commercial availability, as compared to other variants of absorption chillers. Due to these advantages, it is the most commonly used type in solar absorption cooling systems. Usually, the system COP is around 0.7–0.8. The system design depends on the working fluid types. In practice, the system shows better performance with a nonvolatile working pair, such as water-LiBr. If a volatile working pair is used, such as ammonia-water, an extra rectifier should be used before the condenser to provide pure refrigerant (Sarbu and Sebarchievici 2017b). A systematic diagram of a single-effect water-LiBr absorption cycle is shown in Figure 57. Solid lines represent liquid flow and dashed lines represent vapor flow. Here, a solution heat exchanger (SHX) is used to improve the cycle efficiency and chiller performance. The strong solution from the generator is cooled by the weak solution from the absorber. This configuration recovers the sensible heat of the strong solution and therefore reduces the heat input required for the generation process. A low-cost, nonconcentrating flat-plate or evacuated-tube solar collector is normally sufficient to obtain the required temperature for the generator. However, this comes with a lower COP. A higher COP can be obtained by multi-effect systems such as double-effect and triple-effect absorption chillers, which are run by steam produced from concentrating solar collectors.

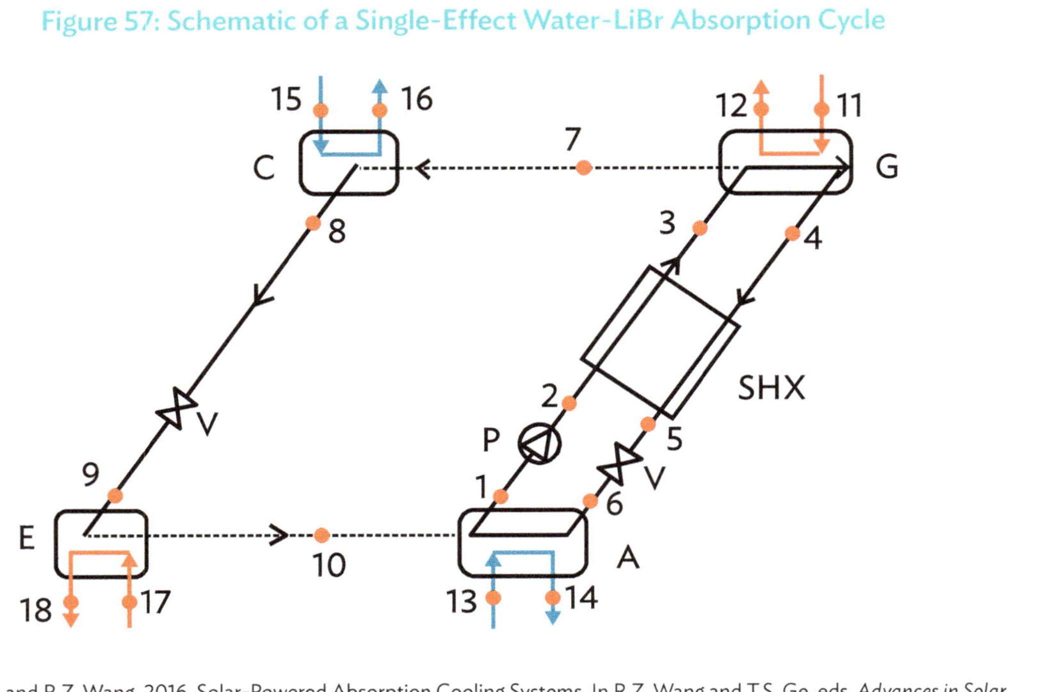

Figure 57: Schematic of a Single-Effect Water-LiBr Absorption Cycle

Source: Xu, Z.Y. and R.Z. Wang. 2016. Solar-Powered Absorption Cooling Systems. In R.Z. Wang and T.S. Ge, eds. *Advances in Solar Heating and Cooling*. Elsevier Science. pp. 251–298.

The half-effect absorption cycle, also called the double-life cycle or two-stage cycle, is another configuration that utilizes low-temperature solar collectors, providing a heat source lower than 100°C. This type of chiller has two solution loops and one refrigerant loop. The refrigerant and solution loops are the same as those of the single-effect absorption chiller, but the connections are different. A half-effect absorption chiller can operate at a driving temperature lower than a single-effect absorption chiller. This advantage, however, comes at the cost of a lower COP of around 0.3–0.4, which is about half that of the single-effect chiller.

The relatively low COP of half-effect and single-effect systems can be improved by double-effect systems, which use heat sources at higher temperatures. A double-effect water-LiBr absorption chiller can achieve a COP of 1.1 to 1.2 with a driving temperature of about 140°C (Ullah et al. 2013). Using gas as the heat source for the chiller is a mature technology. When using solar thermal energy as the heat source, medium-temperature solar collectors, such as concentrating parabolic collectors or parabolic trough collectors, can produce heat at a temperature between 100°C and 200°C, enough to drive a double-effect water-LiBr absorption chiller. Figure 58 illustrates a double-effect cycle using a water-LiBr working pair. Essentially, the cycle can be seen as a combination of two single-effect sub-cycles. The two sub-cycles are coupled by recovering the condensation heat of the high-pressure vapor, leaving the high-pressure condenser to heat the low-pressure generator to boiling.

Triple-effect absorption cooling takes two forms, i.e., a single-loop cycle or a dual-loop cycle. Single-loop triple-effect cycles are basically double-effect cycles with an additional generator and condenser. The system, with three generators and three condensers, operates on principles similar to those of double-effect systems. Primary heat concentrates the absorbent solution in a first-stage generator at a temperature of approximately 200°C to 230°C. A working pair other than water-LiBr is required for the high-temperature cycle. The refrigerant vapor produced is then used to concentrate additional absorbent solutions in a second-stage generator at approximately 150°C. Finally, the refrigerant vapor produced in the second-stage generator concentrates additional absorbent solution in a third-stage generator at approximately 93°C. Theoretically, these triple-effect cycles can obtain COPs of approximately 1.7.

Figure 58: Schematic of a Double-Effect Water-LiBr Absorption Cycle

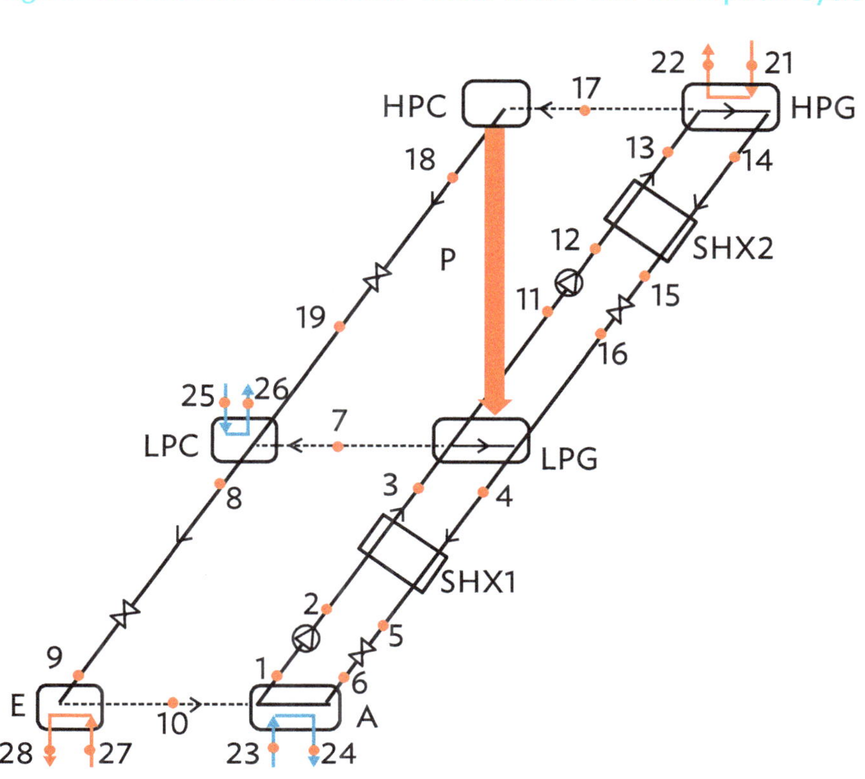

Source: Xu, Z.Y. and R.Z. Wang. 2016. Solar-Powered Absorption Cooling Systems. In R.Z. Wang and T.S. Ge, eds. *Advances in Solar Heating and Cooling*. Elsevier Science. pp. 251–298.

A double-loop triple-effect cycle consists of two cascaded single-effect cycles. One cycle operates at normal single-effect operating temperatures, and the other at higher temperatures. The smaller high-temperature topping cycle has a generator temperature of approximately 200°C to 230°C. A working pair other than water-LiBr is required for the high-temperature cycle. Heat rejected from the high-temperature cycle at 93°C is used as the energy input for the conventional single-effect bottoming cycle. The theoretical COP of this triple-effect cycle can reach approximately 1.8.

Multi-effect absorption cycles are costlier but more energy-efficient due to the use of additional generators and heat exchangers with decreased heat input. The requirements on high temperature and high pressure result in limited choices of working pairs and more stringent technical specifications of generators. The available solar intensity, cooling load, overall performance, and cost considerations collectively determine a particular configuration of the system.

Adsorption Cooling

Like absorption cooling and refrigeration, adsorption cooling and refrigeration is also a thermal-driven technology that can use low-grade heat, such as solar heat or industrial waste heat. Adsorption involves a solid sorbent that attracts refrigerant molecules onto its surface by physical or chemical force and does not change its form in the process. Physical adsorption occurs when the molecules of the refrigerant (adsorbate) fix themselves to the surface of a porous solid adsorbent. Through heating, the refrigerant molecules can be released, i.e., desorption. Chemical adsorption results from the ionic or covalent bonds formed between the refrigerant molecules and the solid adsorbent. The bonding forces are much greater than those of physical adsorption, thereby requiring more heat for desorption.

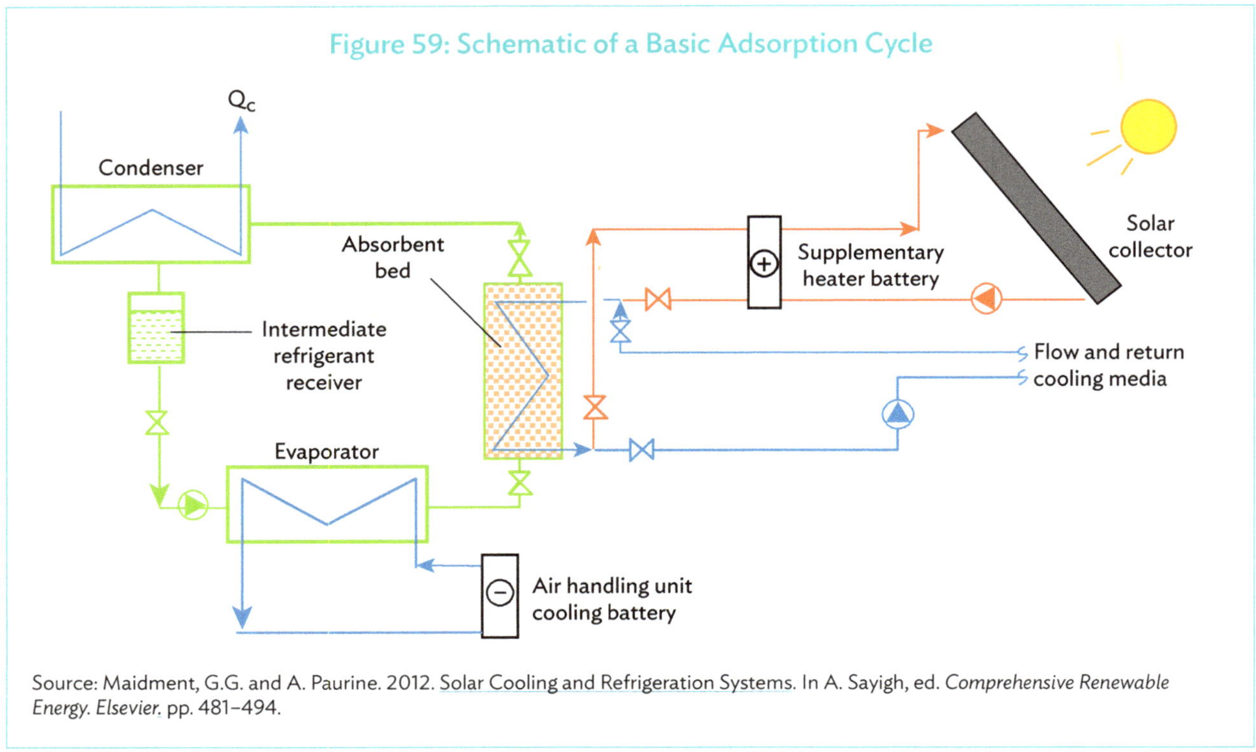

Figure 59: Schematic of a Basic Adsorption Cycle

Source: Maidment, G.G. and A. Paurine. 2012. Solar Cooling and Refrigeration Systems. In A. Sayigh, ed. *Comprehensive Renewable Energy*. Elsevier. pp. 481–494.

A basic solar adsorption system is a closed system consisting mainly of an adsorber powered by a heat source such as solar collectors, a condenser, a refrigerant receiver, and an evaporator (Figure 59).

The adsorption cycle consists of four stage processes (two isobaric and two isosteric processes), as delineated in the Clapeyron diagram in Figure 60.

- **Process A to B (heating and pressurization).** The process starts at point A, when the adsorbent is at adsorption temperature and low evaporation pressure. The mass of the refrigerant is at its maximum level. The adsorber is closed, and the adsorbent is heated. This increases the temperature and pressure along the isosteric line A-B. The mass of the refrigerant remains constant at its maximum.

- **Process B to C (desorption and condensation).** From point B, the desorption process starts along the isobaric line B-C. Progressive heating of the adsorbent causes the refrigerant vapor to be released. The vapor is liquefied in the condenser, and condensation heat is released at the condensation temperature. The liquid refrigerant is then collected in the receiver. The heat consumed to regenerate the adsorbent is supplied by a low-grade heat source, such as solar heat or industrial waste heat.

- **Process C to D (cooling and depressurization).** The previous process ends when the adsorbent reaches its maximum regeneration temperature, and the refrigerant content drops to its minimum value. From point C, the adsorbent cools down, along the isosteric line C-D. Meanwhile, the refrigerant mass remains constant at is minimum value. In this process, the control valve is opened so that the refrigerant flows into the evaporator, and the system pressure decreases until the evaporating pressure is reached.

- **Process D to A (evaporation and adsorption).** The refrigerant evaporates in the evaporator and produces the cooling effect, i.e., heat of evaporation is drawn from the space, which is usually air or water in the case of air-conditioning. The control valve is opened so that the vaporized refrigerant flows from the evaporator to the adsorber, where it is adsorbed until its maximum content is reached. Along the isobaric line D-A, the adsorbent temperature continues to drop, and the heat of adsorption is released into the ambient. At the end of this process, the control value is closed, and the cycle starts over.

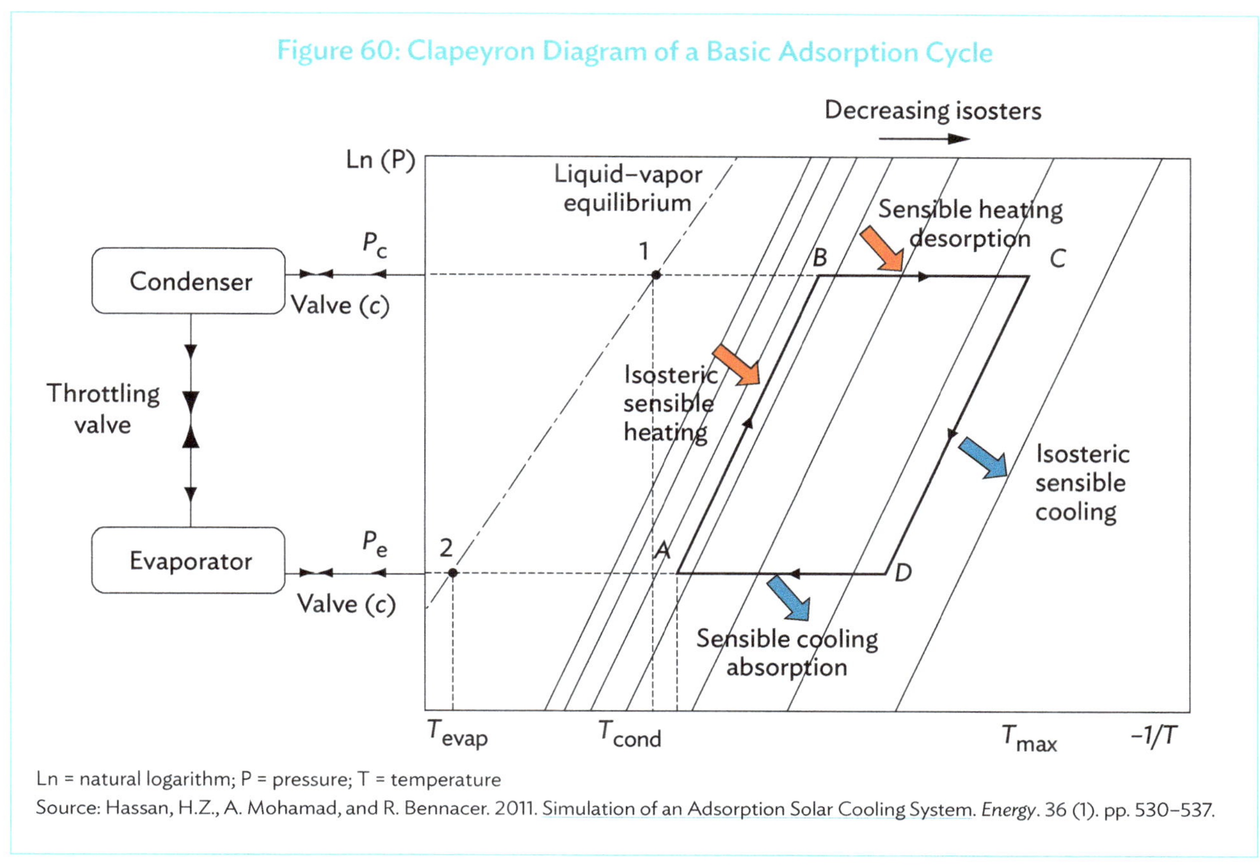

Ln = natural logarithm; P = pressure; T = temperature

Source: Hassan, H.Z., A. Mohamad, and R. Bennacer. 2011. Simulation of an Adsorption Solar Cooling System. *Energy*. 36 (1). pp. 530–537.

This basic cycle is intermittent, which means the cooling effect produced by the cycle is discontinuous. The adsorber's function changes between adsorption and desorption. This decreases the system's COP. To enable the system to operate continuously, two or more adsorbers need to be designed and operated. A common type of continuously working adsorption cooling system is the dual-absorber system, as shown in Figure 61. The heating and cooling processes of the two adsorbers are arranged complementarily. The heating and cooling water media interchange between the two adsorbent beds. While adsorbent bed 1 is heated, adsorbent bed 2 is cooled, and vice versa. This design ensures the desorption and adsorption processes work simultaneously, and the cooling effect can be delivered continuously.

The thermal properties of the adsorbent or refrigerant working pair play a critical role in determining the performance of an adsorption cooling system. The common working pairs include activated carbon-methanol, activated carbon-ammonia, activated carbon fiber-methanol, zeolite-water, silica gel-water, metal hydrides-hydrogen, calcium chloride-ammonia, etc. Among these working pairs, silica gel-water is the most used working pair for adsorption air-conditioning. The advantage of using water as the refrigerant is water's nontoxicity and large latent heat of vaporization. In general, hot water of 55°C to 90°C is good for driving silica gel-water adsorption chillers to produce chilled water of 5°C to 15°C. This low operating temperature range is suitable for commonly used solar collectors such as flat plates and evacuated tubes, which produce hot water below 100°C. For a middle-temperature solar heat source at a temperature between 100°C and 250°C, adsorption cooling with a zeolite-water working pair is still a good option for air-conditioning purposes.

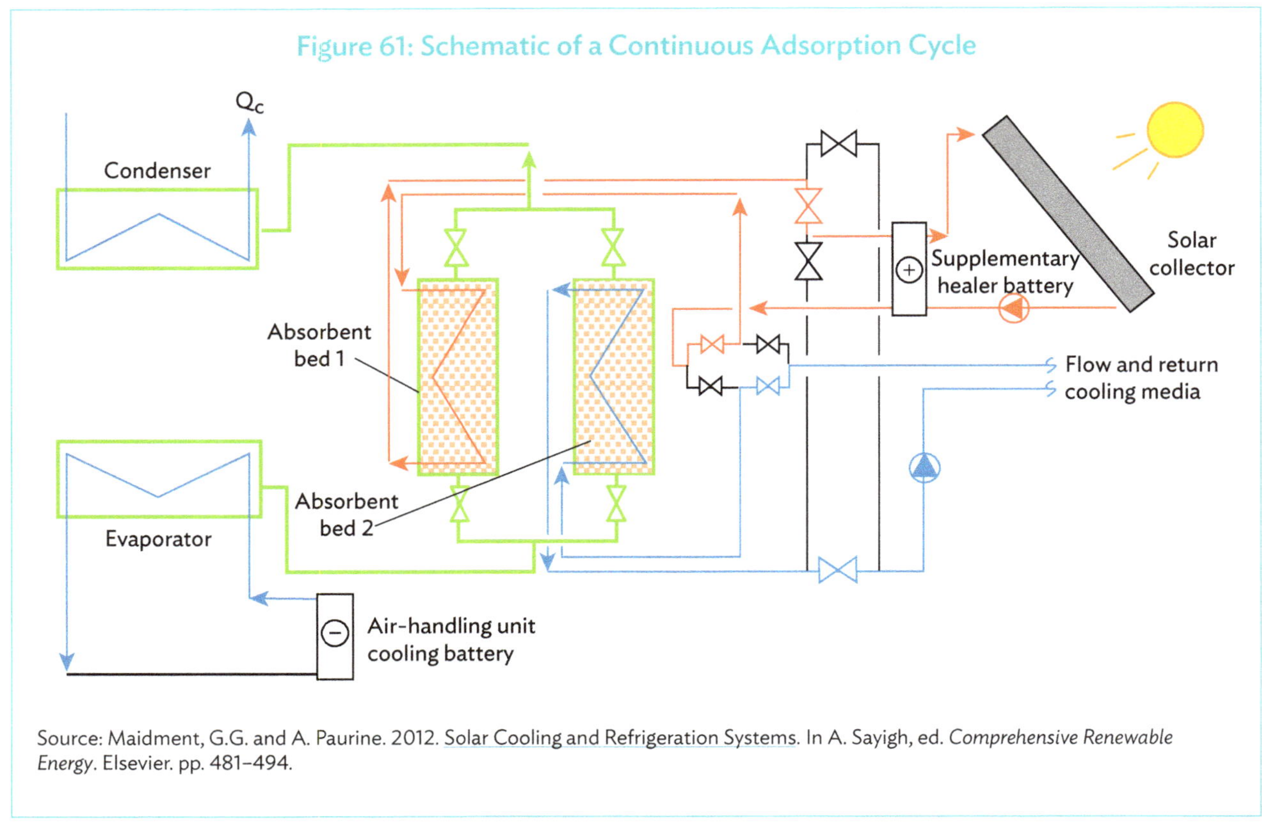

Source: Maidment, G.G. and A. Paurine. 2012. Solar Cooling and Refrigeration Systems. In A. Sayigh, ed. *Comprehensive Renewable Energy*. Elsevier. pp. 481–494.

Desiccant Cooling

Open sorption systems, also commonly known as desiccant cooling systems, are based on the principles of sorption dehumidification and evaporative cooling. In a desiccant cooling system, moisture is transferred from one air stream to another through a sorption process and a desorption process. The sorption process is for dehumidifying air through the difference in water vapor pressure between the humid air and the dry and cold desiccant material. Moisture in the air is absorbed into the desiccant material. In the desorption process, also known as the regeneration process, the desiccant is heated, and moisture previously captured by the desiccant is desorbed. This way, the desiccant is "regenerated" and reusable, and thus ready to absorb moisture again. By going through these processes in parallel, the desiccant system cyclically changes the desiccant's surface vapor pressure and thereby enables the moisture to be transferred continuously (Figure 62). In addition to removing the latent load of the processed air through this dehumidification process, a desiccant cooling system is also equipped with an evaporative cooler, which is used to handle the sensible load of the dehumidified air. With a lowered relative humidity and temperature, the processed air is supplied to the conditioned space for cooling purposes.

Compared with the dehumidification realized by conventional vapor-compression cooling, the sorption desiccant dehumidification system is more energy-efficient because there is no need to first cool the to-be-conditioned air down to its dew-point temperature for condensation of water vapor and then reheat the dehumidified air to the desired temperature (Wang, Xu, and Ge 2016; Chen et al. 2020). Mixed solvent desiccants have become research hot topics recently. Various types of dehumidifiers and their integration with liquid desiccant dehumidification systems have been reviewed. The combination of a liquid desiccant dehumidification system with a solar collector, a vapor compression system, a heat pump system, a combined heat and power system, etc. has been grouped and compared. It is shown that the majority of the recent research work for liquid desiccant dehumidification systems has concentrated on numerical simulations, but a considerable amount of works is still required for the practical investigations of innovative material (mixed solvents). While the desorption process for regenerating desiccant material requires heat, low-grade

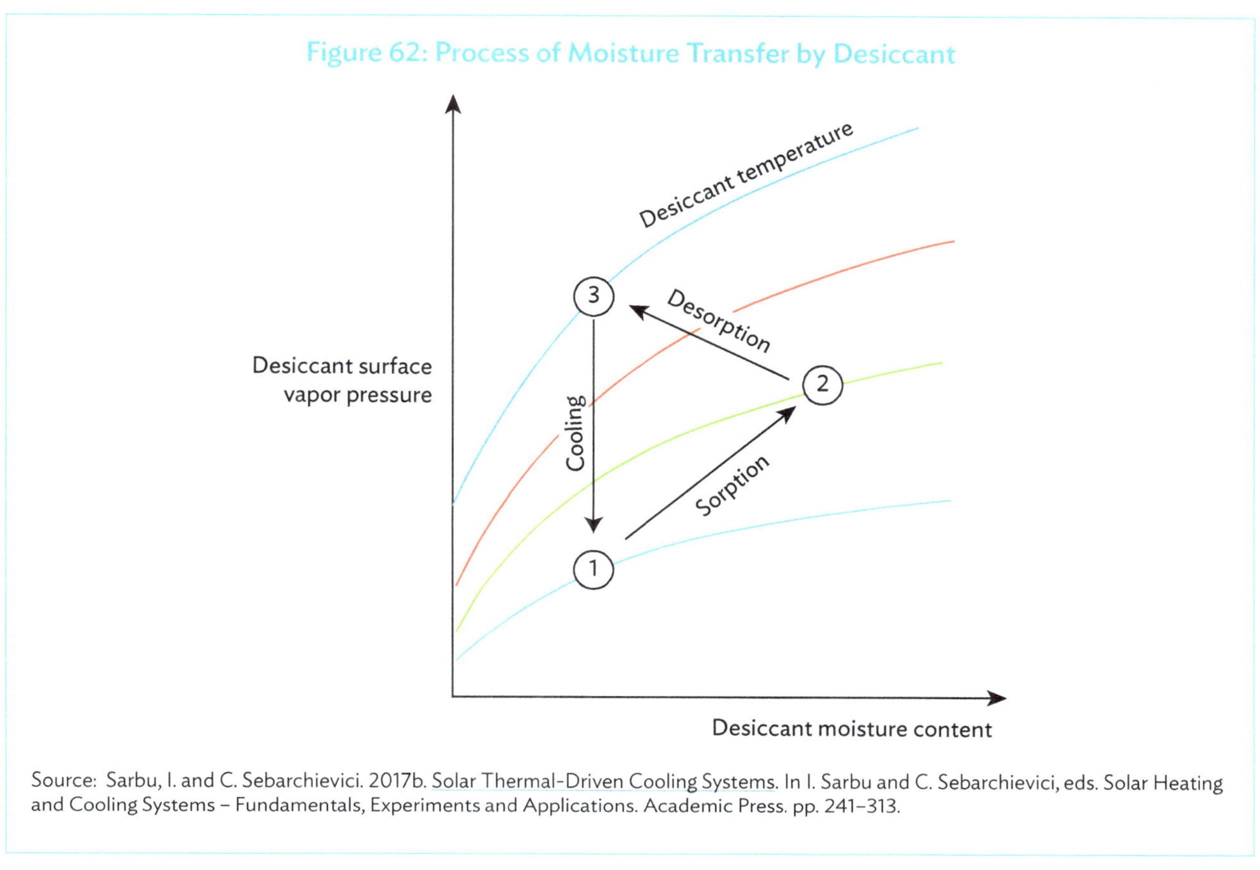

Source: Sarbu, I. and C. Sebarchievici. 2017b. Solar Thermal-Driven Cooling Systems. In I. Sarbu and C. Sebarchievici, eds. Solar Heating and Cooling Systems – Fundamentals, Experiments and Applications. Academic Press. pp. 241–313.

heat at a temperature of about 60°C–95°C is usually sufficient for this purpose (Dincer and Rosen 2007). Therefore, renewable energy such as solar, geothermal, or even waste heat from conventional fossil fuel systems may be used to provide the heat. Owing to these advantages, desiccant cooling is viewed as a good alternative to conventional vapor-compression systems (Dincer and Rosen 2007; Ge and Xu 2016).

There are mainly two types of desiccant cooling systems, namely solid desiccant cooling and liquid desiccant cooling, as introduced in the following paragraphs. Various desiccants are available in liquid or solid phases. In general, all water-absorbing sorbents can be used as desiccant (Sarbu and Sebarchievici 2017b).

Solid Desiccant Cooling

Solid desiccant usually takes the form of a rotary desiccant wheel into which a desiccant material (such as silica gel) is filled. Figure 63 shows a typical example of a solar thermal-driven desiccant wheel evaporative cooling system. The system consists of a desiccant cooling subsystem and a solar subsystem. Within the desiccant cooling subsystem, there are a slowly revolving desiccant wheel, an air heater, a heat exchanger, evaporative coolers, and several other components. Through the two air streams (return air and supply air), flows from and to a conditioned space. The ambient air for conditioning first enters the desiccant wheel for dehumidification. The dehumidified and heated air leaving the wheel is then sensibly cooled down by a heat exchanger, which can be either a plate-type heat exchanger or a heat exchanger wheel. Depending upon its temperature, the conditioned air may either be directly supplied to the conditioned space or be further cooled through an evaporative cooler. In parallel, the return air from the conditioned space goes through the process in a reverse order. It first passes through an evaporative cooler and then enters the heat exchanger, where heat is transferred from the supply air to the return air. The resultant warm and humid stream of return air leaving the heat exchanger is then further heated by the air heater. Afterward, the hot return air enters the desiccant wheel to regenerate part of it before finally being exhausted to the ambient.

The hot water in the air heater for heating the return air is supplied by the solar subsystem, which is largely like a conventional solar hot water system. The solar collector absorbs solar radiation and heats up the water, which is then pumped to the air heater. Common types of solar collectors include flat plates, vacuum tubes, and concentrated collectors. A water tank is part of the subsystem for thermal storage and adjusting the water flux. Depending on the availability of solar radiation, dehumidification, and cooling demand, an auxiliary heater using gas or electricity may be installed for the solar subsystem or the return air side (before regeneration of the desiccant wheel) to provide supplementary heating capacity to ensure continuous operation of the system.

Thermal COP is the most commonly used parameter to evaluate the performance of a solid desiccant cooling system. It is defined as the ratio of the enthalpy change of conditioned air to the thermal energy for the regeneration of desiccant material. In principle, the thermal COP of a solid desiccant cooling system, as an open sorption system, is similar to that of closed sorption systems. Another important indicator to evaluate the performance is the solar fraction, which is defined as the ratio of energy provided by the solar collector for regeneration of the desiccant material to the total energy required for generation, including the auxiliary heater.

Figure 63: Schematic of a Solar Thermal-Driven Desiccant Wheel Evaporative Cooling System

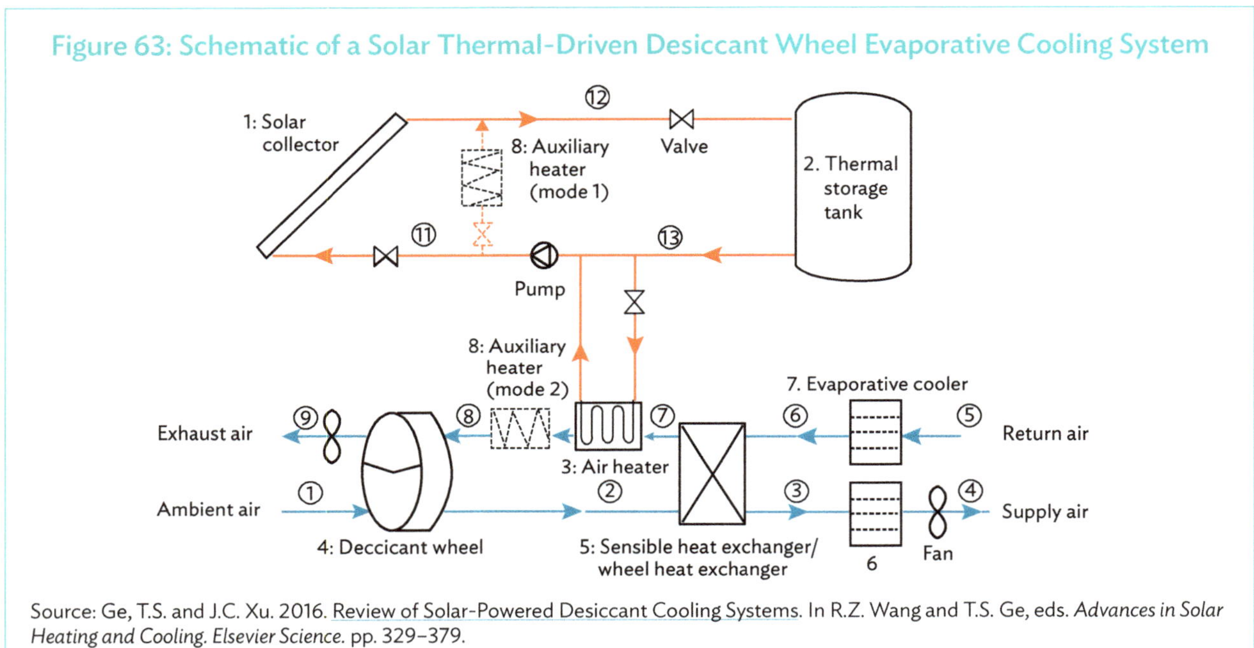

Source: Ge, T.S. and J.C. Xu. 2016. Review of Solar-Powered Desiccant Cooling Systems. In R.Z. Wang and T.S. Ge, eds. *Advances in Solar Heating and Cooling. Elsevier Science.* pp. 329–379.

Box 11: Case Studies of Solar Thermal-Driven Solid Desiccant Cooling Systems

Henning et al. (2001) investigated the performance of a solar-assisted solid desiccant cooling system installed at Riesa, Germany. The system featured a flat-plate collector of 20 square meters (m^2), a hot water buffer storage of 2 cubic meters (m^3), and a nominal air volume flow of 2,700 m^3/hour. On-site measurement results indicated that the system achieved a solar fraction of 76%, an overall collector efficiency of 54%, and a cooling coefficient of performance (COP) of 0.6 under typical summer conditions. Based on further system simulation under climatic conditions in Denmark, Germany, and Italy, they found that combining the solar-assisted solid desiccant cooling system with a backup conventional vapor-compression chiller would be feasible from not only an energy perspective but also an economic perspective, especially in warm and humid climates. Such a feasibility could be maintained even under the requirement that indoor standard comfort criteria had to be met for every hour of the year. The system was found to be able to achieve 50% savings of primary energy at low incremental overall costs.

The performance of solar-assisted desiccant cooling systems using two different desiccant materials: silica gel (SiO_2) and titanium dioxide (TiO_2), with various specifications under East Asian climatic conditions was numerically studied by Enteria et al. (2016). The system was applied in temperate climate (Beijing and Tokyo), subtropical climate (Taipei,China and Hong Kong, China) and tropical climate (Manila and Singapore). In general, it was found that the system could provide the required indoor temperature and humidity ratio in these cities. TiO_2 tended to perform better than SiO_2 in supporting lower indoor temperature and humidity ratio. For hot and humid climate such as in Manila or Singapore, larger size of the solar thermal system and higher volumetric flow of air to support the high cooling load would be required.

In a similar study by Ali et al. (2018)environment-friendly, and reliable water conditioning systems has led to the introduction of several standalone and/or hybrid alternatives. The technology of desiccant evaporative cooling (DEC, five configurations of solar-assisted desiccant cooling systems under five different climatic conditions, covering Karachi in Pakistan, Sao Paulo in Brazil, Shanghai in the People's Republic of China, Adelaide in Australia, and Vienna in Austria, were evaluated and compared. It was observed that the systems were more suitable to be applied in Karachi and Shanghai due to their semiarid and humid climatic conditions. The minimum overall average COP of basic operation cycles (ventilated, recirculated, and dunkle[a]) was greater than one for both climatic zones. In addition, the average cooling capacity for these regions was twice as much as for cities in continental climate conditions, e.g., Vienna.

To improve the performance of solar thermal-driven rotary desiccant wheel cooling system, Ge et al. (2010) proposed a two-stage rotary wheel system and compared its performance with a conventional vapor-compression system (Box Figure 11.1). Both systems were designed to meet the cooling demand of one floor in a commercial office building in two locations under different climatic conditions, namely Berlin and Shanghai. The required regeneration temperatures were 55°C in Berlin and 85°C in Shanghai. The system in Shanghai achieved a higher thermal COP than that in Berlin due to higher-humidity ratio in Shanghai. Compared with the conventional vapor-compression system, the desiccant cooling system had the merits of a higher fresh air ratio and less electricity consumption. While the initial cost of the desiccant cooling system was higher than that of the vapor-compression system, the operating cost was significantly reduced due to the utilization of low-grade solar energy. The dynamic investment payback periods were estimated to be only 4.7 years in Berlin and 7.2 years in Shanghai.

continued on next page

Box 11 *continued*

Box Figure 11.1: Schematic of a Solar Thermal-Driven Two-Stage Rotary Desiccant Cooling System

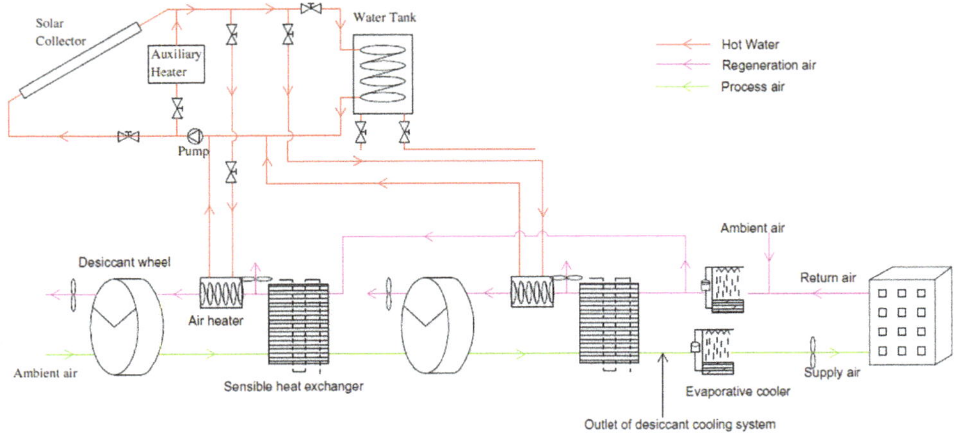

a Dunkle is one of the operating cycles of the solid desiccant cooling system with an additional heat exchanger. It is named after Dunkle who proposed such an operating cycle.

Source: Ge, T.S. et al. 2010. Performance Comparison Between a Solar Driven Rotary Desiccant Cooling System and Conventional Vapor Compression System (Performance Study of Desiccant Cooling). *Applied Thermal Engineering.* 30 (6–7). pp. 724–731.

A variant of the solar thermal-driven desiccant wheel evaporative cooling system is a hybrid solar thermal-driven desiccant cooling and vapor-compression system (Figure 64). The hybrid system brings together the good capacity of the desiccant wheel system in handling latent load and that of the conventional vapor-compression system in handling sensible load. Compared with a conventional vapor compression, a hybrid system usually consumes less energy to provide the same cooling capacity.

Figure 64: Schematic of a Hybrid Solar Thermal-Driven Desiccant Cooling and Vapor-Compression System

Source: Ge, T.S. and J.C. Xu. 2016. Review of Solar-Powered Desiccant Cooling Systems. In R.Z. Wang and T.S. Ge, eds. *Advances in Solar Heating and Cooling. Elsevier Science.* pp. 329–379.

Box 12: Case Studies of Hybrid Solar Thermal-Driven Desiccant Cooling and Vapor-Compression Systems

A study in subtropical Hong Kong, China, found that solar-assisted hybrid desiccant cooling systems had the potential to be more effective in handling commercial premises with high latent cooling loadsin hot and humid climatic conditions compared to conventional air-conditioning systems (Fong et al. 2011). Their hybrid system used an electricity-driven vapor-compression chiller to provide chilled water to a supply air cooling coil to handle the sensible cooling load (Box Figure 12.1). No evaporative coolers were included. The conditioned spaces included a Chinese restaurant and a wet market, which are typical commercial premises in Hong Kong, China, and Guangdong province in the south of the People's Republic of China. The designed zone sensible and latent loads were 19 kilowatts (kW) and 13 kW for the restaurant and 27 kW and 13 kW for the wet market, respectively. Results showed that the hybrid system was able to maintain indoor temperature and humidity ratios much more steadily compared with a conventional system under different loading and climatic conditions throughout a year. At a designed zone relative humidity of 60%, the hybrid system could achieve an annual primary energy savings of 49% for the restaurant and 13% for the wet market. If the humidity control requirement was loosened to 70% for the restaurant, the hybrid system still outperformed the conventional system with an annual primary energy saving as much as 22.7%.

Box Figure 12.1: Schematic of a Hybrid System for Commercial Premises with High Latent Cooling Load

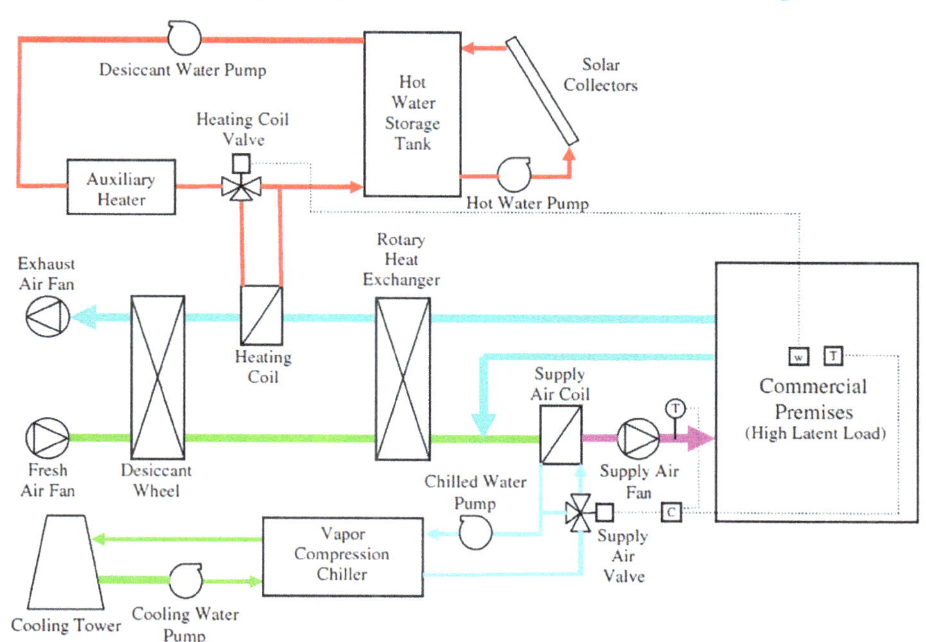

Source: Fong, K.F. et al. 2011. Investigation on Solar Hybrid Desiccant Cooling System for Commercial Premises with High Latent Cooling Load in Subtropical Hong Kong. *Applied Thermal Engineering*. 31 (16). pp. 3393–3401.

Khalid et al. (2009) developed and evaluated a hybrid solar-assisted desiccant cooling and vapor-compression system in Karachi, Pakistan. The system was characterized by using two indirect evaporative coolers instead of a direct evaporative cooler on the processed air side. One was installed before the desiccant wheel to precool the ambient air, and the other was used between the sensible heat exchanger and the cooling coil of the vapor-compression subsystem. Measurements found that due to the climatic conditions of Karachi, the solar-assisted desiccant cooling system alone was not able to handle the high latent and sensible loads and treat the air to the desired comfort conditions. Therefore, the auxiliary vapor-compression cooling unit had an important role to play. It was also demonstrated from measurements that by using indirect evaporative coolers, the regeneration temperature was brought down by 15% while the decrease in the dehumidification was only 6%, thereby saving thermal energy.

continued on next page

Box 12 *continued*

A solar hybrid desiccant air-conditioning system was configured, experimentally investigated, and theoretically analyzed (La et al. 2011). The system, installed at an office building in Jiangsu Province, People's Republic of China, was mainly composed of a solar collection subsystem with a 72-square meter (m²) flat-plate solar collector array and a 4 m² hot water tank, a two-stage desiccant cooling subsystem with a design cooling capacity of 10 kW, an air-source vapor-compression air-conditioning subsystem with a nominal cooling capacity of 20 kW, and a cooling tower that supplied cooling water to the desiccant cooling subsystem (Box Figure 12.2).

Box Figure 12.2: Schematic of a Hybrid Two-Stage Desiccant Cooling and Vapor-Compression System

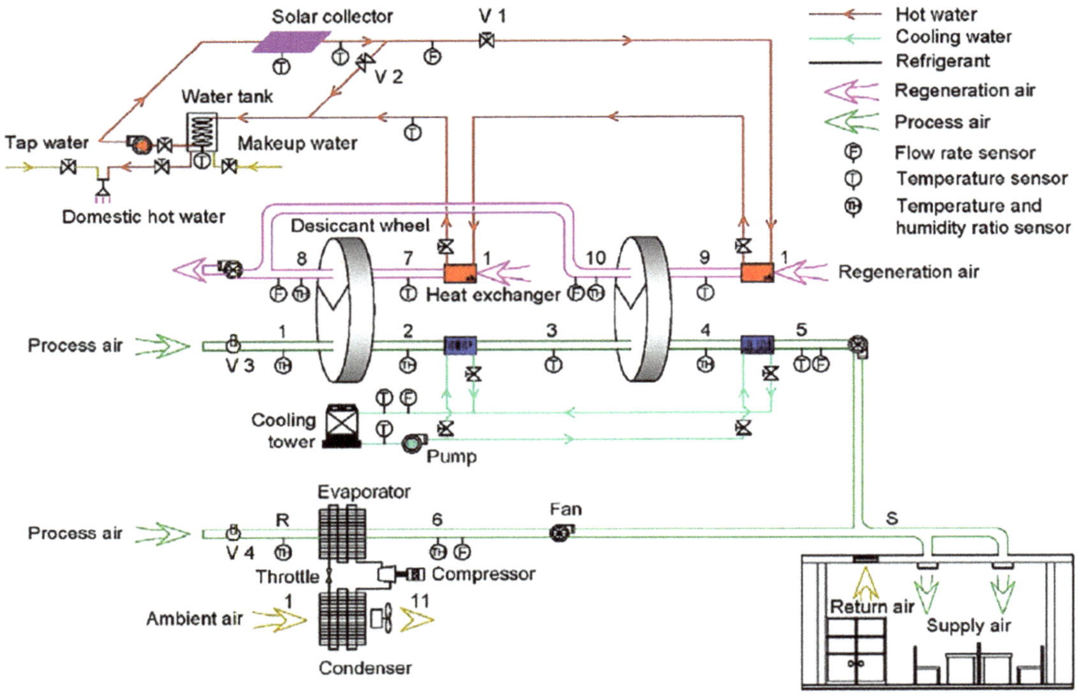

Source: La, D. et al. 2011. Case Study and Theoretical Analysis of a Solar Driven Two-Stage Rotary Desiccant Cooling System Assisted by Vapor Compression Air-Conditioning. *Solar Energy*. 85 (11). pp. 2997–3009.

On-site operation measurements showed that under the local climatic conditions of an average ambient air temperature of 34.6°C, a humidity ratio of 21.54 grams (g)/kilogram (kg), and a solar radiant intensity fluctuating between 122.2 watts per square meter ((W)/m²) and 539.9 W/m², the average efficiency of the solar collector was 0.32 and the two-stage desiccant cooling subsystem achieved an average cooling capacity of 10.9 kW, with the thermal and electric coefficient of performance (COPs) reaching 1.24 and 11.48, respectively. For the whole hybrid system, its electric COP was 4.41, about 34% higher than the conventional vapor-compression subsystem. The system's performance was further analyzed through simulation under different climatic conditions in Beijing (temperate), Shanghai (humid), and Hong Kong, China (extremely humid). The results demonstrated the feasibility of the hybrid system for a wide range of operating conditions. Compared with conventional systems, the hybrid system could achieve a seasonal electricity saving rate of 22% to 34%.

Source: La, D. et al. 2011. Case Study and Theoretical Analysis of a Solar Driven Two-Stage Rotary Desiccant Cooling System Assisted by Vapor Compression Air-Conditioning. *Solar Energy*. 85 (11). pp. 2997–3009.

Liquid Desiccant Cooling

Liquid desiccant systems usually consist of an absorber for dehumidifying incoming air, a regenerator for regenerating the solution of liquid desiccant, an evaporative cooler for sensibly cooling air, and a series of heat exchangers. Key to the systems are dehumidification and regeneration. Depending upon specific climatic conditions and application, a liquid desiccant system can either be used as a stand-alone system or in conjunction with a vapor-compression system or a vapor absorption system, i.e., as a hybrid system. In practice, hybrid systems are used more commonly than stand-alone systems (Ge and Xu 2016).

Commonly used liquid desiccant materials include lithium chloride (LiCl), calcium chloride (CaCl2), and lithium bromide (LiBr) (Sarbu and Sebarchievici 2017b). Compared to solid desiccants, liquid desiccants have a generally greater ability to hold moisture (Chen et al. 2020). Due to the low pressure drop across various system components, the regeneration temperatures of liquid desiccants are mostly between 40°C and 70°C, much lower than those of solid desiccants (between 60°C and 115°C), thereby requiring a lower temperature for regeneration (Elsarrag 2008; La et al. 2010). Thus, as in the case of solid desiccant, low-grade energy such as solar energy is well-placed to meet part of the energy demand of a liquid desiccant system. Also, with the ability to pump liquid desiccant, a liquid desiccant cooling system can be made small and compact (Sahlot and Riffat 2016). Given the relatively lower cost of liquid desiccant materials, liquid desiccant systems generally have a shorter payback period (Chen et al. 2020). However, liquid desiccants do have some disadvantages, including potential corrosion and damage to the system, requiring larger pumps to handle large volumes of liquid and the problem of crystallization (Sahlot and Riffat 2016).

It is worth noting that the liquid desiccant in a liquid desiccant cooling system circulates between an absorber and a regenerator in a way largely similar to that in an absorption system. The major difference is that the equilibrium temperature of a liquid desiccant is determined by the partial pressure of water in the humid air in contact with the desiccant solution, rather than the total pressure (Kim and Infante Ferreira 2008; Sarbu and Sebarchievici 2017b).

Figure 65 shows a schematic of a basic solar-assisted liquid desiccant dehumidification and cooling system and the principles under which the system operates. The system is characterized by two circulations—an air circulation and a liquid desiccant circulation.

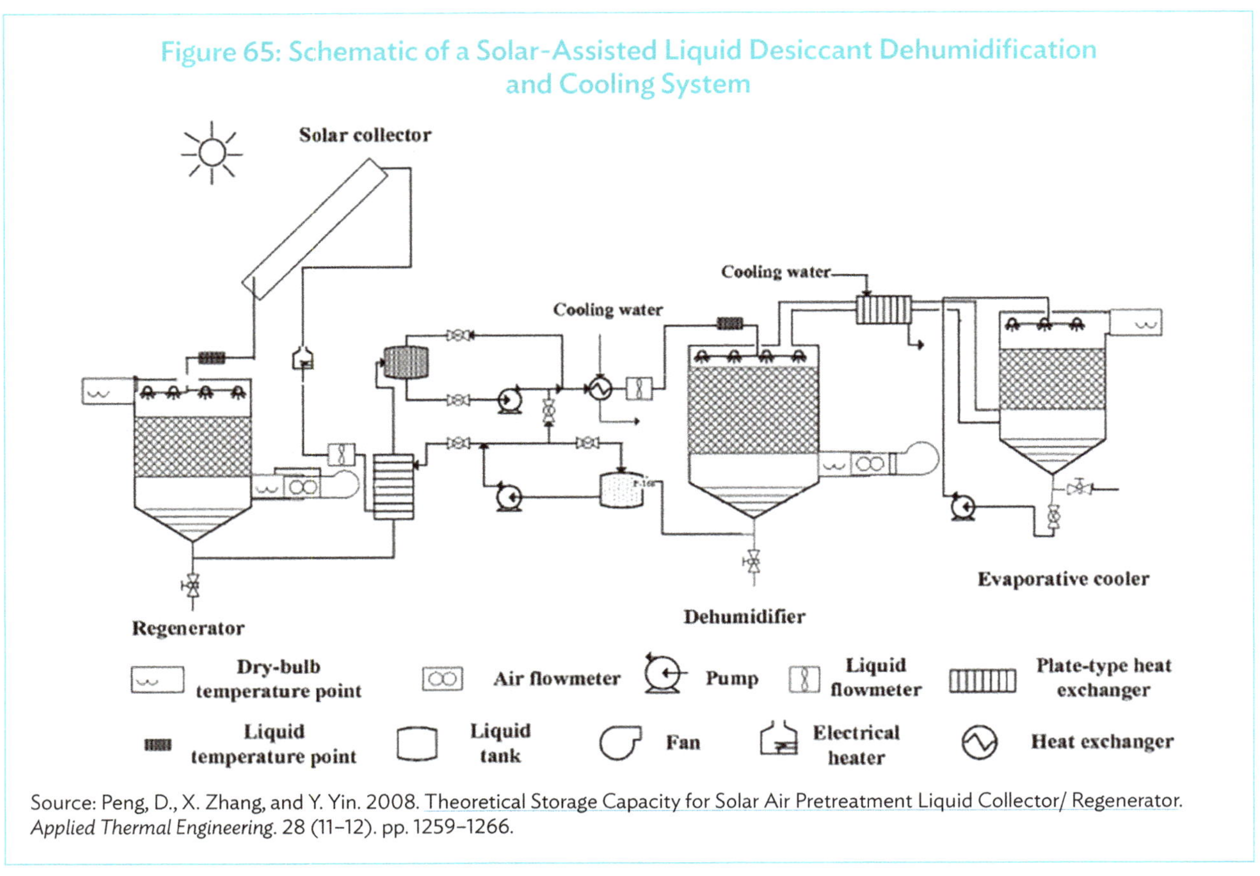

Figure 65: Schematic of a Solar-Assisted Liquid Desiccant Dehumidification
and Cooling System

Source: Peng, D., X. Zhang, and Y. Yin. 2008. Theoretical Storage Capacity for Solar Air Pretreatment Liquid Collector/ Regenerator. *Applied Thermal Engineering*. 28 (11–12). pp. 1259–1266.

In the air circulation, the humid and hot air, which may be ambient air or a mix of ambient air and return air from the conditioned space, is dehumidified in the liquid dehumidifier where the air is brought into contact with a sprinkled concentrated solution of the liquid desiccant. Water vapor is absorbed by the solution. The dehumidified air is then cooled through heat exchange with cooling water and further cooled by an evaporative cooler to the desired condition for supply. In the conditioned space, the humidity and temperature of the supply air increase as a result of removing latent and sensible cooling loads. Part of the return air leaving the conditioned space and some ambient air, as make-up air is supplied as the regenerative air to the regenerator where it gets humidified and heated before being exhausted to the ambient. The other part of the return air is mixed with fresh ambient air and then supplied to the dehumidifier to repeat the dehumidification process. This completes the air circulation.

For the liquid circulation, mass transfer takes place in the dehumidifier. Water vapor in the air is transferred into the strong solution of liquid desiccant because the partial pressure of water vapor in the liquid film is much lower than that of the processed air. The resultant diluted weak solution of liquid desiccant is then pumped to a liquid-liquid heat exchanger, in which it receives heat from the hot concentrated solution of desiccant from the regenerator. Subsequently, it passes through an electrical heater. The diluted solution of desiccant with a raised temperature then flows into the solar collector for further heating to the temperature of the regeneration point. After entering the regenerator, the hot, diluted solution is exposed to the regenerative air. Driven by the difference in vapor pressure, moisture is transferred from the solution to the regenerative air—a reverse of the process in the dehumidifier. This increases the concentration of the solution to an acceptable level near the initial concentration. The regenerated liquid desiccant then passes through the liquid-liquid heat exchanger and transfers heat to the to-be-regenerated diluted solution. Before entering the dehumidifier, the precooled regenerated liquid proceeds to get cooled more deeply by a cooling coil to gain the desired dehumidification capability. This circulation maintains the continual operation of the system.

The flow pattern and various configurations of the dehumidifier have a strong impact on the performance of the dehumidification process. In the humidifier, the incoming humid air gets in contact with the liquid desiccant and has its moisture absorbed. The patterns in which the air-desiccant contact occurs can be broadly categorized into parallel flow, counter flow, cross flow, and counter-cross flow. Depending on whether an internal cooling source is provided, a dehumidifier can either be an adiabatic one or an internally cooled one. The former is a simple unit in which mass is transferred from incoming air to the liquid desiccant, while heat transfer also takes place due to temperature differences and the latent heat of condensation of water vapor. Heat discharge reduces the efficiency of dehumidification. This can be avoided by an internally cooled dehumidifier, which uses an embedded cooling coil to provide an extra cooling source to cool the desiccant solution and effectively remove the heat produced during the dehumidification process. This way, the vapor pressure of the desiccant solution is maintained at a lower level favorable for the dehumidification process, thereby improving system efficiency. The internally cooled dehumidifier can further be categorized into three types, including parallel plate, fin coil, and packed tower with tubes. A comprehensive review of the technical characteristics can be found in (Sahlot and Riffat 2016; Chen et al. 2020).

Also, in single-stage dehumidification, an adverse temperature increase in the desiccant solution will lower the driving force for mass transfer between desiccant solution and processed air. This can be addressed by using a multistage dehumidification with an array of single dehumidifier modules. The liquid desiccant flows through the modules and is cooled separately in each single module. This configuration keeps the vapor pressure difference between the desiccant solution and the processed air at a moderate level with drastic change, and therefore ensures the dehumidification performance (Mei and Dai 2008; Ge and Xu 2016).

As already shown in Figure 66, a liquid desiccant dehumidification system can be combined with an evaporative cooling system to achieve a cooling effect. A liquid the desiccant dehumidification system can also be combined with a vapor-compression system to provide a subcooling effect from the condenser outlet to the compressor inlet and therefore improve the cooling capacity and COP of the vapor-compression system (Chen et al. 2020). Li, Lu, and Yang (2010) modeled the performance of a solar-assisted hybrid system in the hot, wet weather in Hong Kong, China. With the use of a solar-assisted desiccant dehumidification unit, the capacity of the vapor-compression unit could be reduced from 28 kW to 19 kW. The energy-saving potential was significant due to a higher COP resulting from a higher supply of chilled water from the chiller. The annual operation energy savings for the hybrid system reached 6.8 megawatt-hours (MWh). With an extra initial cost of HK$6,360, the payback period of the hybrid system was around 7 years. These results demonstrated the technical and economic viability of hybrid systems in subtropical areas. Khalil (2012) investigated the potential of a hybrid system, which had a total cooling capacity of 6.2 kW. The liquid desiccant was lithium chloride. As shown in Figure 66, a strong solution from the tank was pumped and sprayed uniformly over the evaporator surface area. Process air to be dehumidified passed through the evaporator in the crossflow direction. The evaporator and desiccant enabled simultaneously cooling and dehumidifying the process air while the diluted desiccant solution was collected in the weak solution tank. After that, the diluted solution was pumped to absorb heat from the heat exchanger, which used the waste heat rejected from the condenser of the vapor-compression unit to preheat the diluted solution. A heating coil in the regenerator tank provided the additional heat required by the solution for regeneration. The results showed that, compared with a conventional vapor-compression system with a reheat configuration, the investigated system increased the COP by 68% and attained an annual energy savings of 53%.

There are also merits to combining a liquid desiccant dehumidification system with a heat pump system. Heat released from the condenser of the heat pump can be utilized to regenerate solution in the liquid desiccant cycle (Niu, Xiao, and Ma 2012). Separate control of indoor temperature and humidity can be achieved by a hybrid liquid desiccant and heat pump system, with a system COP higher than conventional systems, as found in applications in the PRC (Zhao et al. 2011; Chen, Yin, and Zhang 2014).

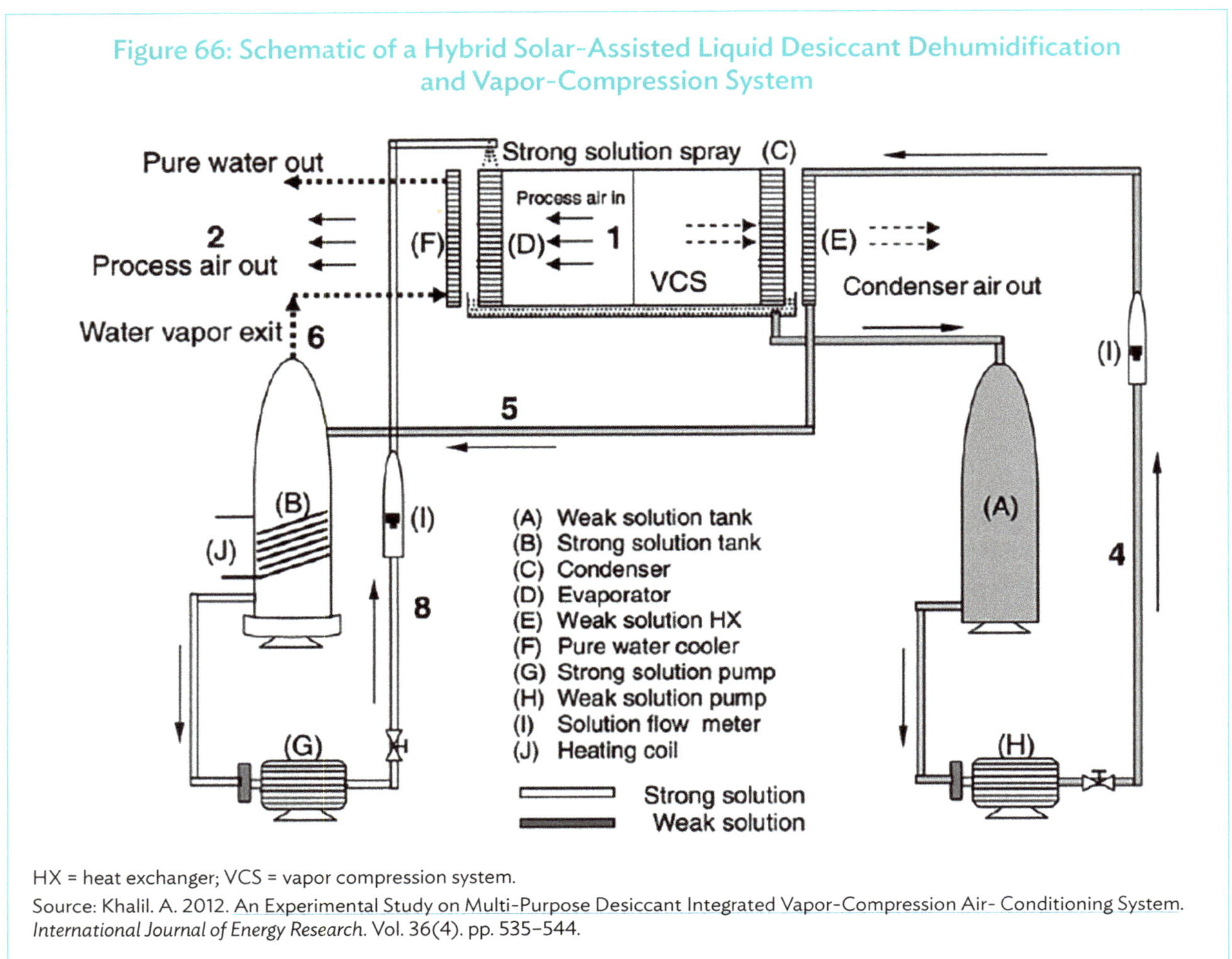

Figure 66: Schematic of a Hybrid Solar-Assisted Liquid Desiccant Dehumidification and Vapor-Compression System

HX = heat exchanger; VCS = vapor compression system.

Source: Khalil. A. 2012. An Experimental Study on Multi-Purpose Desiccant Integrated Vapor-Compression Air- Conditioning System. *International Journal of Energy Research*. Vol. 36(4). pp. 535–544.

Thermal Storage—Phase-Change Materials

By storing and utilizing energy, thermal storage techniques can smooth out the peaks and troughs in cooling demand, thus improving the loading and efficiency of chillers, and reducing energy consumption. The use of phase-change materials (PCMs) is an established method of storing energy because of the large latent storage capacity of PCMs during an almost isothermal process. This means PCMs can provide a large storage density at small temperature intervals.

Materials suitable for thermal storage purposes should have large latent heat and a high thermal conductivity. Their melting temperature should be adaptable to a practical range of operations. They should be chemically stable, flammable, low- or noncorrosive, and nontoxic. Their costs should be low. Some of the widely used PCMs include paraffin, salt hydrates, eutectics, and fatty acids. The main advantages and disadvantages of PCMs are summarized in Table 9. When selecting a PCM, it is important to focus on the following properties: the melting point that assures the storage and extraction of heat; the heat of fusion required to obtain a high storage density; the thermal conductivity required to perform a suitable charging and discharging process; the material density; the need for congruent melting; minimal volume change following phase change; and a high specific heat to ensure a significant sensible storage capacity, among others.

Table 9: Main Advantages and Disadvantages of Organic and Inorganic Phase-Change Materials

	Advantages	Disadvantages
Organic PCM	• Large temperature range • Low super-cooling • Noncorrosive • High congruency in melting • Highly compatible with building materials • Chemically stable • High fusion heat • Recyclable	• Low thermal conductivity • Low volumetric latent heat storage capacity • Flammability
Inorganic PCM	• High volumetric latent heat storage capacity • Low cost and easy availability • High thermal conductivity • Nonflammability	• High volume change • Super-cooling

PCM = phase-change material.

Sources: Cabeza, L.F. et al. 2011. Materials Used as PCM in Thermal Energy Storage in Buildings: A Review. *Renewable and Sustainable Energy Reviews.* 15 (3). pp. 1675–1695; Cui, Y. et al. 2017. A Review on Phase Change Material Application in Building. *Advances in Mechanical Engineering.* 9 (6). Sage Journals; Kuznik, F. et al. 2011. A Review on Phase Change Materials Integrated in Building Walls. *Renewable and Sustainable Energy Reviews.* 15 (1). pp. 379–391.

There are various techniques to integrate PCMs into buildings, such as micro, macro, composite, and structural encapsulation. In the microencapsulation process, the small PCM particles are encapsulated in a thin, high-molecular-weight polymeric film. This technique has the advantage of improved heat transfer and stability. For larger-scale encapsulation, the PCM is enclosed in a container that can be used to build products. This technique avoids the issue of phase-change volume adjustments in building applications. Building-integrated PCMs can also be manufactured as composite materials to improve their properties. Structural encapsulation of PCMs may be applied to various building materials such as plaster, wood, concrete, cement, or vermiculite, which are used in buildings' ceilings, walls, or flooring for cooling and/or heating purposes.

Figure 67 shows an example of a PCM-based ventilative cooling system. It is the "cool phase" system for natural cooling, manufactured by Monodraught. It is a modular device, either ceiling-mounted or integrated within the suspended ceiling, that guides ventilation air through thermal battery modules filled with paraffin PCM. During the night, cool air is circulated and is stored in the form of latent heat in the thermal battery. During the following day, the high-temperature ambient air is circulated through the thermal battery, where the material cools the air by changing its phase.

The efficiency of a PCM-based ventilation and cooling system hinges primarily on the phase-change temperature of the material, the temperature of the ambient air during the night period, and the airflow rate. As found by previous applications, a PCM system using fatty acid, if carefully designed and applied, had the potential to further decrease the maximum room temperature of the following day by almost 2°C, on top of what could be achieved by a common night ventilation approach. Also, using PCM for night ventilation could offer the opportunity for downsizing the mechanical ventilation system in climates with outdoor air temperatures below 18°C at night. It was also found that using a PCM wallboard could reduce peak cooling loads by 28%.

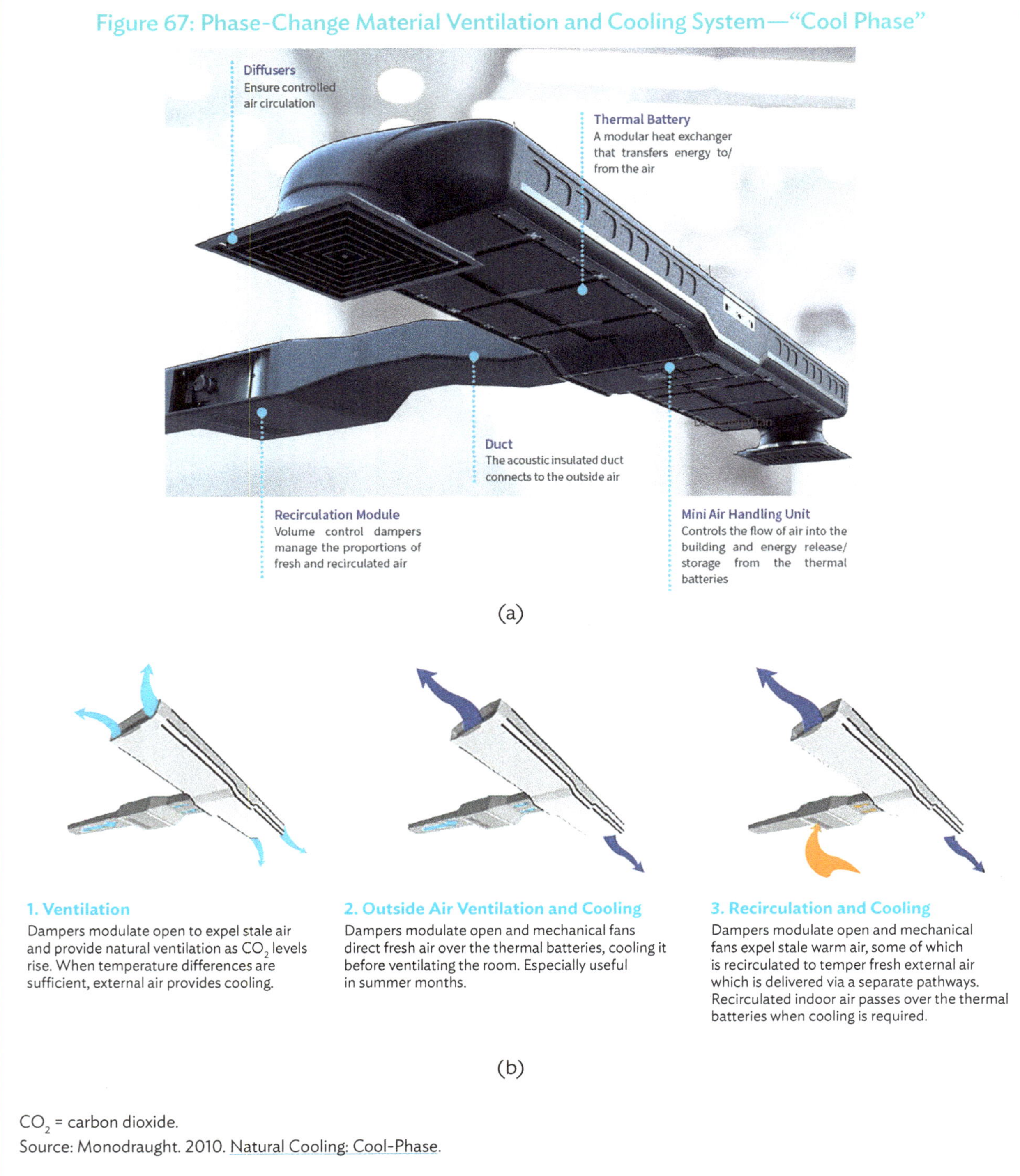

CO_2 = carbon dioxide.
Source: Monodraught. 2010. Natural Cooling: Cool-Phase.

Renewable Cooling Sources

Ground Cooling (Air)

The concept of ground cooling through heat exchange between air and earth is simple and straightforward. The system uses an underground duct, or a network of them, buried into the ground at around two to four meters of depth, at which level the temperature of the ground is constant. This temperature, known as the earth's undisturbed temperature, remains lower than the ambient air temperature during summer, and therefore can serve as a cooling source. As shown in Figure 68, the ambient air is drawn through the duct, cooled through heat exchange with the ground, and then supplied to the building. In some applications, the cooled air is not supplied to the rooms directly, but used to cool the indoor air through an air-to-air heat exchanger (Figure 69).

Figure 68: Ground Cooling (Air) System for Direct Supply

Source: BEBuilding Energy Efficiency Project (BEEP). 2020a. Earth Air Tunnel; Bisoniya, T.S. 2015. Design of Earth–Air Heat Exchanger System. *Geothermal Energy*. 3 (18).

Figure 69: Ground Cooling (Air) System with Air-to-Air Heat Exchanger

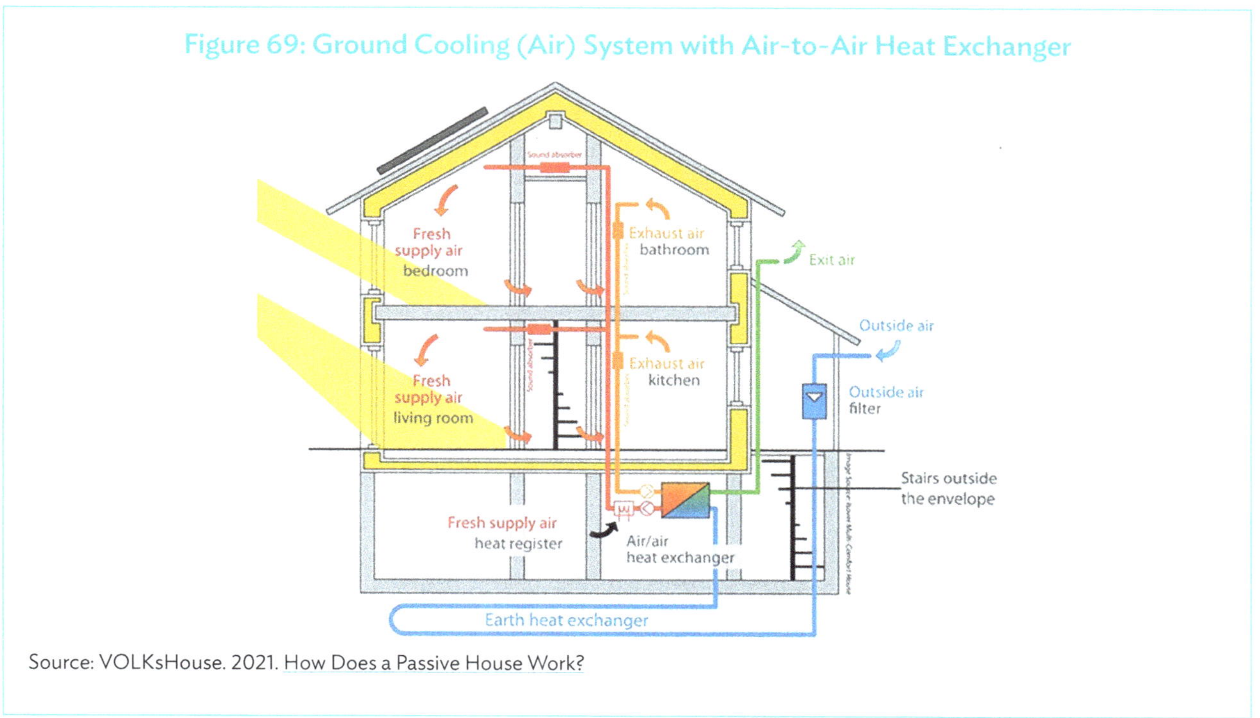

Source: VOLKsHouse. 2021. How Does a Passive House Work?

Ground Cooling (Water)

Ground-source cooling uses the relatively stable temperature of the ground or bodies of water as a source for providing cooling in buildings. There are two main types of ground-source cooling: groundwater cooling and ground-source heat pumps.

Groundwater Cooling

Groundwater cooling uses the relatively stable temperature of water in aquifers to provide space cooling within buildings. It works because, in summer, the groundwater temperature is generally lower than the indoor air temperature and therefore can be used as a cooling source. Groundwater cooling can be either a closed loop or an open loop, with the open loop being more common. Closed-loop applications are normally equipped with heat pumps, i.e., ground-source heat pump systems. Open-loop applications are provided cooled water directly to buildings. In open-loop, direct groundwater cooling systems, water is abstracted from the ground, which is typically an aquifer, a ground or surface water source such as a lake or river, and then passed through a heat exchanger to transfer heat to the ventilation and air-conditioning systems. Finally, the water is either returned to the ground or discharged into a river or sewer.

Open loop, direct groundwater cooling systems have the potential to meet larger cooling loads than closed-loop systems. The available volume of water that can be extracted and the temperature change that is allowed are key factors to determine the system capacity. As power is only required for driving water circulation pumps, the system efficiency is generally high, with a COP of 15 to 20 achievable. There are, however, a series of disadvantages of open-loop groundwater cooling systems, such as resource and regulatory uncertainties, design complexity, thermal interference, maintenance, and so on (Feuvre and Cox 2009).

Box 13: Case Study of Groundwater Cooling

Portcullis House, London

Located in Westminster, London, Portcullis House is a seven-story building housing the United Kingdom's 650 members of Parliament with a number of offices, conference rooms, and committee rooms.

The house was designed to integrate building services. It features a highly active construction of a facade system, which consists of a triple plane glazing with argon filled and low-emissivity coating. A ventilative cavity serves as air distribution ducts. During winter, louvers are used to block lower sun angles. During summer, a light shelf is used to block higher sun angles. Glass prism surfaces act as a reflector of daylight upon the interior space. This brings the benefit of doubling the daylight, especially in north-facing offices, which have their sky views obstructed by adjacent buildings. The facade with a low U-value helps to block heat from outside to inside, and vice versa. Thermal mass materials with high thermal resistance are used at the interior finishes and at the ceiling to absorb heat.

For ventilation strategy, the house uses a low velocity displacement ventilation system, which, owing to buoyancy force and vertical temperature gradient, is more energy efficient as compared to a conventional system. The plan is organized around a central courtyard. Along the building perimeter, a total of 14 chimney stacks provide ventilation for the building. Air is drawn at the base of the chimneys and distributed through the duct system integrated into the facade. Each floor has a ventilation plenum at the floor level where air circulates before entering the occupied spaces. Air is exhausted at ceiling level

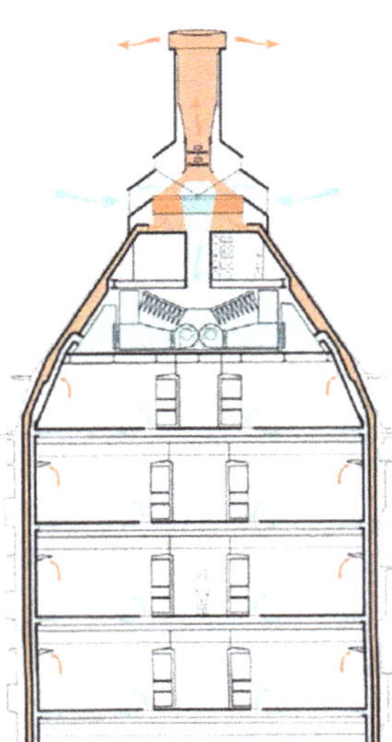

Box Figure 13.1: Ventilation Strategy of Portcullis House, London

Buchanan, P. 2005. *Ten Shades of Green: Architecture and the Natural World*. Architectural League of New York.

continued on next page

Box 13 *continued*

and distributed through the duct system leading to the chimneys. The exhaust air is not recirculated but used for heat recovery through a roof-mounted rotary hygroscopic thermal wheel. The recovered heat in the exhaust air comes from solar heat captured at the facade, internal heat gains from occupants and devices, and heat emitted by radiators.

The cooling of the building is provided by a groundwater cooling system, which extracts groundwater from two boreholes sunk 150 meters into a chalk aquifer below the building. When the outdoor temperature goes high, the groundwater of around 12.5°C is pumped and stored in two 16,500-liter buffer tanks located in the basement. The groundwater is then pumped through plate heat exchangers connected to cooling coils in air-handling units to cool outdoor fresh air down to room temperature, around 19°C. The cooled air is sent to occupied spaces through the displacement ventilation system. Upon leaving the heat exchanger, the groundwater is discharged at a temperature of about 21°C. Some of the groundwater is fed to a gray water system for flushing toilets, whereas the rest is discharged into the river. Groundwater cooling is supplemented by nighttime ventilation when needed. The night ventilation helps to enhance the building's cooling strategy and avoid overheating. It facilitates the removal of internal heat gains absorbed by the thermal mass materials during the daytime. The ventilation rate during night ventilation is about half of the rate during daytime, contributing to reducing the overall energy consumption of the building.

Box Figure 13.2: Schematic of Groundwater Cooling System of Portcullis House, London

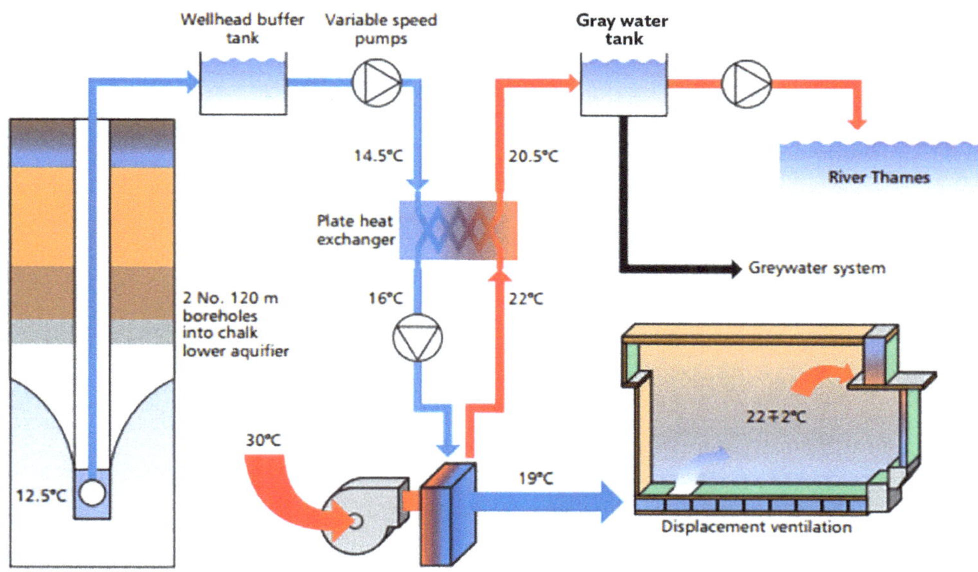

m = meter.
Source: Ampofo, F., Maidment, G.G. and Missenden, J.F. 2006. Review of Groundwater Cooling Systems in London. *Applied Thermal Engineering*. 26 (17–18). pp. 2055–2062; Feuvre, P. Le and C.S.J. Cox. 2009. Ground Source Heating and Cooling Pumps – State of Play and Future Trends; Wincott, N. 2019. Low Carbon Heating and Cooling from Groundwater for Commercial and Residential Developments and District Networks.

Ground-Source Heat Pump for Space Cooling

Reversible ground-source heat pumps can be an effective way to provide either cooling or heating. The basic principle is to make use of the relatively stable temperature of the earth at a certain depth from the surface, of groundwater from wells, or of surface water from a lake, pond, or stream. With the temperature between this heat source or sink and the air in the conditioned space, the refrigerant of the heat pump is driven by an electric compressor to go through the conventional vapor-compression cycle to produce a heating or cooling effect for the conditioned space.

When operating in cooling mode, a ground-source heat pump works like an air-conditioning system, by using the ground, groundwater, or surface water as the heat sink. The heat pump extracts heat from the air in the conditioned space and transfers it to its circulating refrigerant. When the heat pump is of a water-to-water type, this heat transfer takes place between the refrigerant and the chilled water of an air-handling unit (Figure 70). By undergoing the continual vapor-compression cycle, the refrigerant transfers the absorbed heat to the ground through the working fluid (water, or antifreeze solution) in the ground connection subsystem, which takes the form of tubes vertically placed in boreholes, or horizontally placed in trenches, or submerged in a body of water such as a pond or lake.

More details on heat pump classification and description are given in section 5.1 of this handbook.

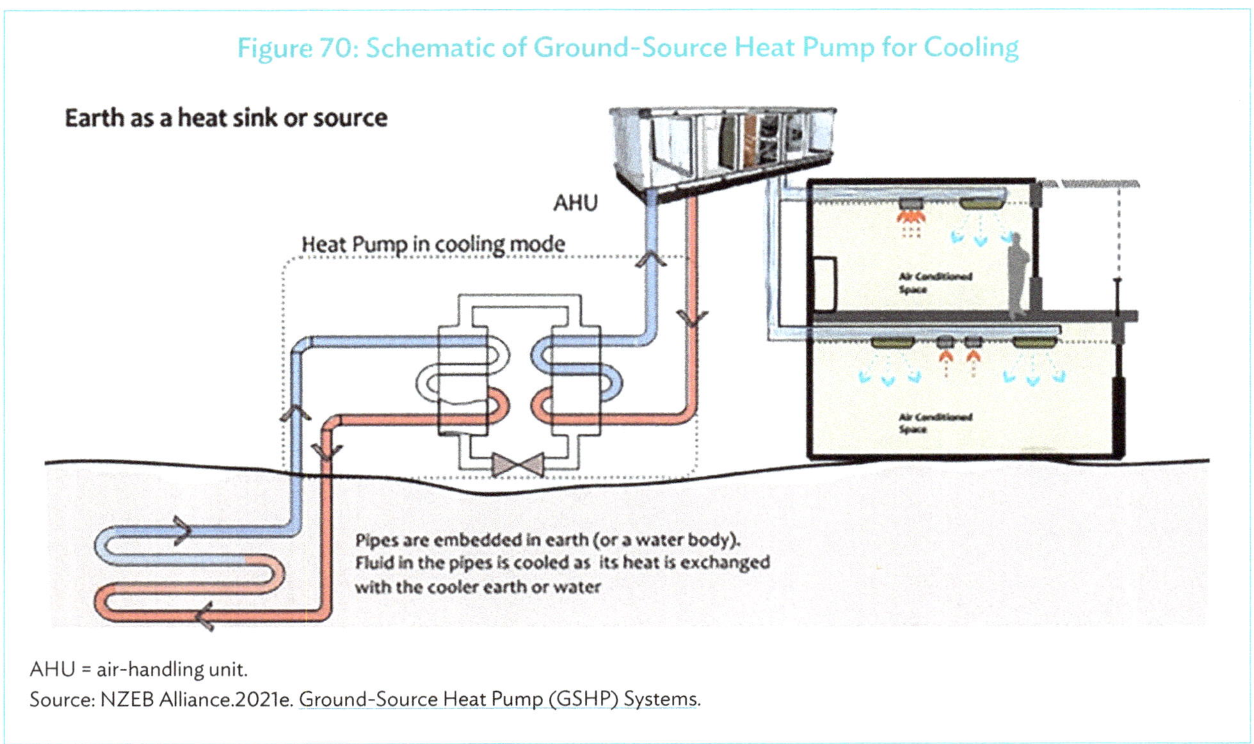

Figure 70: Schematic of Ground-Source Heat Pump for Cooling

AHU = air-handling unit.
Source: NZEB Alliance.2021e. Ground-Source Heat Pump (GSHP) Systems.

Box 14: Case Studies of Ground-Source Heat Pump for Space Cooling

CASE 1: CENTRAL UNIVERSITY OF RAJASTHAN, INDIA

The Central University of Rajasthan is in Ajmer, Rajasthan, India, where the climatic condition is hot and dry. A geothermal hybrid cooling system was installed at the university's student hostel. Water from the boreholes is used to cool the air before being introduced into indoor spaces. Heat exchange between water from the boreholes and supply air takes place inside specially designed air-handling units. After this, air is further cooled through a two-stage evaporative cooling system. The system uses full fresh air and is designed to reduce the temperature of the supply air by around 10°C. The design temperature for the hostel rooms is 29°C. The average outlet water temperature from the boreholes is recorded at 25°C. Outlet temperatures in winter are around 19°C. The energy performance index of the system was measured to be approximately 78 kilowatt-hour/square meter/year in 2012.

Box Figure 14.1: Boreholes of Ground-Source Heat Pump System at Central University of Rajasthan

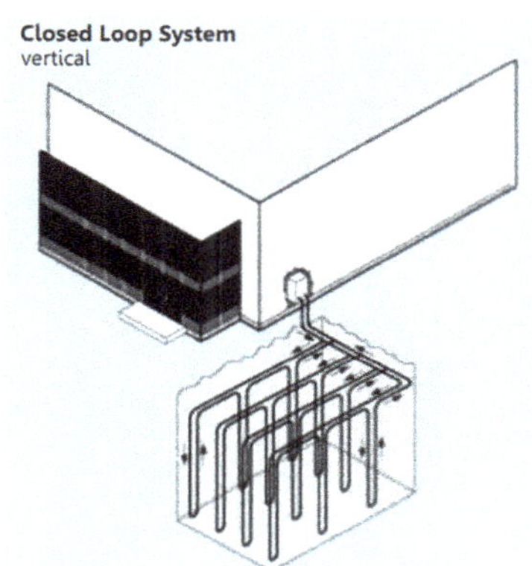

Source: United States Agency for International Development. 2014. HVAC Market Assessment and Transformation Approach for India.

CASE 2: SPACE COOLING BY GROUND-SOURCE HEAT PUMP IN TROPICAL SOUTHEAST ASIA (THAILAND, VIET NAM)

A major concern regarding the applicability of ground-source heat pumps (GSHP) for space cooling in Southeast Asian countries is that the subsurface temperature is very close to the average atmospheric temperature and therefore may not be appropriate to be used as a cold source (heat sink). A solution is to use the effect of cold groundwater, which infiltrates in the underground in high mountain regions and flows toward the nearshore regions with the densely populated Southeast Asian cities.

Yasukawa and Uchida (2018) conducted groundwater temperature surveys at Chao Phraya Plain, Thailand, and the Red River Plain, Viet Nam. The surveys found that (i) in Bangkok, Ayutthaya, Phitsanulok, and Nakhon Sawan of Thailand, subsurface temperature was lower than monthly mean maximum atmospheric temperature throughout the year; and (ii) in Ha Noi, Viet Nam, subsurface temperature was lower than monthly mean maximum atmospheric temperature from May to October by 2–7 degrees Kelvin. These results suggested the applicability of GSHPs in these places.

continued on next page

Box 14 *continued*

Several experimental and demonstrational GSHP systems in Thailand, Viet Nam, and Indonesia were installed and operated by the authors (Table 14.1). All were closed-loop systems. The first four systems were only experimental and were removed after a couple of years' operation. The remaining ones were demonstrational systems for longer periods of operation. Improvements have been applied to these installations for better cost performance, including drilling and well completion technologies for a higher heat exchange rate, a combination of horizontal and vertical heat exchangers, and a controlling system for the heat pump operation.

Table 14.1: Experimental and Demonstrational Ground-Source Heat Pump Systems in Thailand and Viet Nam

No.	Location	Period	Subsurface Heat Exchanger	Surface System	Note
1	Kamphaengphet (DGR), Thailand	October 2006 to March 2008	57-meter (m) deep borehole with double U-tube	Water-water chiller, fan coil	First experiment in tropics. Mostly made in Japan.
2	Chiang Mai (DGR), Thailand	March 2008 to July 2010	80-m deep borehole with single U-tube + 60-m horizontal tube	Same as above	Moving the above system to another site.
3	Bangkok (Kasetsart University), Thailand	July 2010 to 2012	200-m horizontal tube	Same as above	Moving the above system to another site.
4	Bandung (ITB), Indonesia	July 2013 to 2015	200-m horizontal tube	Remodel from air-conditioner	Cooling efficiency 25% up. Done by Akita University.
5	Bangkok (Chulalongkorn University), Thailand	May 2014 to present	50-m deep borehole with single U-tube × 3 (150 m)	Combined chiller and fan unit	Cooling efficiency 30% higher than normal air-conditioner.
6	Saraburi (Chulalongkorn University), Thailand	June 2015 to present November 2016 to present	300-m carpet style (horizontal) 300-m coil style (horizontal)	Combined chiller and fan unit	Machine made in Thailand Remodel from air-conditioner.
7	Pathumthani (Geology Museum, DMR) Thailand	March 2015 to present	50-m deep borehole with double U-tube × 2 (400 m)	Combined chiller and fan unit	Mostly made in Japan. No cementing borehole for higher heat exchange.
8	Ha Noi (VIGMR), Viet Nam	October 2016 to present	50-m deep borehole with double U-tube × 2 (400 m)	Combined chiller and fan unit	Mostly made in Japan. No cementing borehole for higher heat exchange.

Source: Yasukawa, K. and Y. Uchida. 2018. Space Cooling by Ground Source Heat Pump in Tropical Asia. In *Renewable Geothermal Energy Explorations*. IntechOpen.

The installation and operation of these GSHP systems offered the following findings:

- Application of horizontal subsurface heat exchangers is effective in reducing the installation cost of heat exchangers. Combination of horizontal and vertical ones is effective to maintain higher heat exchange rate while reducing installation costs.
- Shorter piping and thermal insulation of the surface pipe are important for effective cooling. The subsurface pipe for the horizontal system should not be bended to maintain high fluid circulation with a small water pump.
- Application of local materials, such as domestic heat pumps and domestic tubes is effective in reducing total cost. Electrofusion welding is useful to connect the head part of the U-tube and local tubes so that import of whole U-tube is not necessary.

continued on next page

Box 14 *continued*

- For higher performance and cost reduction of the vertical heat exchanger, well completion without cementing is essential. Polymer mud is effective for such drilling, and local crews may handle it without problem.
- A sophisticated heat pump control system with a proper heat exchanger achieved 30% electricity savings compared to a conventional air-conditioner.

Sources: Link, K. 2017. Example of Geothermal Cooling with Ground Source Heat Pumps in South-East Asia; Yasukawa, K. and Y. Uchida. 2018. Space Cooling by Ground Source Heat Pump in Tropical Asia. In *Renewable Geothermal Energy Explorations.* IntechOpen; Widiatmojo, A. et al. 2019. Ground Source Heat Pump Application in Tropical Countries. In *44th Workshop on Geothermal Reservoir Engineering,* Stanford University; Chokchai, S. et al. 2018. A Pilot Study on Geothermal Heat Pump (GHP) Use for Cooling Operations, and on GHP Site Selection in Tropical Regions Based on a Case Study in Thailand. *Energies.* Vol. 11(9).

CASE 3: THE PEOPLE'S REPUBLIC OF CHINA'S FIRST GROUND-SOURCE HEAT PUMP SYSTEM AND ICE STORAGE PROJECT—BEIJING TIANCHUANG SHIYUAN BUILDING

The Tianchuang Shiyuan building is in Chaoyang District, Beijing, which has a hot summer and a cold winter. The large commercial facilities at the bottom of the building have an air-conditioned space totaling to about 40,000 square meters, with a design cooling load of 6.05 megawatts (MW) and a heating load of 4 MW. The project adopts the GSHP and ice storage air-conditioning system designed by the China Academy of Building Sciences. Completed in September 2002, the system was the first-of-its-kind in the People's Republic of China.

Box Figure 14.2: Schematic of a Ground-Source Heat Pump System Coupled with Cold Storage

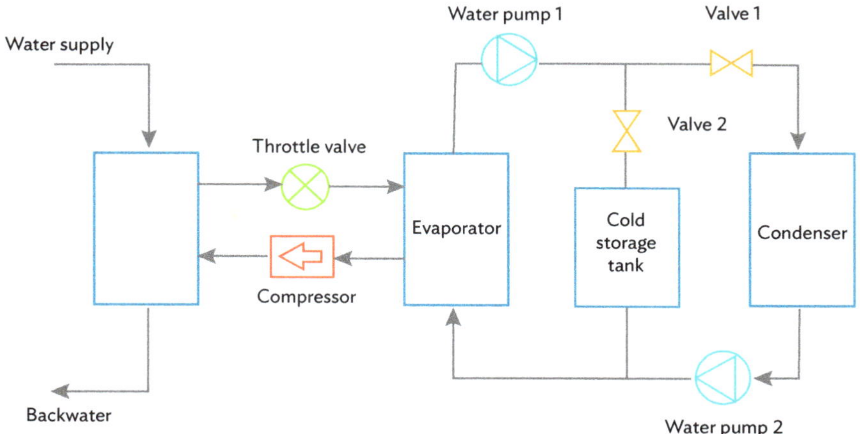

Source: Song, C. et al. 2021. Application and Development of Ground Source Heat Pump Technology in China. *Protection and Control of Modern Power Systems.* 6.

In the hybrid system of GSHP with ice storage equipment, a cold storage tank is added in parallel with the evaporator and end heat exchanger in the conventional heat pump system. This makes possible four operational modes of the system, including conventional cooling cycles, cold storage cycles, combined cooling cycles, and single cold storage and supply cycles. The optimal scheduling among different operation modes can effectively reduce the power consumption of refrigeration equipment, as well as the operational cost of the air-conditioning system during the peak period.

If the heat pump system was used without cold storage, 11 wells would have been needed, and the groundwater consumption would have been quite large. With the hybrid system in operational mode, seven wells are sufficient for energy supply, which reduces system power consumption. The total project investment was about CNY3 million, with a large percentage subsidized by the local government. Compared with conventional systems, the annual operational cost can be reduced by CNY1.58 million, demonstrating a high economic viability.

Source: Song, C. et al. 2021. Application and Development of Ground Source Heat Pump Technology in China. *Protection and Control of Modern Power Systems.* 6.

Sustainable Air Distribution

Chilled Beams and Ceilings

Static cooling devices taking the shape of long rectangular beams, named chilled beams, and rectangular panels, named chilled ceilings, deliver cooling to occupied spaces by using chilled or cooled water passing through coils that exchange heat with ambient air to reduce the air dry-bulb temperature. The water as the cooling medium has a temperature in the range of 13°C to 18°C and can be obtained using natural cold-water storage or free cooling from outside air during cold periods of a year. Compared to all-air ventilation and cooling systems, chilled beams and ceilings, being water-based, are more energy- efficient because of the high specific heat capacity and thermal conductivity of water. Moving heat around a building can be undertaken far more efficiently by pumping cool water in pipes than by moving the same amount of heat in ducted air.

There are a wide range of different types of chilled beams and ceilings, but they fall into three main categories: radiant chilled ceilings, passive chilled beams, and active chilled beams (Figure 71).

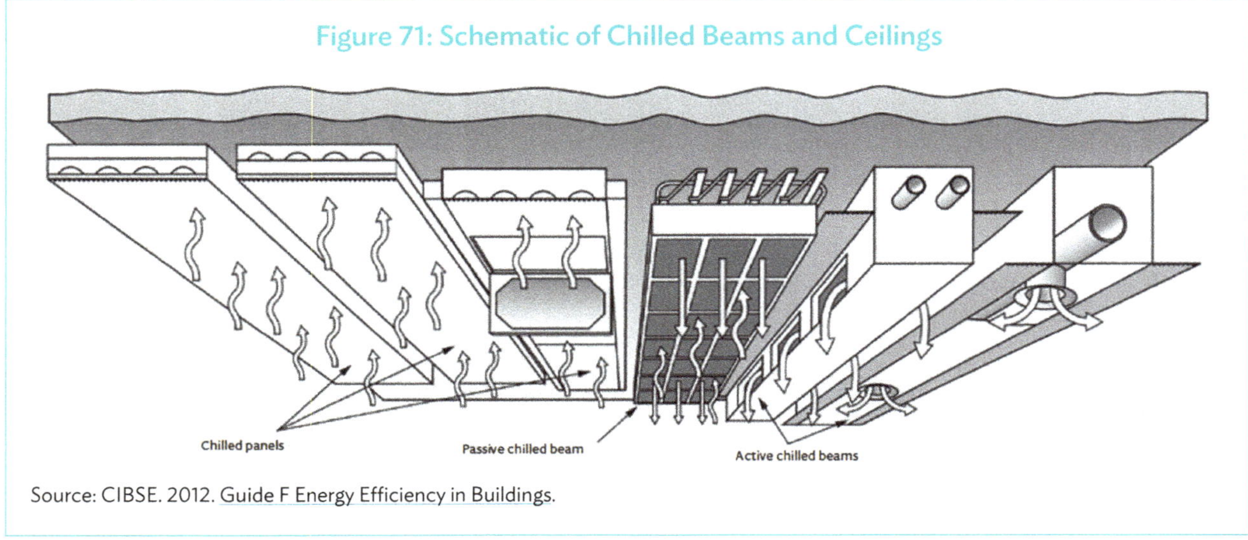
Figure 71: Schematic of Chilled Beams and Ceilings

Source: CIBSE. 2012. Guide F Energy Efficiency in Buildings.

(i)   **Radiant Chilled Ceilings**

Radiant chilled ceilings usually incorporate a chilled water coil into the rear of the ceiling finish material. This means placing a copper pipe matrix on the rear of metal ceiling tiles or panels (Figure 72). Insulation is usually applied to the upper surface of the chilled ceiling. As chilled water passes through the coil, it offers a cool ceiling surface that provides space cooling, mainly by radiation.

Radiant cooling involves the direct absorption of heat radiated from warm surfaces within the room, which occurs when there are cooler surfaces visible to the warmer surfaces. This type of system results in low air velocity and an evenly distributed temperature in the occupied zone, thus ensuring a good level of thermal comfort. As for indoor air quality, a separate ventilation system is required to supply fresh air to the space. Often, chilled ceilings are used with displacement distribution systems.

Figure 72: Radiant Chilled Ceiling Panel

Source: Frenger Systems. 2012. *An Introduction to Chilled Beams and Ceilings*.

(ii)   **Passive Chilled Beams**

Chilled beams are mounted at high levels in the conditioned space and can be fully encased and suspended from a soffit or integrated into a false ceiling. The basic thermal transfer component for chilled beams is a fin and tube heat exchanger, i.e., coils. Chilled beams provide convective cooling. Because of the larger fin surface area, a higher thermal performance can be achieved with chilled beams as opposed to chilled ceilings.

Passive chilled beams respond to natural convection currents in the room. Warm air rising in the space passes over the top and into the passive chilled beam. As the air between the fins is cooled, it becomes denser and drawn downward by gravity, displacing the less dense, warmer air to a higher level (Figure 73).

Passive chilled beams need to be arranged at regular intervals along the ceiling plane to provide uniform cooling to the occupied space. It is important to ensure good air circulation for the operation of passive chilled beams. There need to be sufficiently large openings above the passive beam casing or ceiling system to allow air to circulate properly.

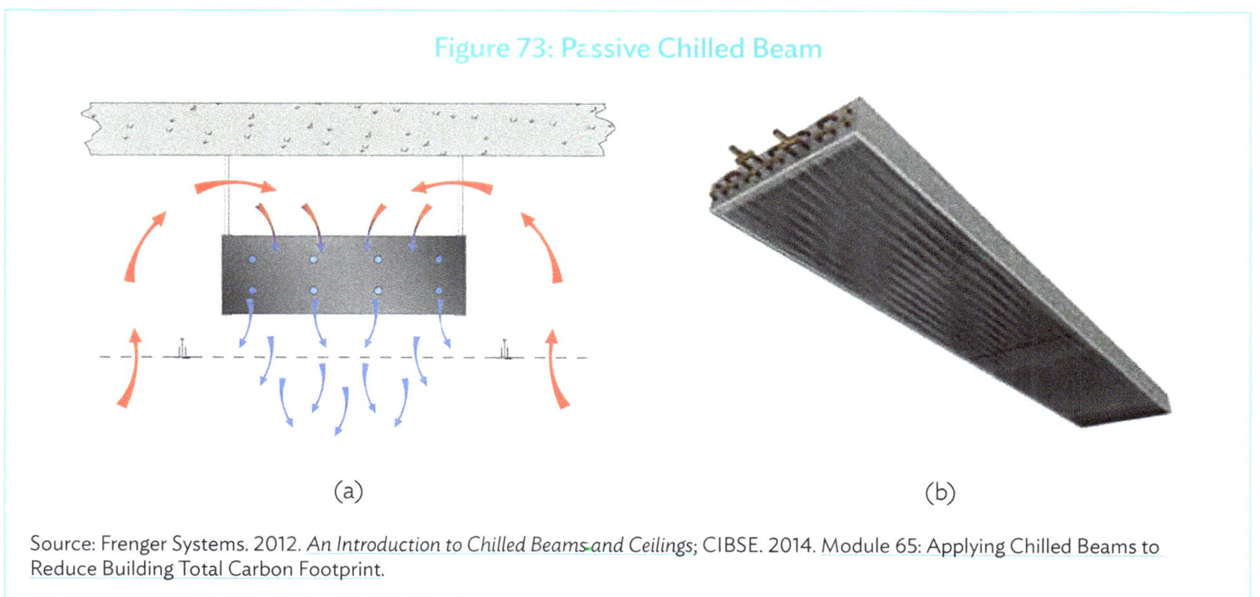

Figure 73: Passive Chilled Beam

(a)

(b)

Source: Frenger Systems. 2012. *An Introduction to Chilled Beams and Ceilings*; CIBSE. 2014. Module 65: Applying Chilled Beams to Reduce Building Total Carbon Footprint.

(iii)   **Active Chilled Beams**

Active chilled beams are supplied with ducted primary air from a centralized system that is often used to meet the fresh air requirement in the space. This air also provides the dehumidification required for the conditioned space by having a supply air moisture content set at a level to offset the latent loads in the space.

The primary air is discharged into the beam through nozzles directed toward the outlet of the beam casing that induce room air up from the room and through the cool coils. The cooled room air is then entrained into the stream of fast-moving supply air, before being discharged into the space through the slots (Figure 74). This enhanced induction helps to significantly increase the velocity of the room air passing across the coils, thereby improving heat transfer. As in the case of passive beams, regular intervals along the ceiling are key to the arrangement of active chilled beams.

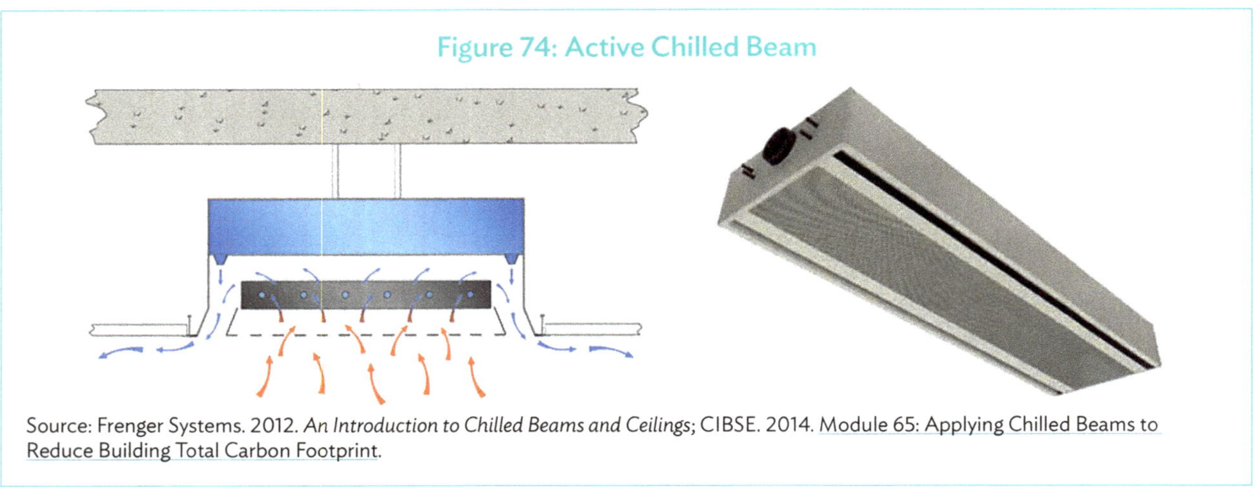

Figure 74: Active Chilled Beam

Source: Frenger Systems. 2012. *An Introduction to Chilled Beams and Ceilings*; CIBSE. 2014. Module 65: Applying Chilled Beams to Reduce Building Total Carbon Footprint.

Displacement Ventilation

Displacement is a proven solution to energy-efficient ventilation and cooling of buildings, particularly those suitable for office buildings. Its basic principle is illustrated in Figure 75. The cool fresh air is supplied at a low velocity, usually 0.25–0.35 meters per second (m/s), and at a lower level, usually through a raised floor. This creates a reservoir of cold air around the level of occupancy. When it gets into contact with an internal heat source, the air is heated and then rises due to the effects of momentum and buoyancy forces. As the air gets gradually warmer while flowing upward, a vertical temperature gradient is formed. The air is extracted through upper-level outlets, usually mounted at the ceiling, where there is a concentrated layer of warm air and pollutants. The movement of the air can be further driven by the heat emitted by the light fittings at the ceiling. Displacement ventilation is more energy-efficient than conventional overhead mixed air systems because the air is supplied near the occupied zone at lower velocity and a higher temperature, which means reduced cooling requirements and fan power consumption. The lower velocity and higher temperature relative to conventional systems also offer the advantage of thermal comfort due to the reduced draft. In some applications, a displacement ventilation has limited cooling capacity and is therefore operated in conjunction with chilled beams or chilled ceilings.

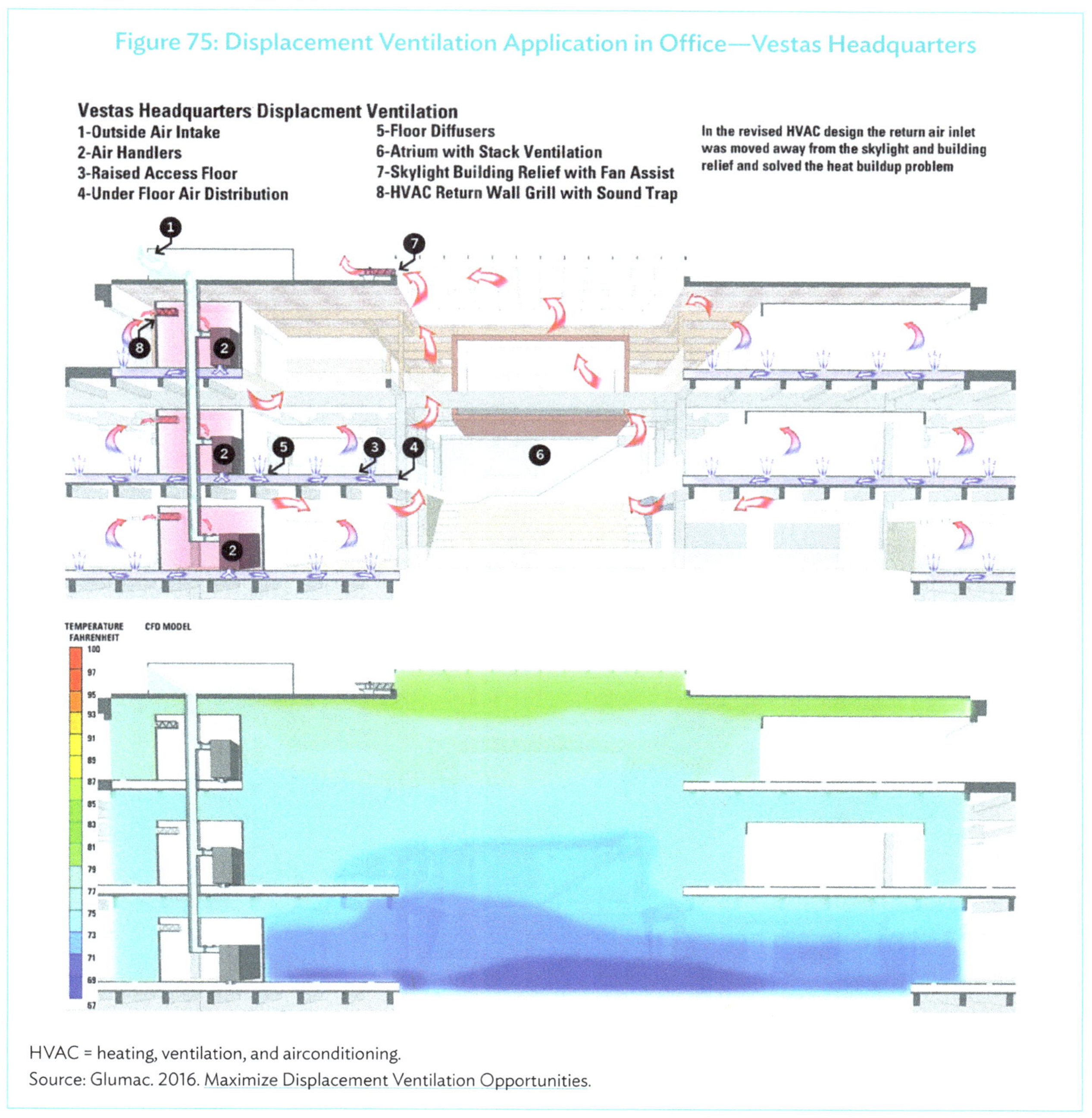

HVAC = heating, ventilation, and airconditioning.
Source: Glumac. 2016. Maximize Displacement Ventilation Opportunities.

Slab Cooling

Slab cooling, also commonly known as a subcategory of radiant cooling, can be realized using air or water. For air applications, slab cooling is often used with night ventilation to maximize the potential of cooling. The slab is constructed in such a way that it allows air to be passed through embedded pathways. During the night, cool air is passed through the slab to enhance the cooling of the slab. During the day, warm air is injected through the cool slab for precooling before being used to condition a space or sent to the air-conditioning system for further cooling.

Often, slab cooling is implemented by using embedded water coils and pipes through which chilled water circulates. The circulating chilled water cools down the slab, which in turn becomes a cold surface acting as a heat sink for sensible loads of internal heat gain (Figure 76). The chilled water can be generated either through conventional electric chiller systems or low-energy systems such as absorption chillers, desiccant chillers, etc.

Figure 76: Slab (Radiant Floor) Cooling and Heating

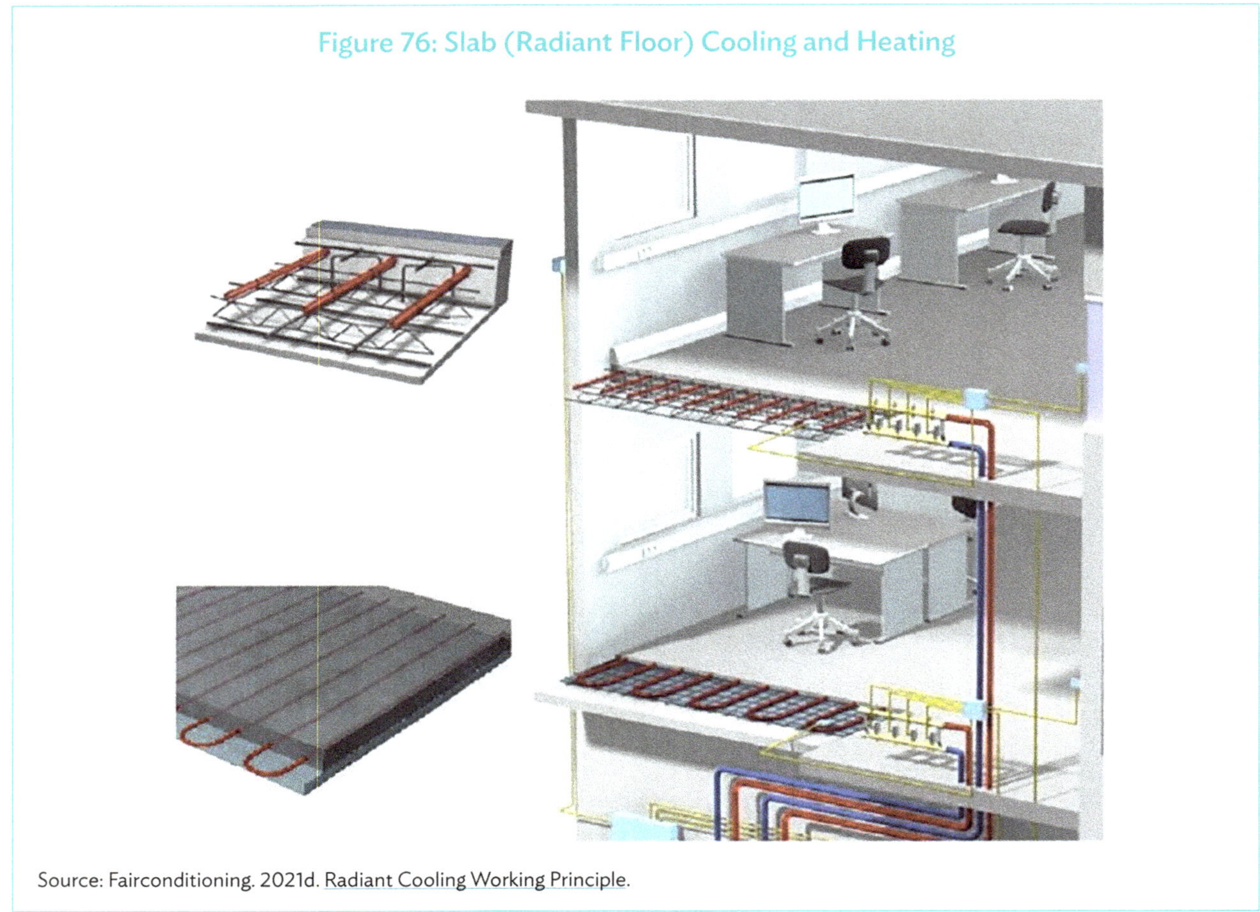

Source: Fairconditioning. 2021d. Radiant Cooling Working Principle.

Concrete structures typically used with slab cooling systems also increase the thermal mass of buildings. This introduces inertia in the structure against temperature fluctuations and allows it to absorb heat from internal spaces. This advantage of the thermal mass can help reduce the need for air-side systems. Thus, the conventional air-conditioning system can be downsized substantially, and operational energy can be saved. Due to potential condensation on building structural elements, slab cooling is more suitable for applications with low latent loads. It is important to control the operation to ensure the surface temperature is always above the dew-point temperature of the surrounding area so that condensation does not occur (Figure 77). As radiant cooling can only address a sensible load, air humidity must be separately handled by an air-conditioning system or a dehumidifier. In general, the capital cost of this system is similar to that of a high-efficiency chilled water system, but the operational cost is lower.

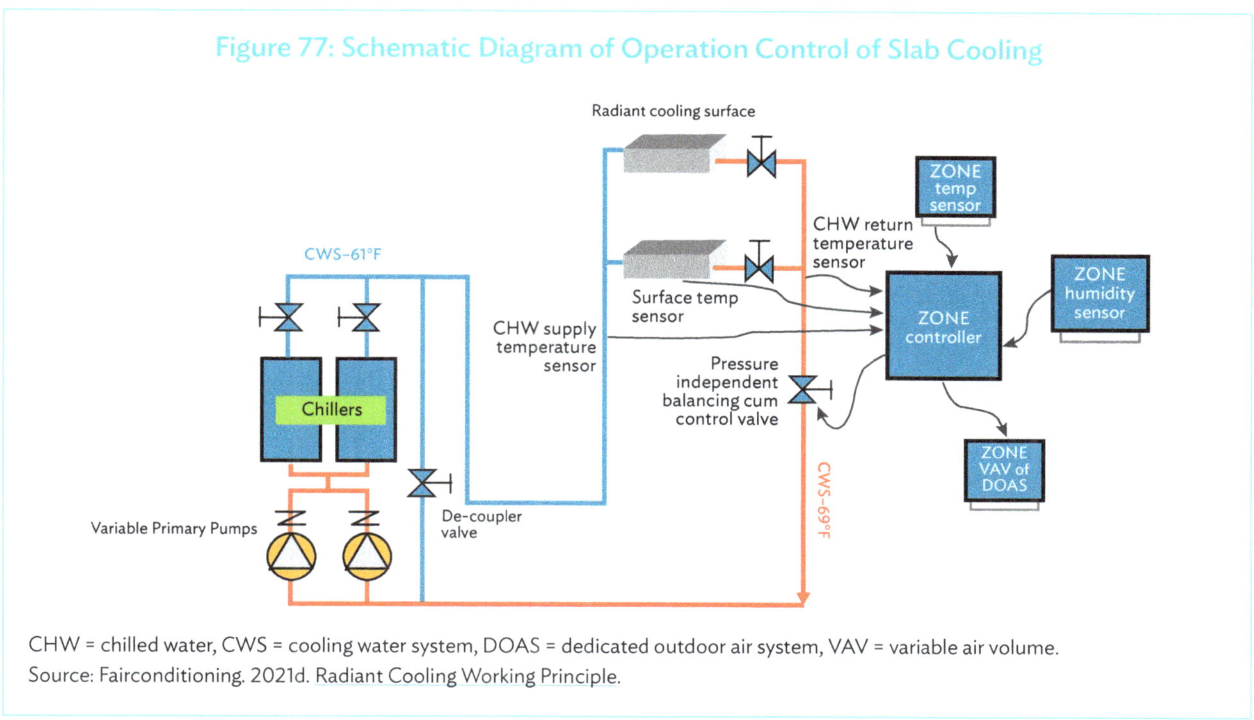

Figure 77: Schematic Diagram of Operation Control of Slab Cooling

CHW = chilled water, CWS = cooling water system, DOAS = dedicated outdoor air system, VAV = variable air volume.
Source: Fairconditioning. 2021d. Radiant Cooling Working Principle.

Box 15: Case Studies of Slab Cooling

CASE 1: INFOSYS HYDERABAD CAMPUS, INDIA

Infosys constructed the first radiant cooled commercial building in India on its campus in Hyderabad, where the climatic conditions are hot and dry. The total built-up area was 24,000 square meters (m^2). The building was split into two symmetric halves: one with conventional air-conditioning (high efficiency and surpassing ASHRAE standards by about 30%), and the other with radiant floor cooling. The radiant system was designed for a cooling output of 75 watts per square meter (W/m^2). Chilled water design temperatures were 14°C for supply and 17°C for return. The cooling tower approach temperature was 2°C. A low-pressure piping and ducting distribution system was installed. An energy recovery system was also installed to provide dehumidified air to offices.

Infosys Campus, Hyderabad, India.
Source: Fairconditioning. 2021c. Radiant Cooling Case Studies.

The energy index for the radiant system was 25.7 kilowatt-hours (kWh)/m^2, about 33% lower than the conventional system (38.7 kWh/m^2). The average chiller plant efficiency was 0.45 kilowatts per ton of refrigeration (kW/TR) for the radiant system and 0.6 kW/TR for the conventional system. The quantity of water required by the radiant system was about 20% of that required by a conventional chiller of similar capacity. As no recirculation was involved, contamination was reduced, and air quality was improved. Comfort conditions measured inside the building were within the permissible limits of ASHRAE 55-2004 and ASHRAE 62.1-2007 for most of the time. The capital costs of the two systems were comparable, with the radiant system costing ₹3,302/m^2 and the conventional system costing ₹3,327/m^2.

Source: NZEB Alliance.2021g. Radiant Cooling Systems; Fairconditioning. 2021d. Radiant Cooling Working Principle; Fairconditioning. 2021c. Radiant Cooling Case Studies.

continued on next page

Box 15 *continued*

CASE 2: UNDER FLOOR INSTALLATION AT L&T HAZIRA, SURAT, INDIA

Design Brief:

- Site location: L&T Hazira Surat, Gujarat
- Climate conditions: hot and humid, with high latent load
- Design conditions: indoor design temperature 25°C, relative humidity 60%

Solutions:

- Radiant cooling was used for the ground floor
- Insulated ground floor slab mitigated loss of cooling effect
- Latent load and residual sensible load were met by conventional air-conditioning system
- Underfloor cooling system was installed in areas containing workstations connected to electrical conduits in the floor

Impact of Radiant Colling on Total Power Requirement

Parameter	Conventional HVAC	Radiant + Conventional HVAC
AHU Power	65 kW	21 kW
Chiller Power	181.3 kW	56.5 kW
Water Pumps		50 kW
TOTAL	246 kW	127.5 kW
SAVINGS		118.5 kW (48%)

AHU = air handling unit; HVAC = heating, ventilation, air-conditioning; kW = kilowatt.
Source: Fairconditioning. 2021c. Radiant Cooling Case Studies.

CASE 3: A SLAB-INTEGRATED RADIANT COOLING SYSTEM IN PUTRAJAYA, MALAYSIA

Kwong et al. (2017) conducted a comprehensive study on the energy conservation potential and cost-effectiveness of a slab-integrated radiant cooling system installed at a green office building in Putrajaya, Malaysia.

The Office Building

Completed in 2010, the eight-story office building has a total area of approximately 23,284 m², of which about 15,225 m² were air-conditioned. The building was north–south oriented, and the center of the building was designed to be hollow to allow penetration of natural daylight. Most of the building facilities were located around the building perimeter. The office building consisted of three main parts, including the base (one and a half stories), the office block (six stories), and the rooftop. The rooftop consisted of an open-air garden built under the concept of "green building." For air- conditioning, chilled water was supplied from a centralized chiller plant, where both electric and gas-fired chillers were installed, to the office building through pre-insulated seamless underground pipes.

The Slab Cooling System

The multi-floor radiant slab cooling system was designed to handle 40% of the sensible cooling loads generated from the internal heat gains of the office building. The 22-millimeter (mm) cross-linked polyethylene (PEX) pipes were embedded at 50 mm above the underside of the 200 mm thick concrete slabs during construction. This turned the building structure into a thermal storage system. With no insulation layers in the slabs, the system acted as both a chilled ceiling and a chilled floor. The chilled water was supplied and circulated on the concrete floors at night to cool the building

continued on next page

Box 15 *continued*

structure down to 18°C–20°C. A timer was installed to control the operating time of the chilled water flow in the hydronic tubes to ensure that the thermal mass was sufficient to absorb 40% of the sensible heat. Typically, the thermal mass charging was from 10:00 p.m. to 6:00 a.m. The system was shut off during office hours to allow the floor slabs to passively absorb heat gained from occupants, office equipment and devices, solar irradiation, and other sources. This effectively reduced the cooling load and the resultant energy consumption of the air-side system. The ceiling surface temperature was maintained at 2°C or higher than the dew-point temperature of the circulated indoor air to avoid condensation. The latent load was handled by overhead air plenums installed on each floor, which supplied filtered and conditioned air to the indoor space. The indoor air relative humidity was maintained in the range of 50%–60% by modulating the intake of treated fresh air, whereby the risk of condensation problems commonly associated with radiant cooling systems was effectively mitigated.

Box Figure 15.1: Graphical Illustration of the Slab Cooling System Used in a Green Office Building in Putrajaya, Malaysia

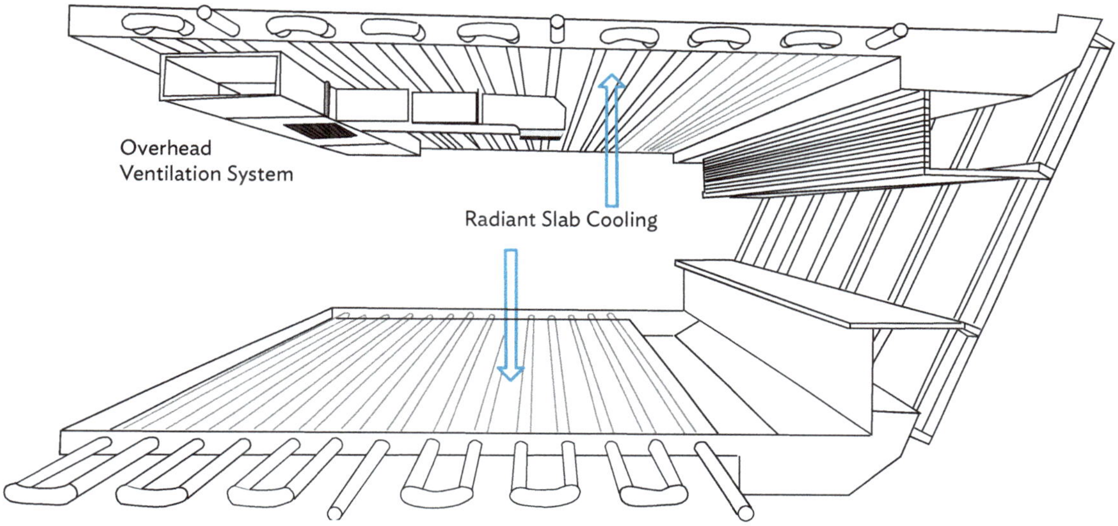

Source: Kwong, Q.J. et al. 2017. Evaluation of Energy Conservation Potential and Complete Cost-Benefit Analysis of the Slab-Integrated Radiant Cooling System: A Malaysian Case Study. Energy and Buildings. 138. pp. 165–174.

Energy Performance

The energy consumption pattern of the system was studied using the operation data collected from the building energy management system. For comparison purposes, a reference conventional air-conditioning system, which was an all-air variable air volume system, was designed for the same building, using the simulation software eQuest. The simulated energy consumption of the reference system was then compared with the actual performance data of the radiant slab cooling system. Results showed that the radiant cooling system consumed 34% less energy per m^2 per year than the conventional system (Box Figure 15.2), leading to a substantially lower yearly electricity bill (Box Table 15.1).

continued on next page

Box 15 *continued*

Box Figure 15.2: Comparison of Building Energy Index* Between the Radiant Slab Cooling System and the Conventional System

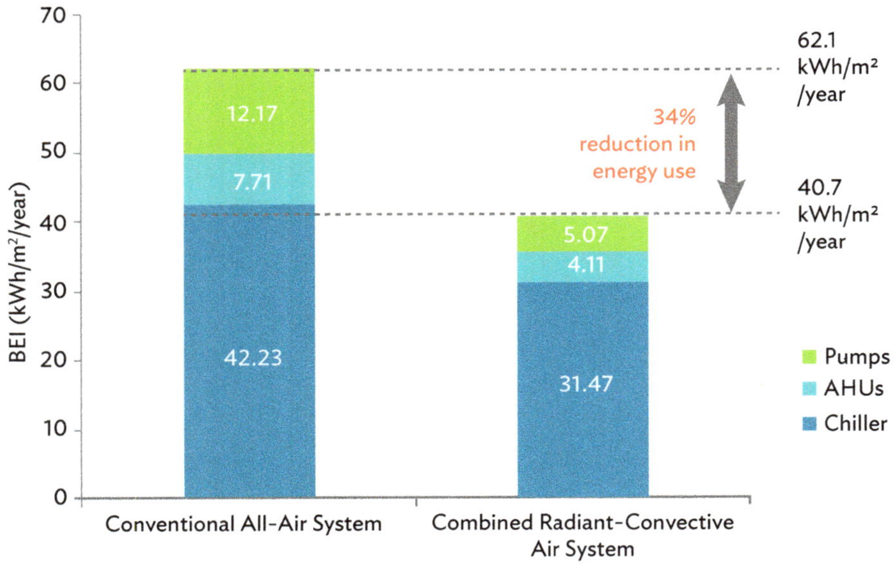

AHU = air-handling unit, BEI = building energy intensity, kWh = kilowatt-hour, m² = square meter.
*(kWh/m²/year).
Source: Kwong, Q.J. et al. 2017. Evaluation of Energy Conservation Potential and Complete Cost-Benefit Analysis of the Slab-Integrated Radiant Cooling System: A Malaysian Case Study. Energy and Buildings. 138. pp. 165–174.

Box Table 15.1: Comparison of Electricity Usage and Bill Between the Radiant Slab Cooling System and the Conventional System

Details	Radiant cooling	Conventional system
BEI (kWh/m²/year)	40.7	62.1
Operating hour (hour/day)	10	10
Monthly usage (kWh/month)	51,638.13	78,789.38
Tariff rates (RM/kWh)	0.192	0.312
Monthly bill (RM)	9,914.21	24,582.29
Yearly bill (RM)	118,974.25	294,987.48

BEI = building energy intensity, kWh = kilowatt-hour, m² = square meter, RM = Malaysian ringgit.

Cost–Benefit Analysis

The total costs for the radiant slab cooling system (together with the integrated air-side system) and the conventional variable air volume system of similar cooling capacity were calculated based on the price quotations and service contract agreements obtained from several internationally renowned air-conditioning product manufacturers and service providers. As shown in Box Table 15.2, while the radiant system had a higher initial cost than the conventional system, its annual operating cost was significantly lower. This led to a lower total cost over the 20-year operational lifespan. The system cost per m² per year of the radiant cooling system was approximately 79% of the conventional system. The simple payback period of the radiant cooling system was as short as 2.5 years.

continued on next page

Box 15 *continued*

Box Table 15.2: Cost-Benefit Analysis of the Radiant Slab Cooling System and the Conventional System

Description	Radiant slab cooling system (RM)	Conventional variable-air-volume system (RM)
Initial cost		
Water-cooled chiller	509,000	674,100
Pumps (cooling water + chilled water)	80,000	80,000
Air-handling units (AHUs)	208,800	744,000
Cooling tower	108,000	126,000
Ductwork	510,500	947,610
Direct digital control	600,000	800,000
Cooling water pump and chilled water pump pipes	20,995	80,844
Slab-integrated pipes and accessories	1,858,532	
Building energy management system	300,000	300,000
Miscellaneous	300,000	300,000
Estimated maintenance cost (20 years)		
Scheduled maintenance	2,202,193	2,202,193
Annual preventive service maintenance	575,733	575,733
Service repair cost (20 years)		
Major overhauls (chiller and cooling tower)	165,330	165,330
Bearings, seal, and belting (pumps and AHU)	174,300	174,300
Coil replacement (AHU)	550,000	550,000
Infill replacement (cooling tower)	120,000	120,000
Miscellaneous repair (all equipment)	727,500	727,500
Operating cost (20 years)		
Operating cost	2,379,485	5,901,648
Total cost	11,390,368 ($3,722,342)	14,469,218 ($4,728,503)
Simple payback period (year)	2.52	
Cost/m²/year	RM 37.41 ($12.22)	RM 47.52 ($15.53)

AHU = air-handling unit, m^2 = square meter, RM = Malaysian ringgit.

Source: Kwong, Q.J. et al. 2017. Evaluation of Energy Conservation Potential and Complete Cost-Benefit Analysis of the Slab-Integrated Radiant Cooling System: A Malaysian Case Study. Energy and Buildings. 138. pp. 165–174.

While many ADB DMCs are geographically located in climate zones that are cooling-dominated or have significant cooling demand during summer, there exists significant heating demand during a year in DMCs in the temperate climate zone, such as Mongolia and the northern PRC in East Asia, and countries in Central and West Asia. For example, Kazakhstan has a heating season lasting more than 6 months, varying from 132 to 231 days per year, with its annual average temperature ranging from 2°C in the north to 13°C in the south. In 2018, 30% of Kazakhstan's households used coal and/or firewood for heating (IEA and EU4Energy 2020). In Mongolia, the share of space heating in the total annual energy consumption of urban households in 2018 was 68% for houses, 75% for *gers* (tents), and 79% for apartments, respectively. Out of Mongolia's total households, 70% used stoves, and 27.5 had access to district heating. The rest used low-pressure boilers or electric heating systems (Economic Research Institute for ASEAN and East Asia 2021).

To bring down energy and emission intensities of space and water heating and accelerate low-carbon transformation of the buildings and building construction sector in these countries, it is important to promote and scale up the applications of higher-efficiency heating technologies and renewable energy as a clean heat source. This chapter introduces various heat pump technologies, including air-source and ground-source, and the utilization of solar energy for space heating and hot water, which have high relevance to and immense potential in the aforementioned DMCs.

Heat Pump

A heat pump is a thermal installation that extracts heat from a source at a low temperature to a sink at a high temperature by consuming drive energy. This produces a thermal effect, which can be applied for space heating or cooling for buildings, domestic hot water heating, heating or cooling for industrial processes, etc. This means all cooling and refrigeration equipment, including air-conditioners and chillers with refrigeration cycles, are heat pumps in light of the fundamental principles of their operations. In fact, a heat pump operating in cooling mode is technically an air-conditioning system. However, in the general context of engineering, the term heat pump is usually reserved for installations that produce heat for process or comfort, such as space heating, rather than those removing heat for cooling purposes only. Reversible heat pumps can operate in dual modes to provide both heating and cooling. In this section, unless otherwise specified, heat pump is discussed in the context of space heating and domestic hot water supply.

There are four basic types of heat pump cycles: closed vapor-compression cycle, open vapor recompression cycle, mechanical vapor recompression cycle with heat exchanger, and heat-driven Rankine cycle. Among them, the closed vapor-compression cycle, driven by electricity, is the most used heat pump cycle for HVAC applications and industrial processes. It uses a conventional, separate refrigeration cycle consisting of an evaporator, a condenser, a compressor, and an expansion valve (Figure 78).

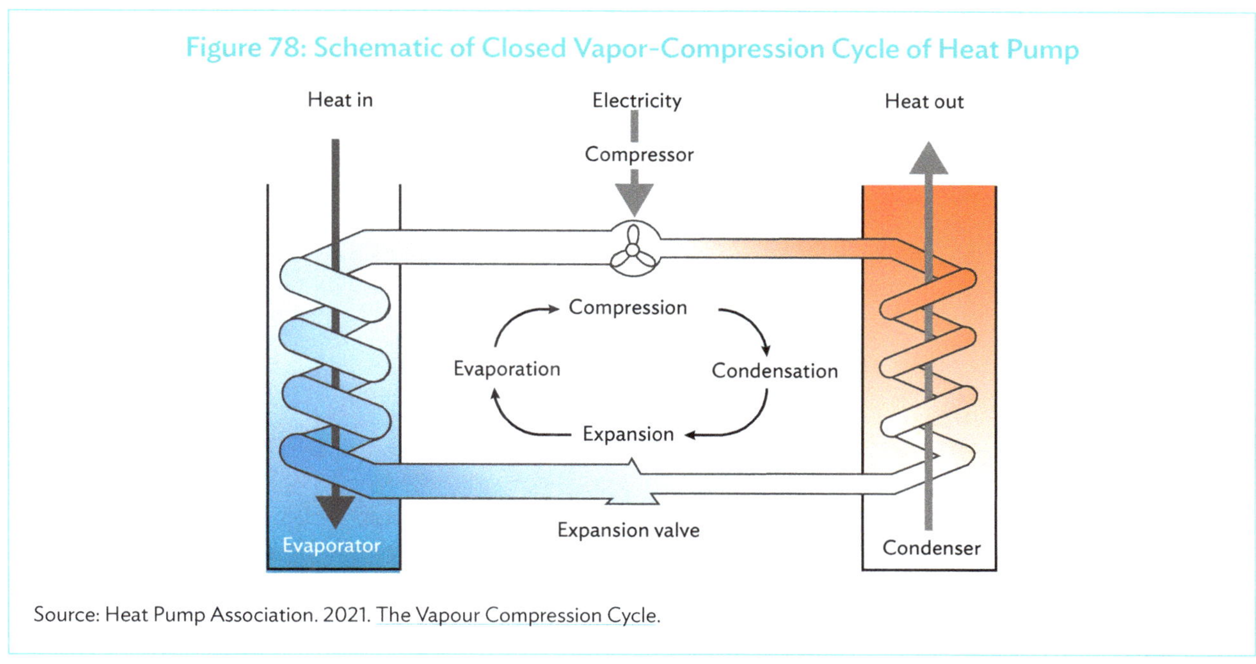

Figure 78: Schematic of Closed Vapor-Compression Cycle of Heat Pump

Source: Heat Pump Association. 2021. The Vapour Compression Cycle.

Various sources can be potentially used as the heat source and sink for a heat pump application, depending upon the climatic condition, geographical location, availability, cost, and type of structure. The main sources are air, water, and the ground. Outdoor air is a universal heat source and sink medium for heat pumps, and is widely used in residential and commercial systems. Exhaust air from ventilation can also be used as a heat source, but it is often insufficient for typical loads and therefore needs supplemental heat. Water from various sources can be used for heat pumps. A good option is groundwater, which has a relatively high and nearly constant temperature. Surface water such as rivers, lakes, open ponds, and so on can also be used, although there is a risk of freeze-up due to the temperature drop across the evaporator during winter. Sewage often has temperatures higher than those of surface water or groundwater and may be an acceptable heat source. Another possibility is wastewater from industrial processes. The ground is used extensively as a heat source and sink. Soil composition and moisture content have a significant effect on thermal properties and heat transfer performance.

Usually, the heat source and the heating and cooling distribution fluid are used as the basis for the classification of heat pumps. Thus, a heat pump can be designated as an air-to-air, air-to-water, water-to-air, or water-to-water heat pump. Alternatively, based on the "system" (air, water, ground) into which a heat pump is installed, a heat pump can be designated as an air-source, water-source, or GSHP. Since groundwater and surface water are more commonly used than sewage or wastewater for space heating and cooling applications, sometimes they are considered as a subset of the general GSHPs in this context. For the specific type of GSHPs that transfer heat from soil through vertical boreholes or horizontal trenches, rather than through groundwater or surface water, it is often referred to as a "ground-coupled" heat pump.

Air-Source Heat Pumps

Air-to-Air Heat Pumps

Air-to-air heat pumps are the most common type, which is particularly suitable for factory-built unitary heat pumps. This type of heat pump is widely used in residential and commercial applications. Multiunit and multi-split system installations are particularly advantageous as they allow zoning and, therefore, heating (or cooling) in different zones on demand. A typical air-to-air heat pump has a full-hermetic compressor, finned heat exchangers for the evaporator and condenser, and an expansion valve. They normally operate in dual modes, i.e., heating and cooling. Under heating mode, the outdoor fan coil unit extracts heat from outdoor air and transfers it, via the refrigerant loop, to indoor space through the indoor fan coil (Figure 79). As the outdoor air temperature decreases, the indoor heat demand increases, and the capacity of the system decreases accordingly.

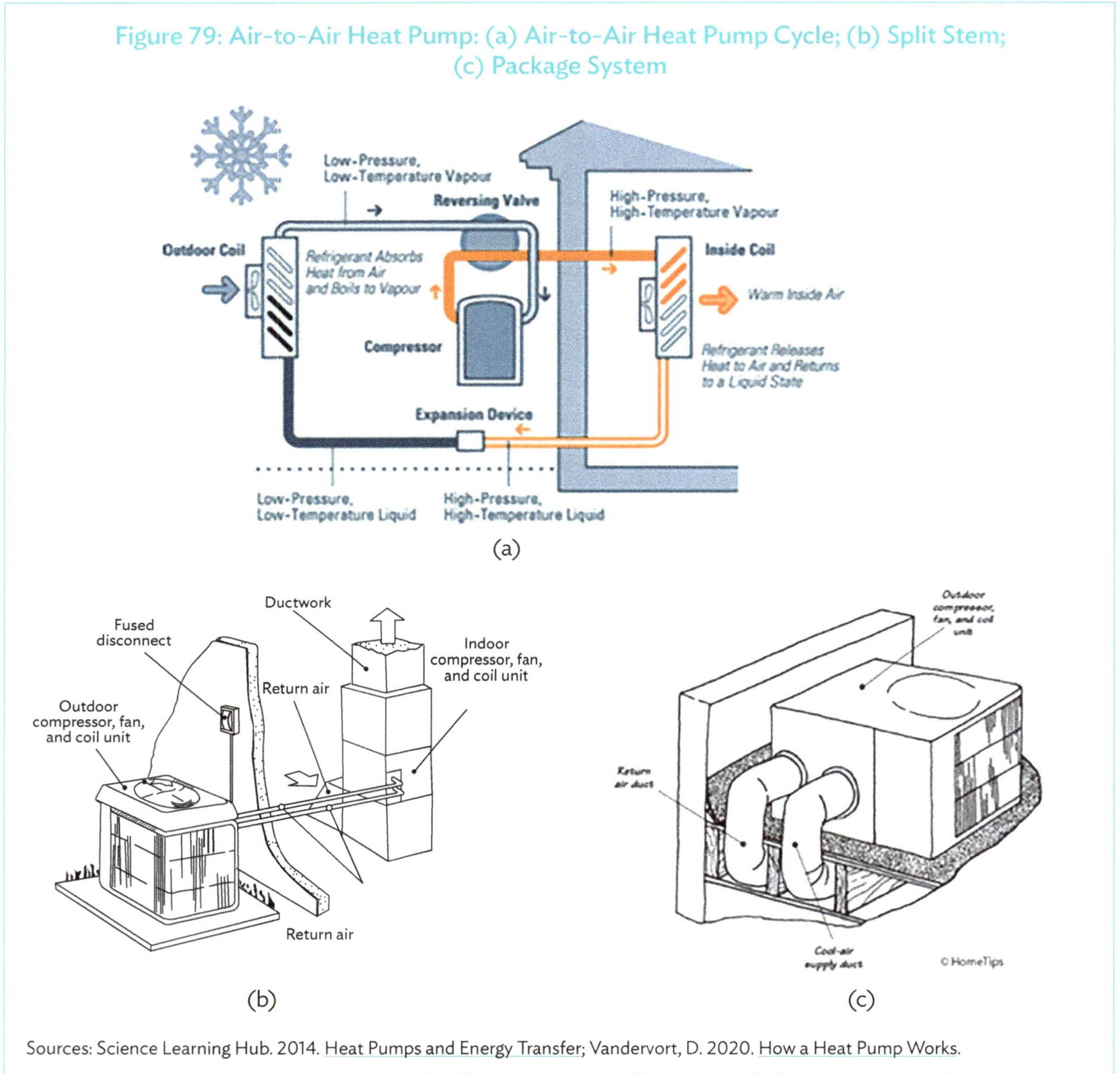

Figure 79: Air-to-Air Heat Pump: (a) Air-to-Air Heat Pump Cycle; (b) Split Stem; (c) Package System

Sources: Science Learning Hub. 2014. Heat Pumps and Energy Transfer; Vandervort, D. 2020. How a Heat Pump Works.

Air-to-Water Heat Pumps

Air-to-water heat pumps use outdoor air as their heat source and are mostly operated in bivalent heating systems (either parallel or alternate), as well as for cooling, heat recovery, and domestic hot water production. The indoor unit contains most of the components and is connected to the outdoor unit via refrigeration lines. In some tropical climates, the sensible and latent heat in ambient air can be used to heat water economically where power systems are costly and local natural gas is scarce, e.g., in small island countries. Thus, air-to-water heat pumps are also sometimes called heat pump water heaters (Figure 80).

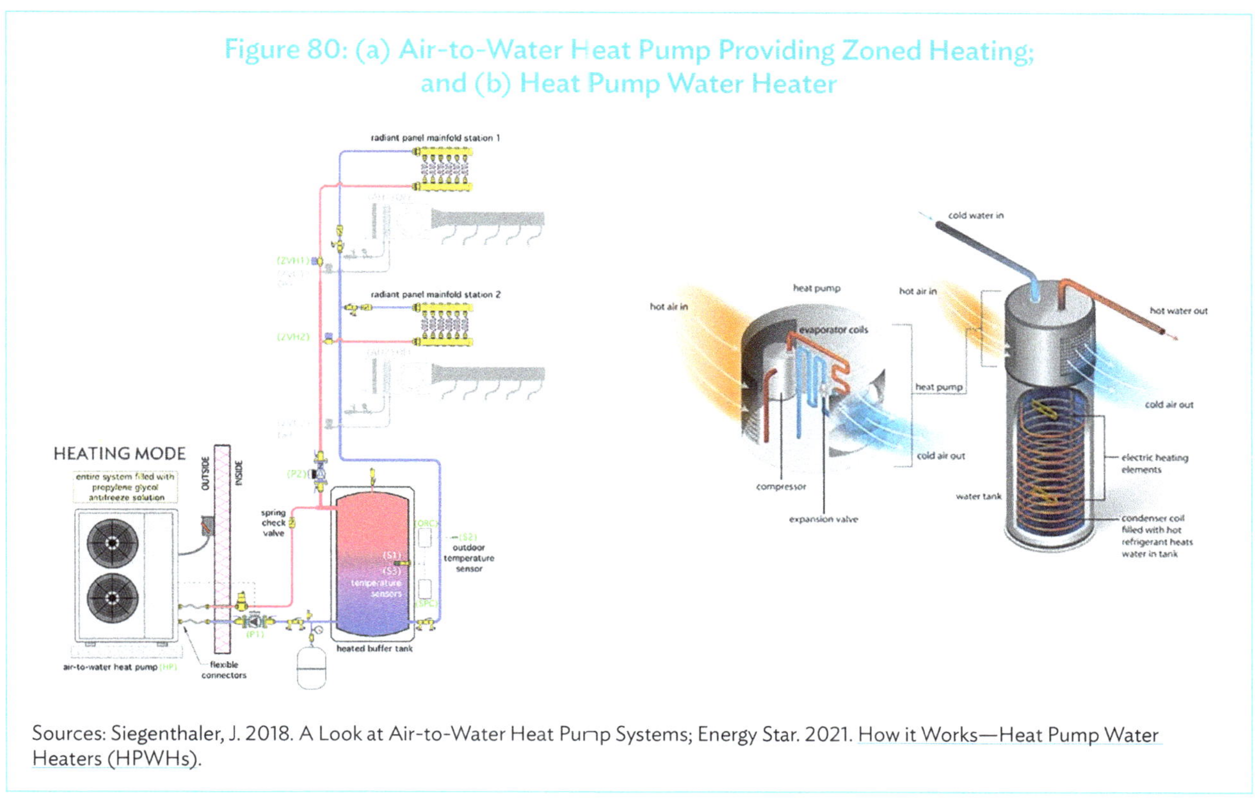

Figure 80: (a) Air-to-Water Heat Pump Providing Zoned Heating; and (b) Heat Pump Water Heater

Sources: Siegenthaler, J. 2018. A Look at Air-to-Water Heat Pump Systems; Energy Star. 2021. How it Works—Heat Pump Water Heaters (HPWHs).

Ground-Source Heat Pumps

Ground-source heat pumps comprise a wide variety of configurations that can use ground, groundwater, or surface water as the heat source or sink (Figure 81). A typical ground-source heat pump is composed of three loops, or subsystems, i.e., a ground connection subsystem, a heat pump subsystem, and a distribution subsystem for space heating, hot water, or cooling. The performance of a ground-source heat pump is usually good when it is in heating mode and optimized to operate at a lower water temperature than radiator, radiant floor, or radiant wall systems. A ground-source heat pump tends to be more cost-effective than conventional systems when: (i) the heat pump is for new construction where the installation and connection are relatively easy to complete, or for replacing an existing heating system reaching the end of its operational lifetime; (ii) the heat pump is operated in climatic conditions characterized by cold winter and hot summer, or geographical locations where natural gas has limited availability and/or a cost higher than electricity.

Figure 81: Various Types of Ground-Source Heat Pumps

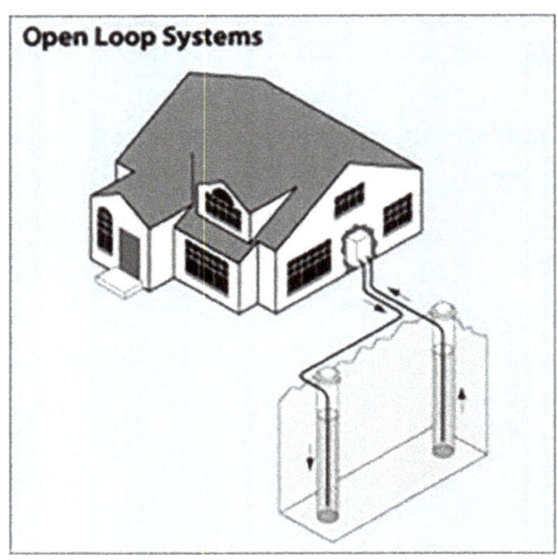

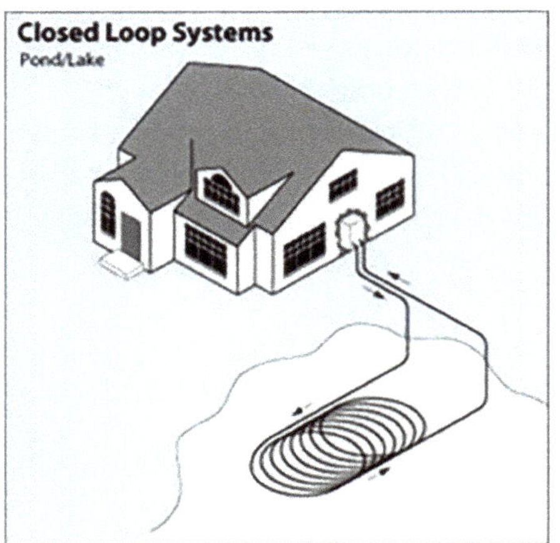

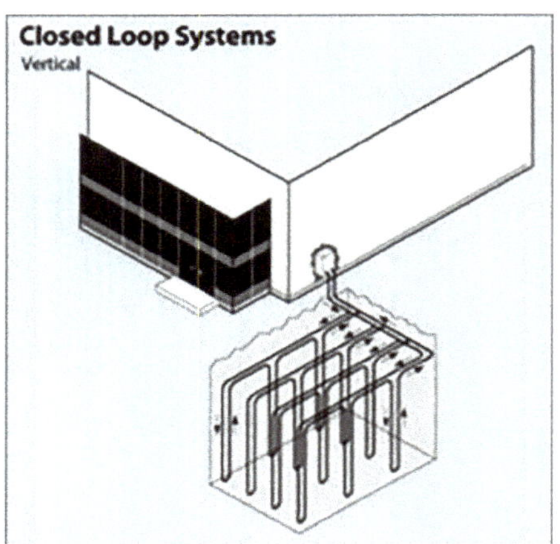

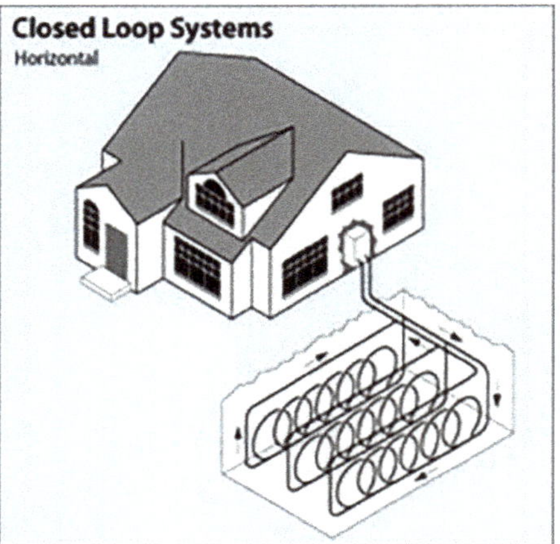

Source: Government of the United States, Department of Energy, Office of Energy Saver. 2021b. Geothermal Heat Pumps.

Groundwater Heat Pumps

Groundwater heat pumps use groundwater from wells as a heat source and/or sink. They can either circulate source water directly to the heat pump or use an intermediate fluid in a closed loop. After it has circulated through the system, the water is either reinjected into the ground through a reinjection well or discharged to a surface system such as a river, lake, or drain (Figure 82). This system can be designed as either a unitary type or a central plant. For the unitary type, a large number of small water-to-air heat pumps are distributed throughout a building. For the central plant, one or a small number of large-capacity chillers are used to supply hot or chilled water to a two- or four-pipe distribution system. The unitary design is more commonly used and more energy-efficient.

Compared with other ground-source heat pumps, groundwater heat pumps have the advantages of a low initial cost and a minimal requirement for ground surface area. However, some key operational factors, such as the limited availability of groundwater resources and the high maintenance cost due to fouling corrosion in pipes and equipment seriously restrict the wide application of these systems. Limited availability of groundwater resources means this type of heat pump is only practical in locations where there is an adequate supply of relatively clean water and the relevant regulations about groundwater utilization and discharge are fully complied with. High maintenance costs due to corrosion issues may be addressed through the installation of an intermediate plate-type heat exchanger to protect the heat pump unit. This issue is site-specific and should be evaluated in a local context.

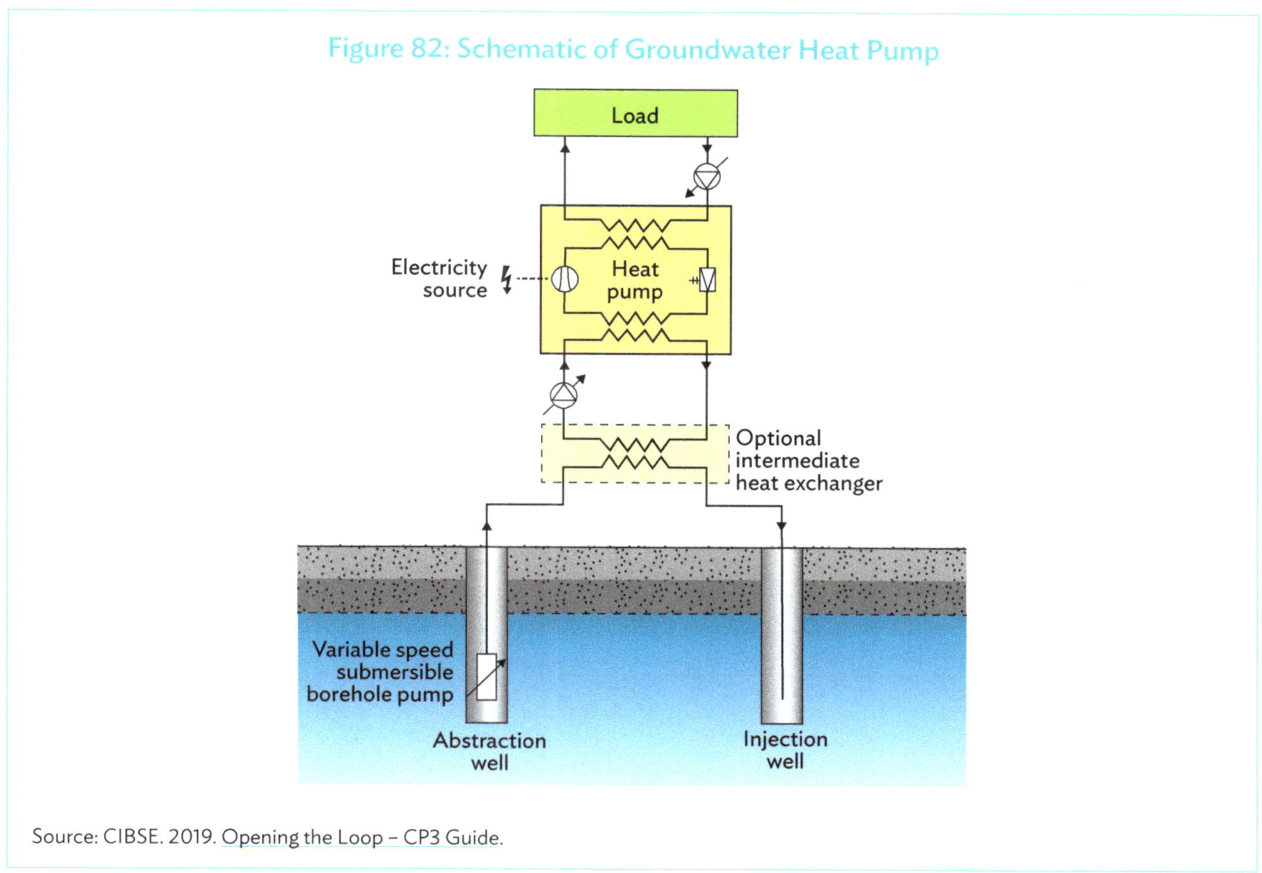

Source: CIBSE. 2019. Opening the Loop – CP3 Guide.

Surface Water Heat Pumps

Surface water can be a good heat source and sink if used properly. Surface water heat pumps can be either closed-loop systems, similar to ground-coupled heat pumps (discussed in the following section), or open-loop systems, similar to typical groundwater heat pumps. For open-loop systems, surface water from a nearby lake, pond, reservoir, or stream is pumped to the heat pump for heat exchange (Figure 83). The heat pump itself can be of either the water-to-air or water-to-water type. After passing through the heat pump, the water is returned to the source or a drain, with its temperature several degrees higher or lower, depending upon the operating mode of the heat pump. Closed-loop systems use a closed loop of working fluid (water or a water-antifreeze solution) that includes pipes or tubing submerged in the surface water (lake, river, large pond, or other open body of water). Usually, the pipes are made of high-density polyethylene (HDPE). A pump circulates the water or water-antifreeze solution through the heat pump water-to-refrigerant heat exchanger and the submerged piping loop for transferring heat from or to the body of water, depending upon the operating mode.

Compared to typical groundwater or ground-coupled heat pump systems, closed-loop surface water heat pumps can cost less due to reduced excavation costs. These systems have reduced pumping energy and operating costs, along with low maintenance requirements. However, there is the possibility of coil damage in public bodies of water. Also, the performance of systems using small and shallow bodies of water tends to vary substantially as a result of the wide fluctuation in water temperature. Thus, when designing a system, it is important to duly consider the adequacy of the total thermal capacity of the body of water under different weather conditions.

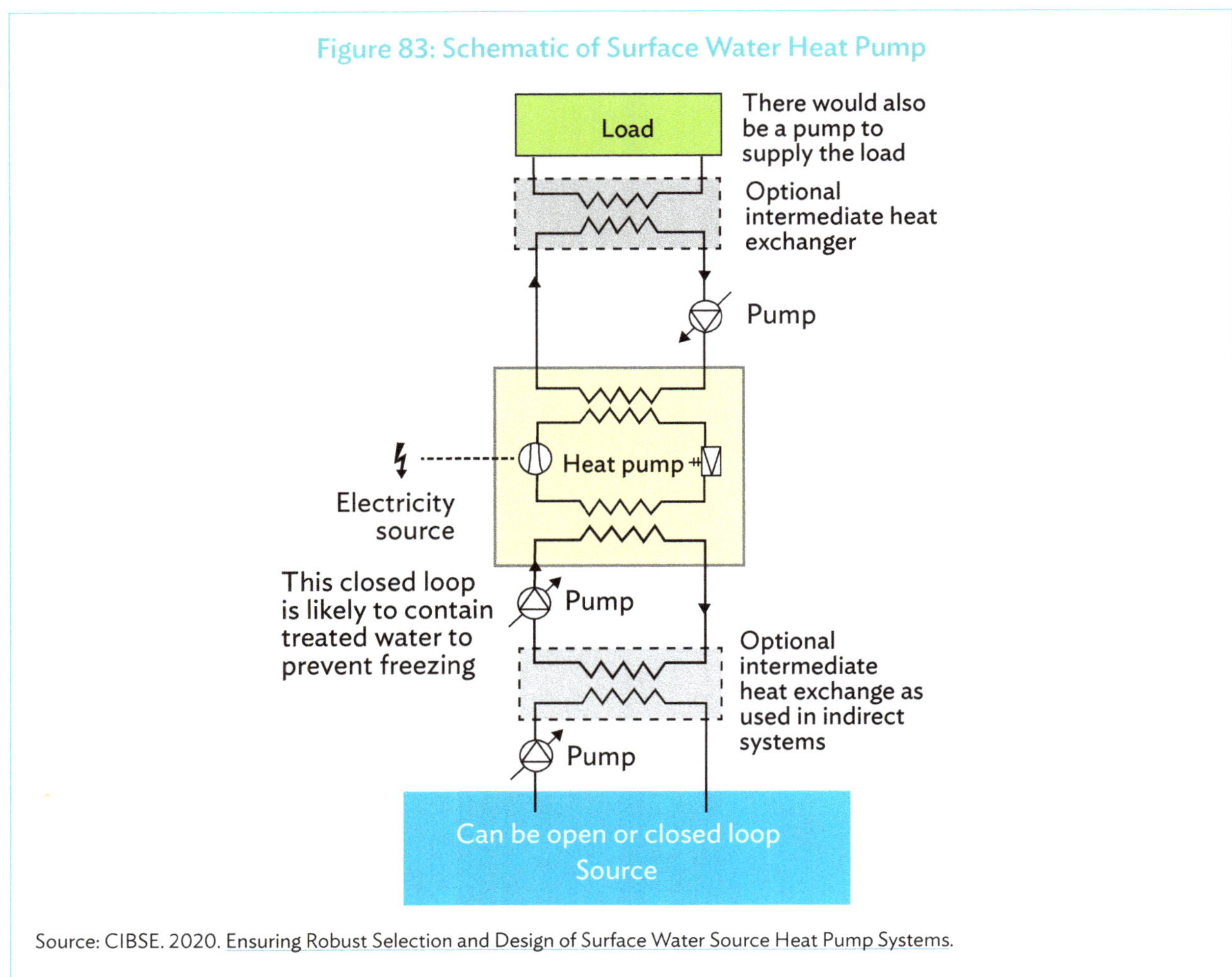

Figure 83: Schematic of Surface Water Heat Pump

Source: CIBSE. 2020. Ensuring Robust Selection and Design of Surface Water Source Heat Pump Systems.

Ground-Coupled Heat Pumps

The ground can serve as an ideal heat source and sink. For ground-coupled heat pumps, a closed pipework loop is buried in the ground, either vertically via boreholes or horizontally via coiled or straight pipes in trenches. This creates a ground heat exchanger (GHE). The working fluid, e.g., water or water-antifreeze solution, is pumped through this GHE to extract heat from the earth, which is then transferred to the heat pump evaporator (under heating mode), or to reject heat to the earth, which is received from the heat pump condenser (under cooling mode). The massive thermal capacity of the earth provides a temperature-stabilizing effect on the circulating working fluid. Soil type, moisture content, composition, density, and uniformity close to the surrounding field areas affect the heat exchange performance. With some piping materials, the material of construction for the pipe and the corrosiveness of the local soil and underground water may affect the heat transfer and service life.

Vertical Ground Heat Exchanger

A vertical GHE consists of one or more vertical boreholes down which a HDPE pipe loop is placed. The borehole is backfilled with a solid medium, either sand or a high-heat transfer material commonly referred to as grout. The individual vertical loops are then joined together to form one or more circuits to be connected to the heat pump. The depth of boreholes varies, as a function of the required energy output, geological conditions, and the space availability. In general, vertical boreholes are suitable for applications that require the installation of sufficient heat exchange capacity under a confined surface area, such as a limited land area, an existing landscape to which disturbance needs to be minimized, or rocky earth close to the surface.

The HDPE pipe loop in a borehole can take two forms of configuration, namely U-tube and coaxial tube. The U-tube configuration can be single- or double-pipe, whereas the coaxial tube can either use a simple configuration of two straight pipes of different diameters or use more complex configurations (Figure 84).

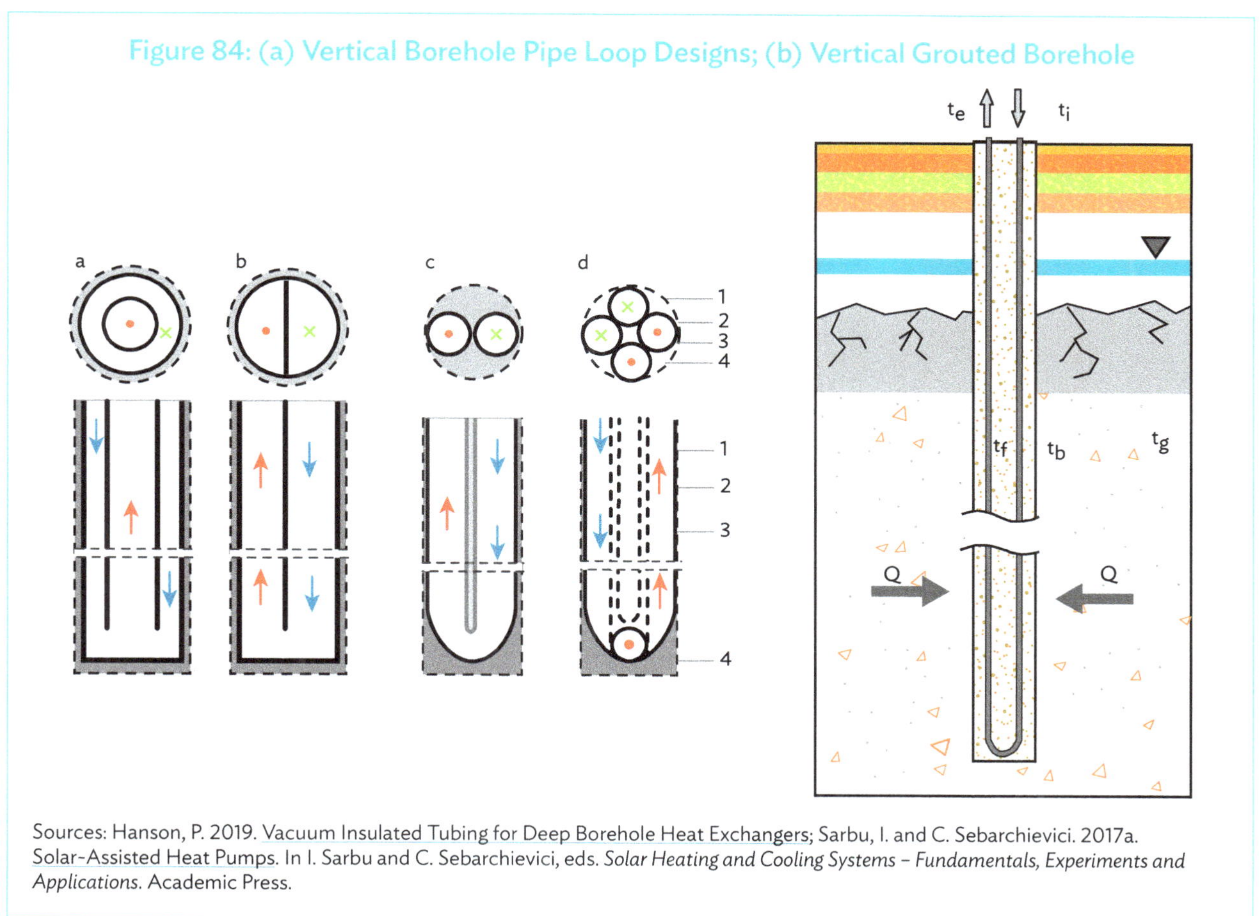

Figure 84: (a) Vertical Borehole Pipe Loop Designs; (b) Vertical Grouted Borehole

Sources: Hanson, P. 2019. Vacuum Insulated Tubing for Deep Borehole Heat Exchangers; Sarbu, I. and C. Sebarchievici. 2017a. Solar-Assisted Heat Pumps. In I. Sarbu and C. Sebarchievici, eds. *Solar Heating and Cooling Systems – Fundamentals, Experiments and Applications*. Academic Press.

A specific configuration of vertical GHE is to integrate the HDPE pipes into the structural foundation piles of the building (Figure 85). The pipes are attached to a reinforcing bar or cage and then placed into the foundation bore before concrete is poured. This method has the benefit of saving the cost and time required for drilling dedicated boreholes and installing conventional vertical loops. The limitation, however, is that its energy output may be limited because foundation piles do not normally go as deep as dedicated vertical boreholes for heat pumps.

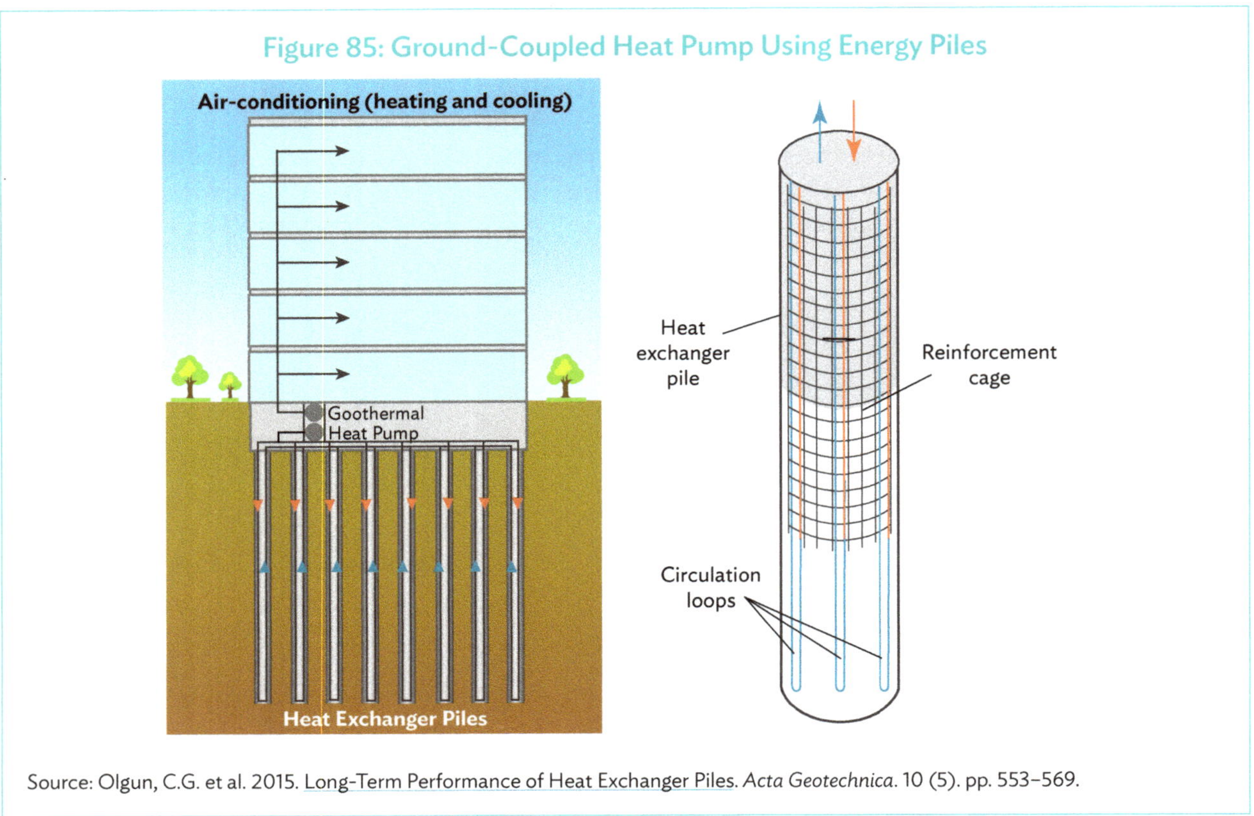

Source: Olgun, C.G. et al. 2015. Long-Term Performance of Heat Exchanger Piles. *Acta Geotechnica*. 10 (5). pp. 553–569.

In general, apart from the advantage of requiring a relatively small ground area, ground-coupled heat pumps using vertical boreholes can achieve a high efficiency of operation thanks to reduced variability in soil temperature and thermal properties. They require a reduced amount of piping and associated pumping energy. However, a key disadvantage is the high cost incurred by drilling the vertical boreholes.

Horizontal Ground Heat Exchanger

Compared to vertical GHEs, horizontal GHEs generally require much more ground area. However, they offer the benefits of installation simplicity and reduced installation costs. Typical configurations include single-pipe, multiple-pipe, and stretched coil (also called spiral slinky) (Figure 86). A single-pipe horizontal GHE consists of a series of parallel pipe arrangements laid out in trenches. It requires more ground area than multiple-pipe and stretched coil. A multiple-pipe configuration usually places two to six pipes in a single trench. This reduces the required ground area, but the total pipe length needs to be increased to overcome the interference from adjacent pipes. The stretched coil configuration consists of pipes unrolled in circular loops in trenches with a horizontal configuration. This can reduce the required ground area substantially. However, sometimes trench lengths may have to be increased for greater thermal performance.

In general, the installation of horizontal ground loops is the easiest for the construction of new buildings. Retrofitting is also feasible and is becoming easier with technological advances. Digging technologies allow drilling horizontal boreholes and thus make possible the installation of these systems into existing buildings with limited disturbance of the topsoil. Options for installing the loops under existing buildings, car parks, or driveways are all possible. However, a disadvantage of horizontal loops is their operation at a reduced efficiency due to their proximity to the ground surface and consequent susceptibility to seasonal soil properties and air temperature fluctuations. The higher pumping energy requirements also affect the efficiency level. For horizontal loops operating in heating-only mode, the main way of thermally recharging the ground is through solar radiation. Therefore, it is necessary to ensure the surface above the loops is not covered.

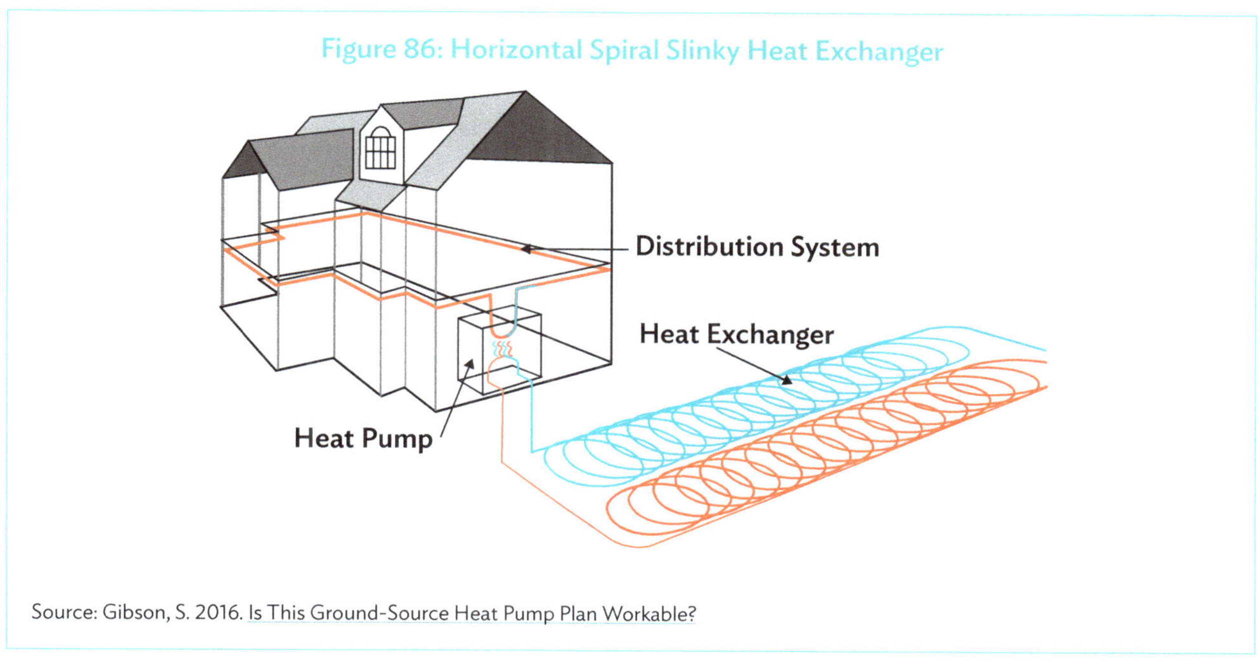

Source: Gibson, S. 2016. Is This Ground-Source Heat Pump Plan Workable?

Box 16: Case Study of Ground-Source Heat Pump for Space Heating

Beijing's First Ground-Source Heat Pump Project—Linked Hybrid MOMA

Beijing Linked Hybrid MOMA is in the Dongcheng District and has a total construction area of 220,000 square meters, comprising apartments, commercial areas, parking, hotels, cinemas, and educational facilities. This mega-scale complex, connected by steel-structured corridors, adopted a ceiling radiant cooling and heating system and displacement ventilation system, coupled with the largest ground-source heat pump system in the world at the time. Approximately 70% of the complex's annual heating and cooling load is handled by this ground-source heat pump system, which consists of 655 vertical ground tube heat exchangers with a depth of 100 meters below the basement foundation. The system received the 2006 American Popular Science Engineering Award for the largest geothermal housing complex.

Beijing's Linked Hybrid MOMA and its buried pipes of ground-source heat pump system.

continued on next page

Box 16 *continued*

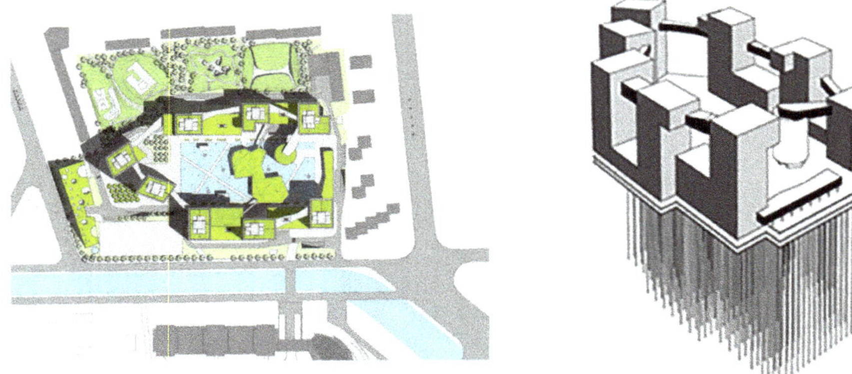

Sources: Cucchiaro, R. 2011. Beijing's Linked Hybrid: one of the largest geothermal cooling and heating systems in the world;
Song, C. et al. 2021. Application and Development of Ground Source Heat Pump Technology in China. *Protection and Control of Modern Power Systems*. 6.

Solar Space Heating

Solar space heating refers to the utilization of solar energy to meet the heating demands of conditioned spaces. In general, solar space-heating systems can be broadly classified into active and passive ones. Active systems use active mechanical and electrical devices to transfer heat from solar collectors. In passive systems, building structure components like the facade or roof are used to absorb and store solar radiation energy, which is then transferred to indoor spaces to meet heating demand.

Passive Solar Space Heating

The architectural design of buildings is at the heart of passive solar space-heating systems. Construction, morphology, and internal environmental and thermal comfort requirements are key factors to be considered. The form and orientation of a building are closely related to the building's exposure to thermal stress, therefore having a fundamental impact on the extent to which solar heat gains result in thermal comfort (or discomfort). Ideally, an optimal form should be able to collect most of the available solar radiation during the heating season while rejecting most of it during the other seasons. Under temperate climatic conditions, a building form that is elongated along the east–west direction is preferred, as it maximizes solar heat gains from the south during winter and limits solar heat gains from the east and west when being close to the horizon during summer. Under hot arid climatic conditions, given the thermal stress due to indoor–outdoor air temperature differences, compact building forms should be the option to limit heat transfer through the building envelope.

The amount of solar radiation absorbed within the building form depends on building envelope characteristics. These may include the glazed areas, overall transmittance, shading devices, insulated shutters, solar reflectors, etc. The heat capacity of a building determines its ability to store the absorbed solar energy before releasing it. To ensure that the heat gains will not result in large temperature fluctuations and thermal discomfort, it is important to balance the design of solar capture and regulating devices (windows, clerestories, skylights, shading devices, reflectors, shutters), heat storage (dark surface, thermal mass, phase-change materials), and heat distribution (conduction, convection, and irradiation).

Based on the design and operation of solar capture and heat storage in a building, passive solar space heating can be broadly classified into direct, indirect, and isolated systems, through which the heat is stored directly inside, in-between, or outside the heated space (Figure 87).

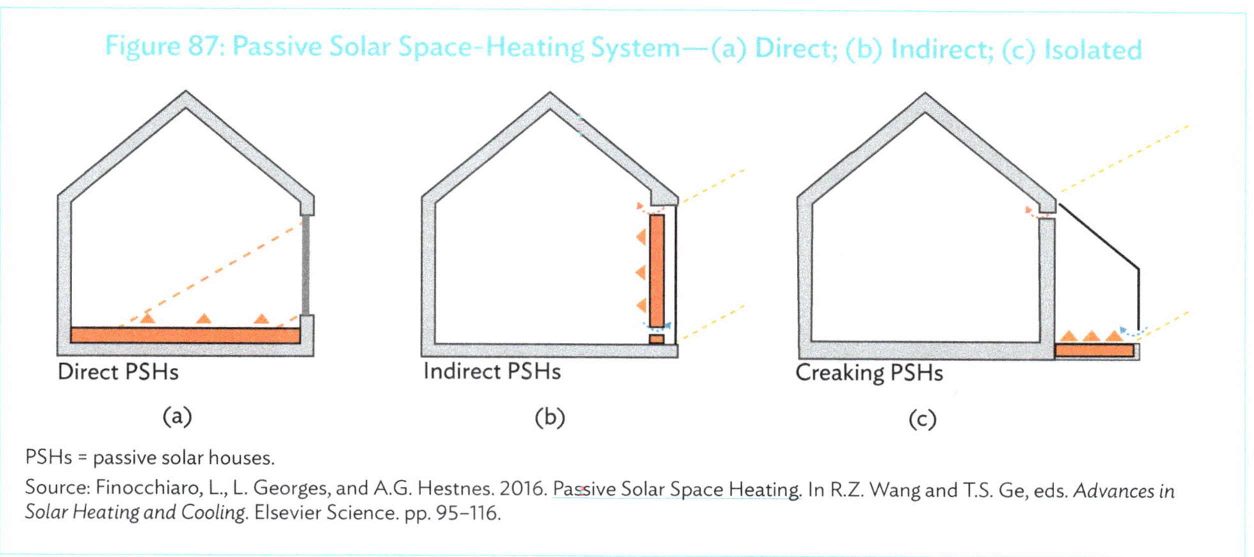

Figure 87: Passive Solar Space-Heating System—(a) Direct; (b) Indirect; (c) Isolated

PSHs = passive solar houses.

Source: Finocchiaro, L., L. Georges, and A.G. Hestnes. 2016. Passive Solar Space Heating. In R.Z. Wang and T.S. Ge, eds. *Advances in Solar Heating and Cooling*. Elsevier Science. pp. 95–116.

Direct Systems

As the simplest passive solar space-heating system, a direct system typically has the highest percentage of the total heating load met by solar energy. Through the solar capture and heat storage components of the building, the solar radiation is admitted and stored directly into the heated space. The heat storage thermal mass absorbs and detains the heat when the room temperature remains high. When there is no sunshine and the room temperature drops, the thermal mass slowly releases the heat due to thermal losses and reduced internal heat gains from occupants and equipment. In this system, the choice of thermal mass materials needs to take into consideration the climatic conditions and environmental space requirements. The solar capture components need to be appropriately sized to ensure that the admitted solar heat can be stored within the space and temperature fluctuations can be stabilized within the comfort zone.

Indirect Systems

Indirect passive solar space-heating systems are characterized by a heat storage mass placed in-between the solar capture components (e.g., glazed areas) and the space to be heated. Solar energy is indirectly transferred into the space through irradiation. Compared with direct systems, indirect systems have much lower risks of overheating.

A typical example of an indirect system is a Trombe wall, which is a combination of a dark masonry wall (often made of bricks, stone, or concrete) and a glazed area, with a narrow air cavity in between (Figure 88). Solar radiation passing through the glass is absorbed by the wall and then slowly released into the heated space. As direct solar radiation has a shorter wavelength, it is easily conducted through glass. The re-emitted heat from the wall takes the form of longer-wavelength radiation, which cannot pass through glass as easily as short-wavelength radiation. As a result, heat is trapped in the air cavity, allowing the wall to effectively absorb heat while limiting its re-emission into the environment. Since the glass panel is only on the exterior of the wall, heat can be freely transferred into the interior of the space. For a wall made of concrete with a thickness of 20–30 centimeters, this heat transfer process typically takes around 6–8 hours. Daily, this means that the wall absorbs heat during the day and slowly re-emits it into the space at night. Therefore, the need for conventional heating can be effectively reduced.

Figure 88: Basic Trombe Wall

Source: Williams, J. 2016. Two Ways to Make a Solar Wall. The Earthbound Report.

Often, the heat transfer through the masonry wall is supplemented with convection. This is realized by placing vents at the bottom and top of the wall (Figure 89). As the air in the cavity is heated, it rises into the top vent, through which it is redirected into the room. At the same time, lower-temperature air from inside the room passes through the lower vent into the cavity, where it is heated and later redirected back into the home through the upper vent. This creates a loop of natural convection. To avoid thermal losses that could reduce the efficiency of the system, or even a reversed heat transfer process in some circumstances, the air vents should be closed during the night when heat stored into the wall can still be slowly released in the space through irradiation.

Figure 89: Schematic of Ventilated Trombe Wall

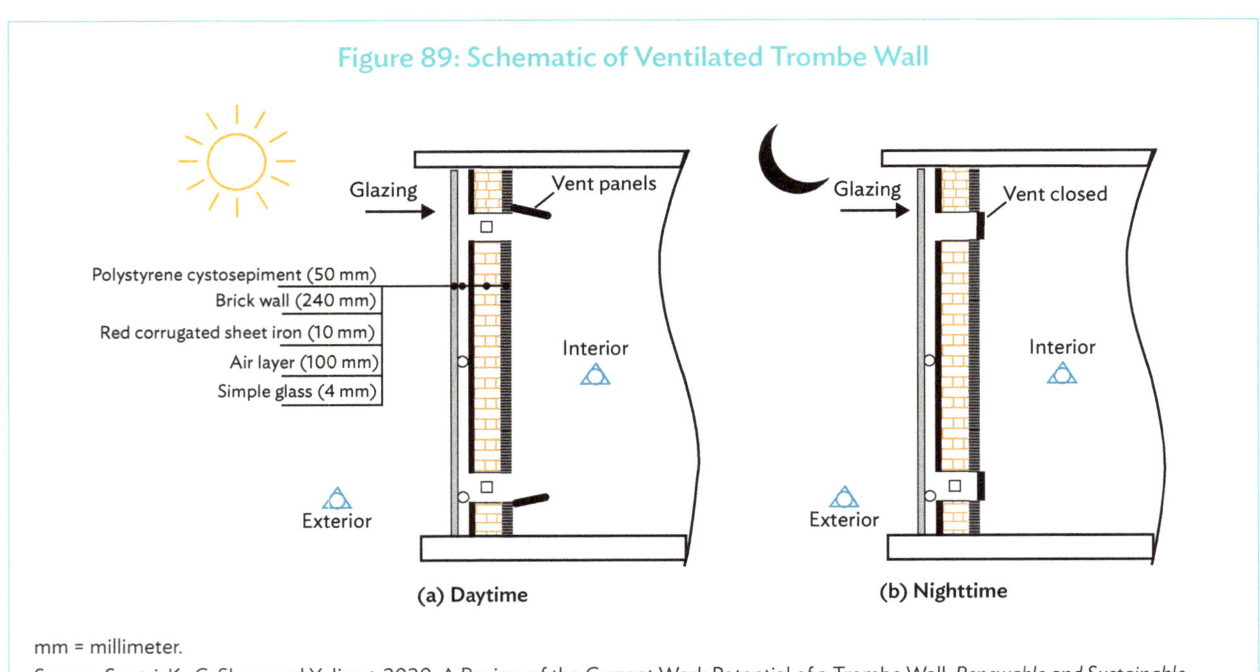

mm = millimeter.

Source: Sergei, K., C. Shen, and Y. Jiang. 2020. A Review of the Current Work Potential of a Trombe Wall. *Renewable and Sustainable Energy Reviews*. 130.

Isolated Systems

An isolated passive solar space-heating system is attached to the living space. It is typically made of a flat panel placed over a storage thermal mass, such as a rock bed. Heat is transferred into the living space through convection by an air-based siphon system, until an equilibrium in the temperature distribution is reached. A common type of isolated system is a greenhouse, which structurally extends the living space and therefore can be used as a functional space when its environmental condition is reasonably comfortable. A greenhouse serves as a thermal buffer space. Usually, it should be elongated along the south side of the building to maximize thermal exchange with the living space while minimizing exposure to the east and west. Another typical example of isolated systems is the Barra–Costantini wall, which is technically a variant of the Trombe wall. The major variation is the use of an insulation layer placed between the thermal mass wall and the heated indoor space. This setup can better manage system heat losses under cold climatic conditions.

Active Solar Space Heating

Active solar space-heating systems use solar energy to heat a working fluid (liquid or air) in the solar collector loop and then transfer the solar heat to meet the heading load of the interior space, or to a storage tank from which the heat can be kept for later use. The key difference from passive systems is that active systems require continuous electricity to power pumps and fans to keep the system running. Liquid systems are more often used when heat storage is included, and they are well-suited for radiant heating systems and boilers with hot water radiators. They can also be combined with heat pumps for space heating. In some applications where the demand for space heating is large, an auxiliary system is integrated into the solar space-heating system to supply additional heat.

Classification of Solar Space-Heating Systems

In general, there are two loops in an active solar space-heating system, i.e., a solar collector loop in which a working fluid is mechanically circulated, and an indoor space-heating loop in which a heating medium is mechanically circulated. By the type of the working fluid and the heating medium, active systems can be broadly categorized into four types, which have different operational characteristics (Table 10).

Table 10: Types of Active Solar Space-Heating Systems by Working Fluid and Heating Medium

	Solar collector loop	Space-heating loop	Operational characteristics
1	Water or antifreeze solution	Water	Water works better than air as a heat transfer medium due to its higher thermal capacity and heat transfer coefficient. Often, space heating is in the form of radiant floor and/or wall heating. Mainly used for high-latitude countries in cold climates.
2	Water or antifreeze solution	Air	Often, space heating is in the form of traditional air space heating coupled with a ventilation system. Mainly used for low-latitude countries in warm climates.
3	Air	Air	Using air as the heat transfer medium has the benefit of avoiding issues of freezing and boiling in solar collectors, as in the case of using water as the medium. Often solar air collectors are integrated with the building envelope (roof or wall). So, an air-space-heating system is sometimes is considered as a semi-passive one.
4	Air	Water	The air-water combination has a low thermal efficiency. The system requires a high flow of air and a large heat exchange surface area for air to transfer heat to water, thus having energy and noise implications.

Source: Author.

By the number of energy sources used, active solar space-heating systems can be categorized into monovalent mode, bivalent mode, and multivalent mode. The monovalent mode uses solar energy as the only source of energy. This is a possible option for applications in low-latitude countries where the heating load is relatively low. In these applications, often the active system is integrated with passive solar heating. In bivalent mode, which is the standard operational mode of active solar space-heating systems, usually solar energy and electricity are the two energy sources used. An electric heater can serve as an auxiliary heater when solar energy cannot meet the heating load on its own. Preferably, the electricity is generated by renewable sources. Sometimes the auxiliary heater can be a fossil-fuel fired boiler, or a biomass boiler. When the availability of solar energy is very low, the auxiliary heater can be operated to serve as the primary energy source to supply heat. As for multivalent mode, the system involves the use of three energy sources. This mode is often found in new technologies and applications of space-heating systems aiming for high efficiency and reliability. Here, the priority of utilization of a specific energy source is dynamically dependent upon which energy source is the most operationally and economically efficient for heat production in real time. Apart from solar energy, the other two energy sources are usually electricity and heat generated by fossil fuel-fired or biomass-fired boilers. The complexity of a multivalent mode of operation requires well-planned, effective, and precise control of system components.

Figure 90 presents two standard active solar space-heating systems that use water or antifreeze solution as the solar collector working fluid, i.e., liquid-based systems. Part of each system is a heat exchanger between the solar collector loop and the storage tank. This is because water is used as a storage medium owing to its thermal properties, which

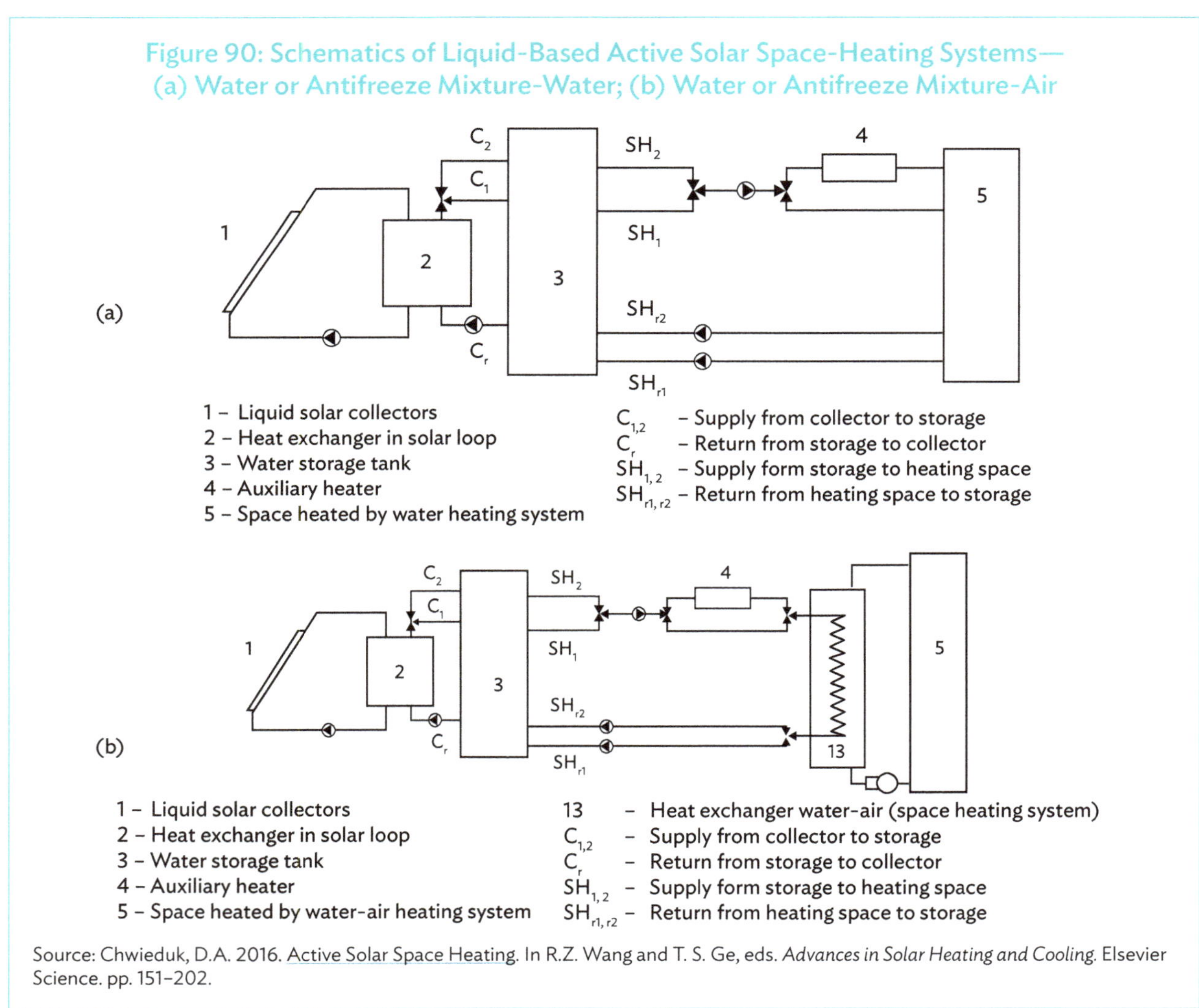

Figure 90: Schematics of Liquid-Based Active Solar Space-Heating Systems—(a) Water or Antifreeze Mixture-Water; (b) Water or Antifreeze Mixture-Air

Source: Chwieduk, D.A. 2016. Active Solar Space Heating. In R.Z. Wang and T. S. Ge, eds. *Advances in Solar Heating and Cooling*. Elsevier Science. pp. 151–202.

outperform those of the antifreeze mixture. In (a), water is used as the heating medium in the space-heating loop, which often uses radiators or a radiant floor heating system. In (b), the space-heating loop uses air as the heating medium for space heating through ventilation system.

Figure 91 presents an air-based system that uses air as the working medium in the solar collector loop. Air is also used as the heating medium for the space-heating loop.

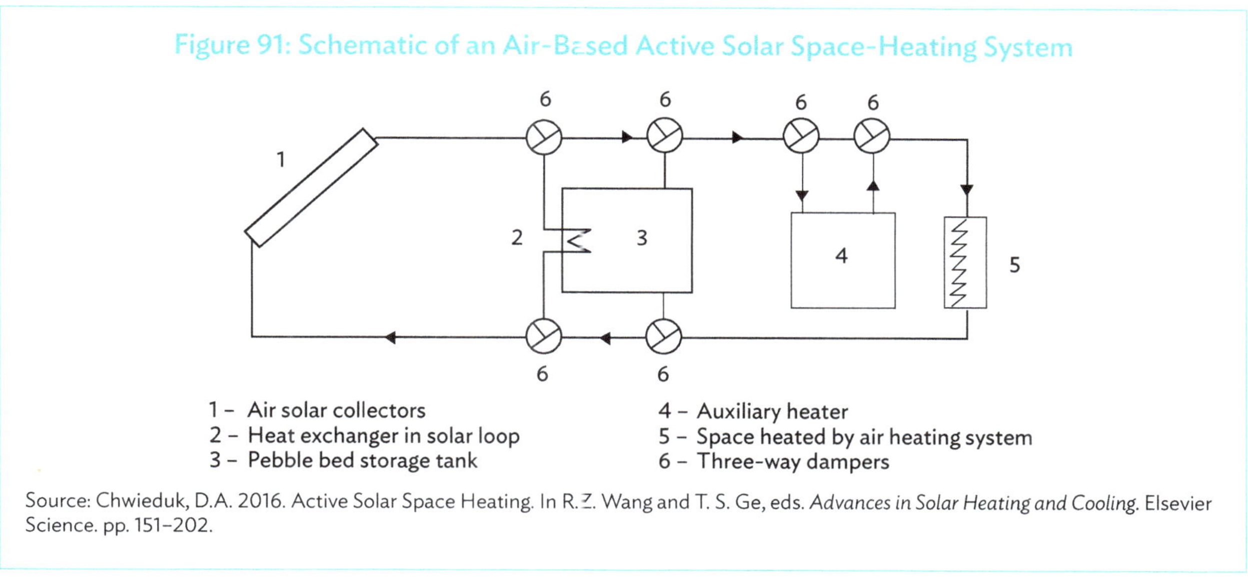

Figure 91: Schematic of an Air-Based Active Solar Space-Heating System

1 – Air solar collectors
2 – Heat exchanger in solar loop
3 – Pebble bed storage tank
4 – Auxiliary heater
5 – Space heated by air heating system
6 – Three-way dampers

Source: Chwieduk, D.A. 2016. Active Solar Space Heating. In R. Z. Wang and T. S. Ge, eds. *Advances in Solar Heating and Cooling*. Elsevier Science. pp. 151–202.

Solar Combi Systems

Solar combi systems provide both space heating and domestic hot water using a common array of solar thermal collectors backed up by an auxiliary non-solar heat source, such as an electric heater, a fossil fuel–fired boiler, or a biomass-fired boiler. The size of solar combi systems can range from small applications for individual properties to large ones serving a large group of buildings as a district heating system. Depending upon system installation, energy demand, and climatic conditions, the annual space heating contribution of a combi system can range from 10% to 60%, or more when the buildings being heated are low-energy buildings. When concentrating solar thermal heat and/or large inter-seasonal thermal storage are used, the contribution can even reach up to 100%.

Solar combi systems are relatively complex in terms of the system structure, configuration, and connections of various components. Compared to basic systems, combi systems require more sophisticated control and regulation of the functioning of all components and their interactions. As combi systems have at least two loops (space heating and domestic hot water), at least two thermal storage tanks are required. So are heat exchangers, which are integral to modern storage tanks. The more complex a system is, the more heat exchangers are required. Storage tanks are the central components that integrate different parts of a combi system. Thus, the design of storage has a fundamental impact on the overall system performance.

A standard active solar combi system is shown in Figure 92. The system uses liquid as the working medium in the solar collector loop, water as the storage medium and working fluid in the domestic hot water loop, and water as the heating medium in the room space heating loop. As there is heat stratification in the storage tank, the tank has various positions of hot water inlets, depending upon the temperature rise of the working medium in the solar collector loop. Likewise, the tank has various positions of hot water outlets to cater to the water temperatures required for domestic water and space-heating applications (floor heating, wall heating, radiators, etc.).

Figure 92: A Standard Active Solar Combi System (Water or Antifreeze Mixture-Water)

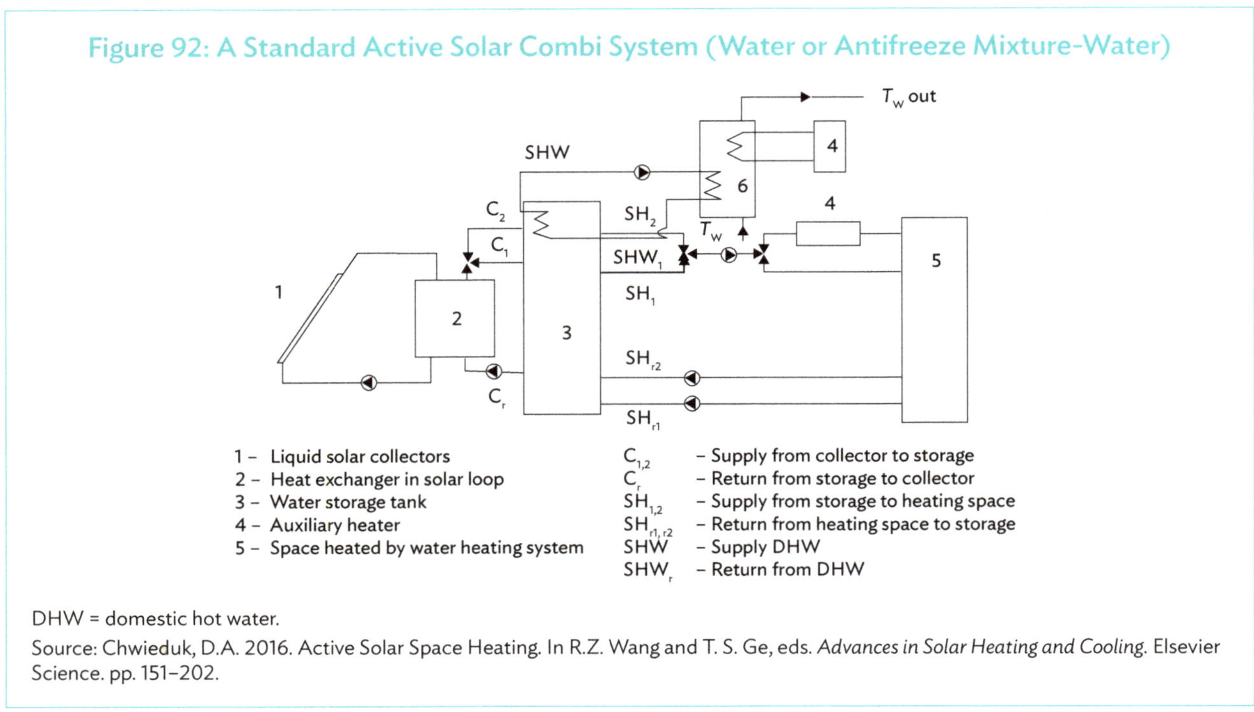

1 – Liquid solar collectors
2 – Heat exchanger in solar loop
3 – Water storage tank
4 – Auxiliary heater
5 – Space heated by water heating system

$C_{1,2}$ – Supply from collector to storage
C_r – Return from storage to collector
$SH_{1,2}$ – Supply from storage to heating space
$SH_{r1,r2}$ – Return from heating space to storage
SHW – Supply DHW
SHW_r – Return from DHW

DHW = domestic hot water.
Source: Chwieduk, D.A. 2016. Active Solar Space Heating. In R.Z. Wang and T. S. Ge, eds. *Advances in Solar Heating and Cooling*. Elsevier Science. pp. 151–202.

Figure 93 presents a radiant floor heating system with a vacuum tube solar collector and a groundwater heat pump. The system is operated at water temperatures of 20°C to 30°C. Heat is transferred from the solar collector and heat pump to a stratified hot water tank. Hot water is reheated by an additional heat pump. For a family house with a heat demand of 8 MWh/year, the solar collector can cover approximately 2 MWh/year. The rest is covered by the heat pump, whose electricity consumption is 1.5 MWh/year, reaching a coefficient of performance (COP) of 4.

Figure 93: A Radiant Floor Heating System with Solar Collector and Heat Pump

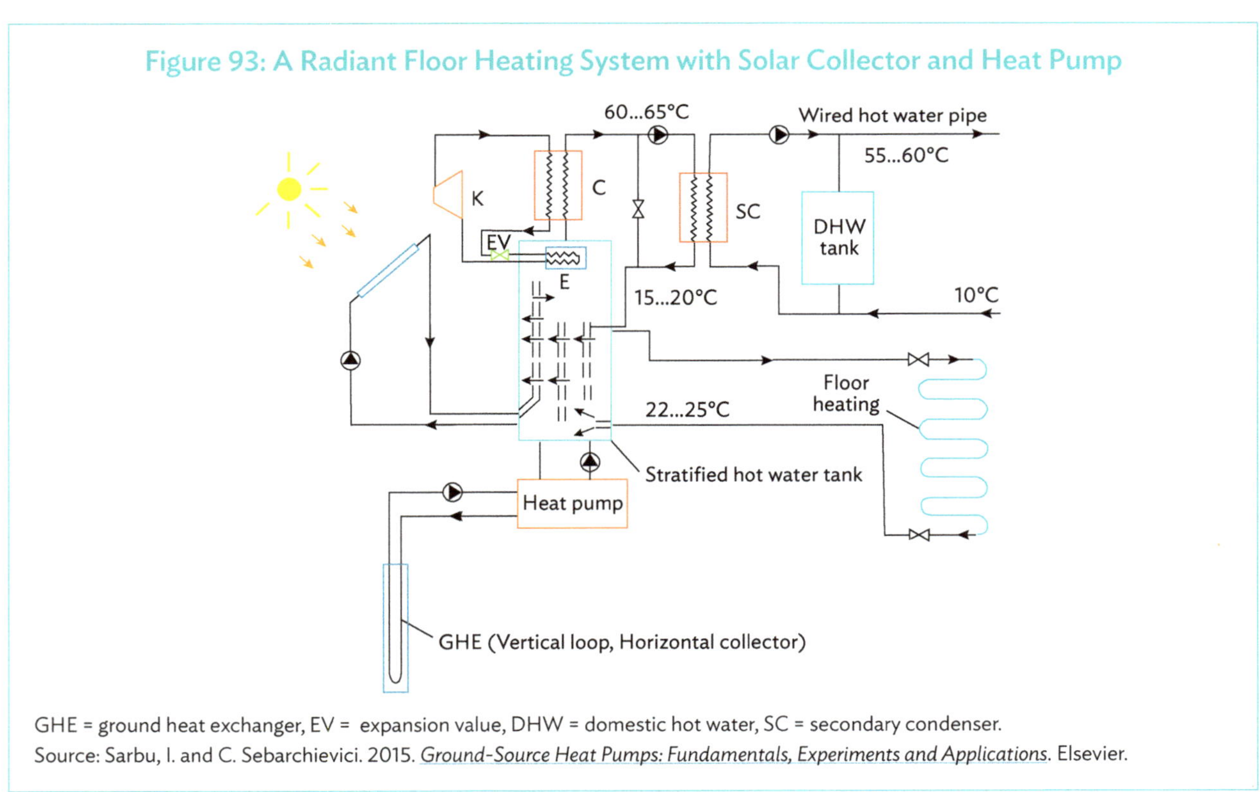

GHE = ground heat exchanger, EV = expansion value, DHW = domestic hot water, SC = secondary condenser.
Source: Sarbu, I. and C. Sebarchievici. 2015. *Ground-Source Heat Pumps: Fundamentals, Experiments and Applications*. Elsevier.

Solar-Assisted Heat Pump

An active solar system can be coupled with a heat pump to supply heat to a building for space heating and domestic hot water provision. The heat pump can achieve a much higher COP when the temperature of the heat source for its evaporator is raised by solar energy. The coupling can take various forms. In general, based on the heat source of the heat pump and the heat pump's connection with the thermal storage unit, solar-assisted heat pump (SAHP) systems can be broadly categorized into three basic configurations, including "series," "parallel," and "dual source."

Series Solar-Assisted Heat Pump System

A solar collector and a heat pump are aligned in series, directly or indirectly, through a storage tank, so that the working fluid heated in the collector loop is used by the evaporator of the heat pump as a heat source. The heat released from the condenser of the heat pump is used for space heating, either through radiant floor heating or a ventilation system. Figure 94 gives two examples of SAHPs with series connections. One is a solar-assisted water-to-air heat pump, and the other is a solar-assisted water-to-water heat pump.

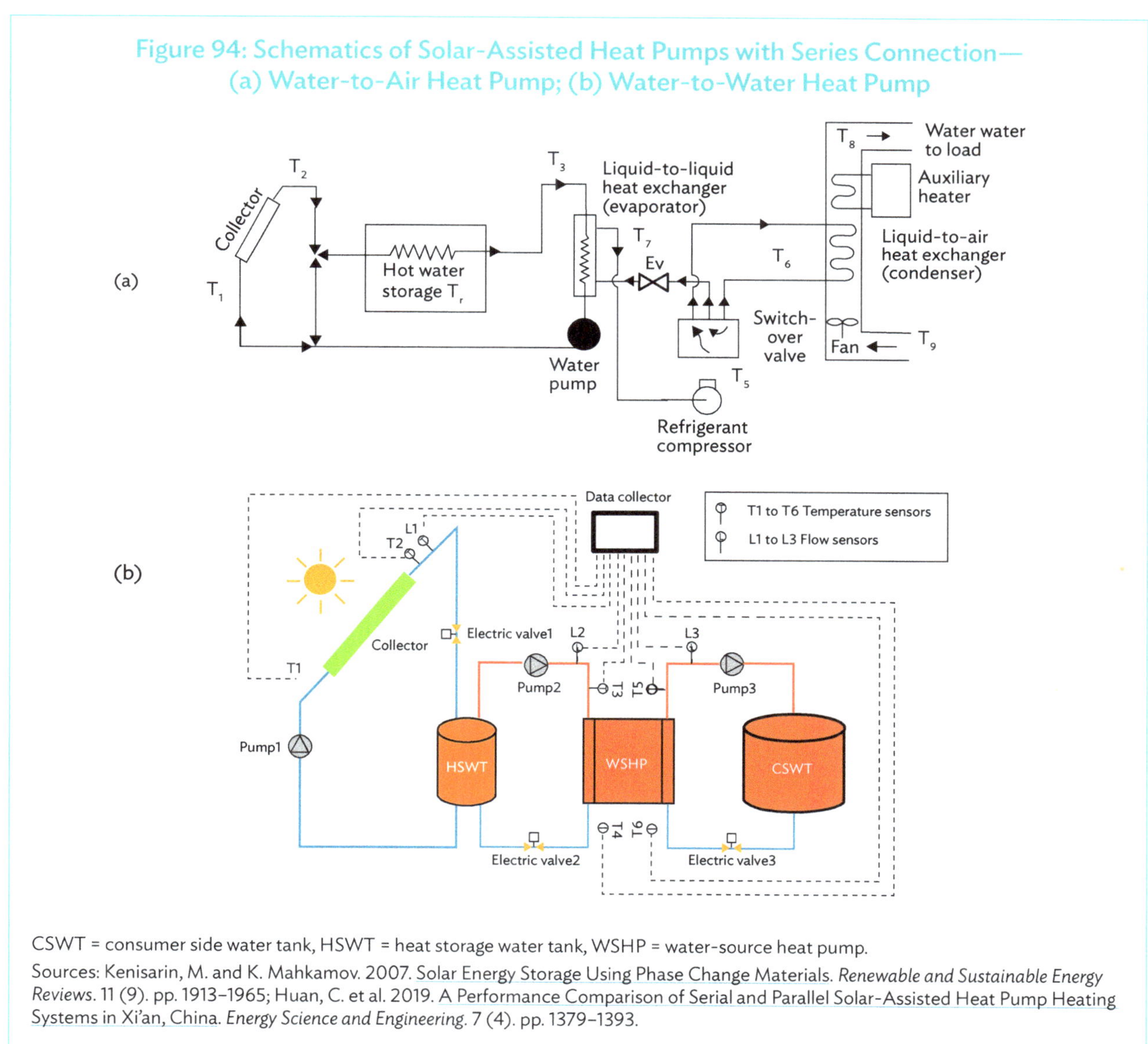

Figure 94: Schematics of Solar-Assisted Heat Pumps with Series Connection— (a) Water-to-Air Heat Pump; (b) Water-to-Water Heat Pump

CSWT = consumer side water tank, HSWT = heat storage water tank, WSHP = water-source heat pump.

Sources: Kenisarin, M. and K. Mahkamov. 2007. Solar Energy Storage Using Phase Change Materials. *Renewable and Sustainable Energy Reviews.* 11 (9). pp. 1913–1965; Huan, C. et al. 2019. A Performance Comparison of Serial and Parallel Solar-Assisted Heat Pump Heating Systems in Xi'an, China. *Energy Science and Engineering.* 7 (4). pp. 1379–1393.

Parallel Solar-Assisted Heat Pump System

A solar collector and a heat pump are connected in parallel, both providing the space heating directly. The heat pump is "separated" from the solar collector by using a heat source other than solar energy. In other words, the heat pump can be air-source, ground-source, etc. Vice versa, the solar collector loop is independent of the heat pump subsystem. An automatic control system decides the subsystem (solar collector or heat pump) to be operated to deliver the best thermal performance. Figure 95 gives two examples of SAHPs with parallel connections. One is a solar-assisted air-to-air heat pump, and the other is a solar-assisted air-to-water heat pump.

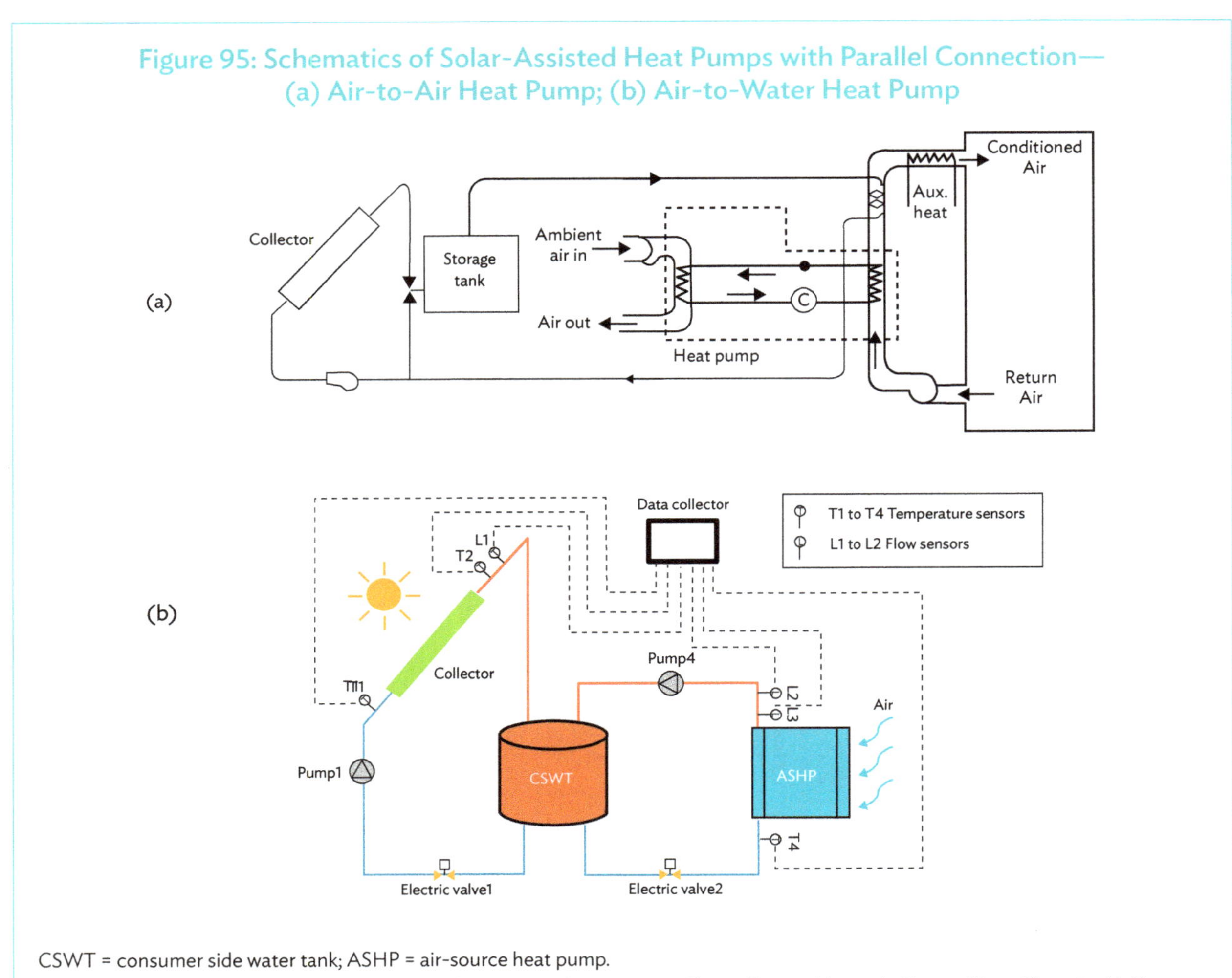

Figure 95: Schematics of Solar-Assisted Heat Pumps with Parallel Connection— (a) Air-to-Air Heat Pump; (b) Air-to-Water Heat Pump

CSWT = consumer side water tank; ASHP = air-source heat pump.
Sources: Kenisarin, M. and K. Mahkamov. 2007. Solar Energy Storage Using Phase Change Materials. *Renewable and Sustainable Energy Reviews*. 11 (9). pp. 1913–1965; Huan, C. et al. 2019. A Performance Comparison of Serial and Parallel Solar-Assisted Heat Pump Heating Systems in Xi'an, China. *Energy Science and Engineering*. 7 (4). pp. 1379–1393..

Dual-Source Solar-Assisted Heat Pump System

A dual-source system can be viewed as a combination of a series system and a parallel system. Solar energy can be used either as the heat source for the heat pump evaporator, or for directly providing heat for space heating. Accordingly, the heat pump can either use the solar energy as a heat source (directly from the solar collector loop or directly through a storage tank) or use an alternative heat source such as air or ground, which is independent of the solar collector loop. The operation of the heat pump is modulated by an automatic control system that dynamically chooses a heat source based on the thermal properties of the two sources (solar and the alternative source), for optimal system performance. Figure 96 shows the schematic of a basic dual-source SAHP.

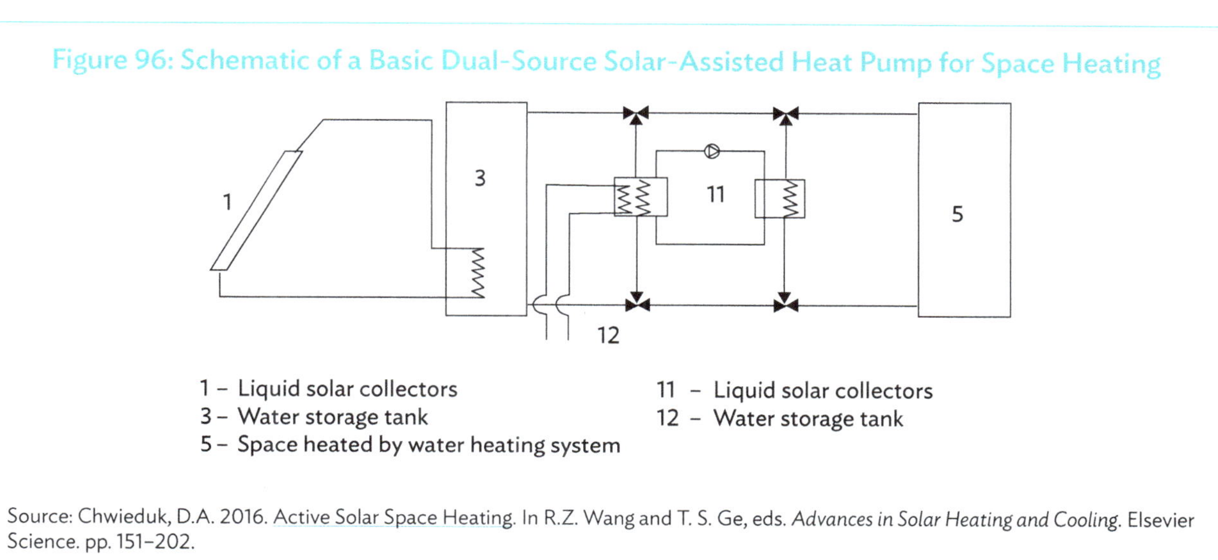

Figure 96: Schematic of a Basic Dual-Source Solar-Assisted Heat Pump for Space Heating

Source: Chwieduk, D.A. 2016. Active Solar Space Heating. In R.Z. Wang and T. S. Ge, eds. *Advances in Solar Heating and Cooling*. Elsevier Science. pp. 151–202.

In many applications, the configuration of an SAHP is often different from, and probably more complex than, the typical ones above, i.e., series, parallel, and dual source. A system may not necessarily be classified as belonging to a specific type, but rather a variant, or an upgraded design for better efficiency of operation in specific application circumstances. For example, Figure 97 shows a water-to-water SAHP system for space heating and domestic hot water supply. Water heated by the solar collector is primarily supplied to the domestic hot water tank. When the water temperature in the buffer storage tank (as the heat source for the heat pump) is low, the water heated by the solar collector is fed into the buffer storage tank.

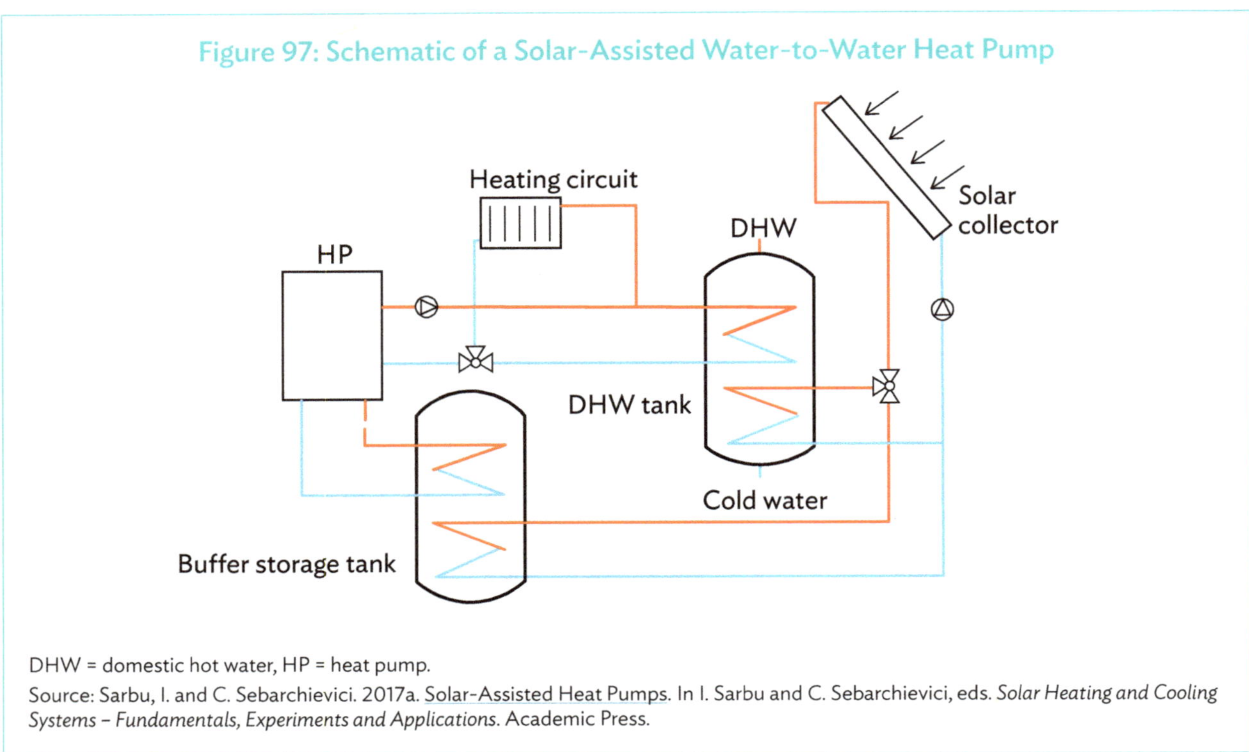

Figure 97: Schematic of a Solar-Assisted Water-to-Water Heat Pump

DHW = domestic hot water, HP = heat pump.
Source: Sarbu, I. and C. Sebarchievici. 2017a. Solar-Assisted Heat Pumps. In I. Sarbu and C. Sebarchievici, eds. *Solar Heating and Cooling Systems – Fundamentals, Experiments and Applications*. Academic Press.

With the increasing importance of thermal storage, the heat pump's condenser, instead of its evaporator as in many cases of traditional configurations, is sometimes connected to the thermal storage tank to store the heat produced from the condensation process. As shown in Figure 98, the thermal storage tank is the key component through which the solar collector loop is coupled with the GSHP loop. The condenser of the heat pump is connected to the storage tank to transfer heat to it. This is a variant of the traditional parallel configuration because the two energy sources (two loops) are not independent of each other. They both supply heat to the same storage, which means there is substantial interplay between their respective performances. In a way, the two loops (the solar collector and heat pump), both serving to provide space heating to the building, are complementary to each other. Therefore, the key to system operation control is to ensure efficient utilization of energy from both sources.

An SAHP system with more modes of operation is shown in Figure 99. The system features the use of an underground heat exchanger, which serves the function of storing extra solar heat during summer and providing space heating during winter. The relatively complex configuration of this system makes possible various closed loops involving the solar collector. The solar collector can supply heat directly to the thermal storage tank. It can also supply heat to the evaporator of the heat pump, which, through its condenser, supplies heat to the storage tank. Either the upper or lower heat exchanger in the storage tank can be used, depending upon the temperature of the working fluid in the loop (the solar collector loop, or the heat pump loop). The solar collector can also be connected to the piping of the vertical GHE loop to store extra solar energy gained during summer. The stored energy can then be extracted and supplied to the evaporator of the heat pump to generate heat for space heating during winter. On the condition that the same working fluid is used in the solar collector loop and the underground heat exchanger, when the working fluid in the solar collector loop has a higher temperature than that in the underground heat exchanger, the system can make the working fluid flow through the underground heat exchangers and the solar collector in sequence so as to gain more heat to raise its temperature before flow into the heat pump evaporator. This way, the working fluid with a higher temperature, used by the heat pump as its heat source, can render a better performance of the heat pump, i.e., a higher COP.

Figure 98: Schematic of a Solar-Assisted Ground-Source Heat Pump

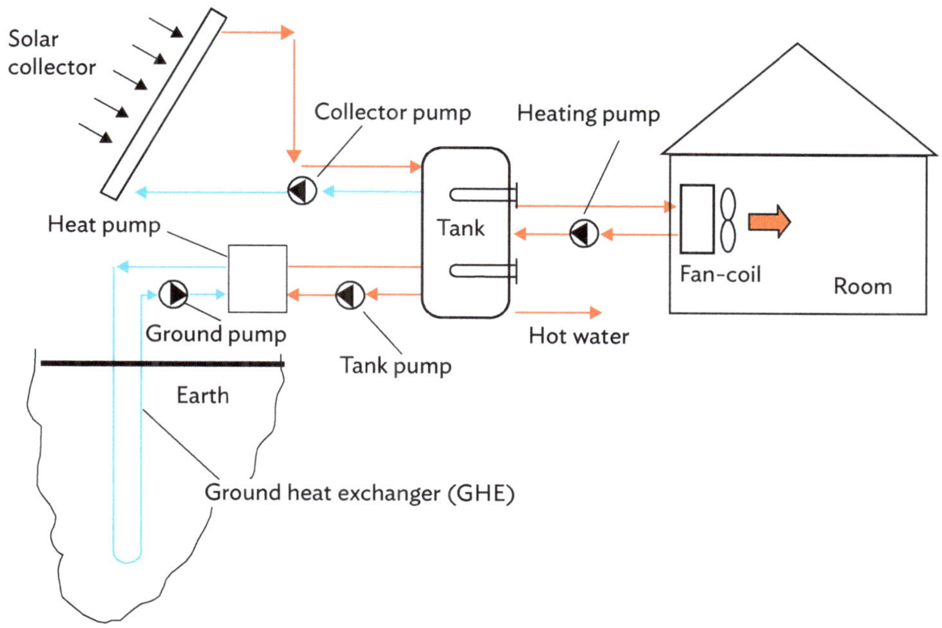

Source: Coskun, S. 2020. Performance analysis of a solar-assisted ground-source heat pump system in climatic conditions of Turkey. *Thermal Science*. 24. pp. 977–989.

Figure 99: Schematic of a Solar-Assisted Heat Pump with Underground Seasonal Thermal Storage

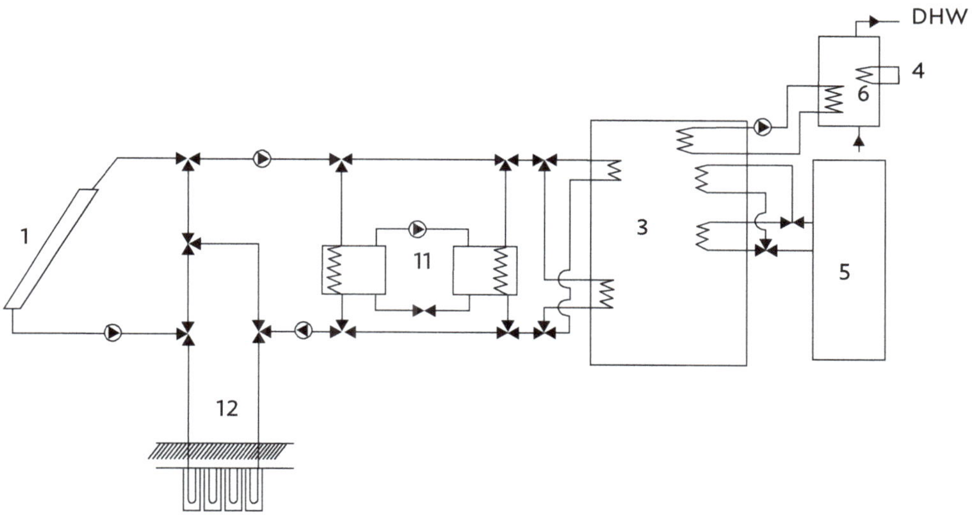

1 – Liquid solar collectors
2 – Water storage tank
4 – Auxiliary heater
5 – Space heated by water heating system
6 – DHW Tank
11 – Heat pump
12 – Ground heat exchangers

DHW = domestic hot water.

Source: Chwieduk, D.A. 2016. Active Solar Space Heating. In R.Z. Wang and T. S. Ge, eds. *Advances in Solar Heating and Cooling*. Elsevier Science. pp. 151–202.

Solar Water Heating

Similar to solar space heating, solar water heating can be broadly classified as passive and active systems. The difference between the two categories is that a passive system transfers heat from a solar collector to a water tank by natural circulation, or thermosiphon, whereas an active system uses an electric pump controlled by temperature sensors and relays, or a differential thermostat. Although passive systems are technologically simple and less costly, active systems have been proven to be more cost-effective and therefore have been gaining popularity in applications. This handbook focuses on active solar water heating systems.

Direct System

Direct systems, also called open-loop systems, use pumps to circulate water through collectors and operate at standard line pressure. These systems are suitable for single applications of domestic hot water supply in areas where the freezing period is short, and the water is not hard or acidic. Owing to their simple design and operation, direct systems cost the least among all active systems. No heat exchangers are used. Water is used as the working fluid and heated within the collector to 50°C–60°C. The simplicity and flexibility of direct systems make easy capacity addition and integration with existing systems possible to meet increased demand for hot water. In general, direct systems can offer higher operating performance than passive systems, which have nighttime heat losses from hot water stored on rooftops, and closed-loop systems, whose efficiency is affected by the heat exchange process.

The major disadvantage of direct systems is their limited freeze protection. A drain-down system is an improved version of direct systems to address this issue (Figure 100). When the risk of freezing is low, an open loop is used where collectors are filled with domestic water under supply pressure. Once the system is filled, a differential controller operates a pump to move water from the tank through collectors. A drain-down valve functions as

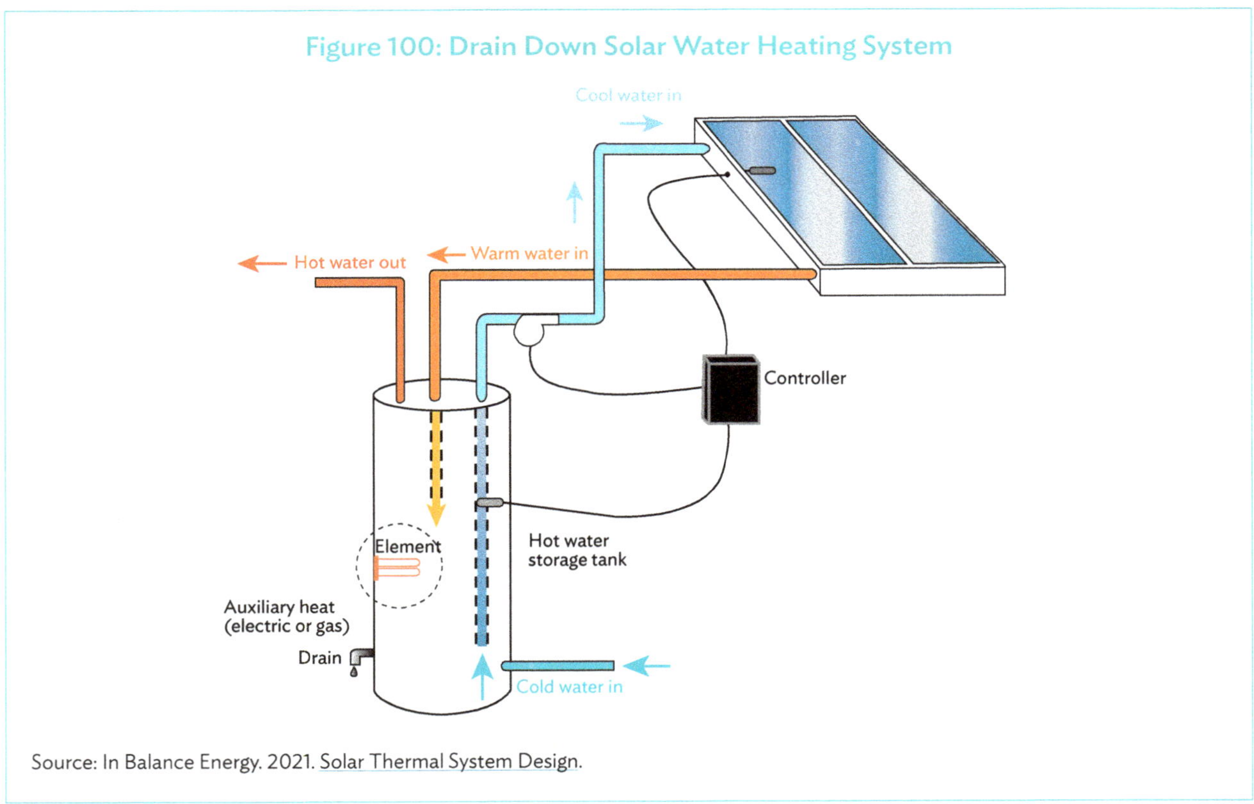

Figure 100: Drain Down Solar Water Heating System

Source: In Balance Energy. 2021. Solar Thermal System Design.

freeze protection. When the drain-down valve is activated by the controller due to the low temperature, it shuts off the water supply to the collector from the tank. The controller simultaneously opens a valve to allow water in the collector to drain away. At the top of the collector, a vacuum breaker is installed to let in air and prevent the backflow of water.

Indirect System

Indirect, closed-loop systems use pumps to circulate the heat transfer fluid through the collectors (Figure 101). Commonly used heat transfer fluids are water and water-antifreeze mixtures (typically propylene glycol). There are two circulation loops in the system: a closed loop of the heat transfer fluid and an open loop of water. The heat transfer fluid circulates within the closed loop and gets heated in the collector. It then transfers the heat through a heat exchanger to the water in the open loop, which flows into the storage tank. The location of the heat exchanger can be either internal or external of the storage tank.

In general, an indirect closed-loop system is more complicated than a direct open-loop system due to the need for a heat exchange. Because of the heat exchanger, the collector loop needs to be pressured and run at slightly higher temperature than a direct open-loop system. With the antifreeze fluid, indirect systems offer the benefit of freeze protection. They can also avoid the risk of overheating which is an issue of direct open-loop systems. Overall, indirect systems are slightly more costly than direct systems. The antifreeze needs to be recharged every 3–5 years, depending on its quality and system operation temperatures.

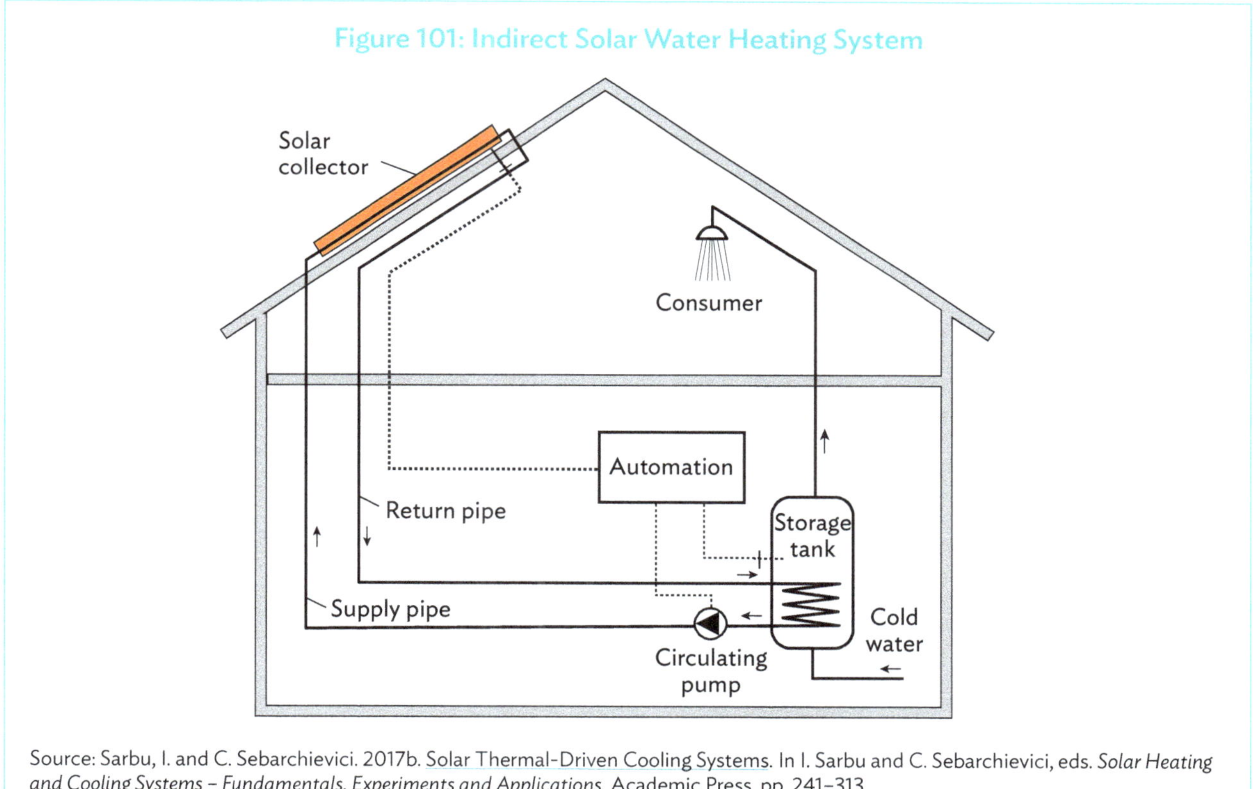

Figure 101: Indirect Solar Water Heating System

Source: Sarbu, I. and C. Sebarchievici. 2017b. Solar Thermal-Driven Cooling Systems. In I. Sarbu and C. Sebarchievici, eds. *Solar Heating and Cooling Systems – Fundamentals, Experiments and Applications*. Academic Press. pp. 241–313.

Drain-Back System

A drain-back system is a variant of indirect closed-loop systems (Figure 102). It can use water as the heat transfer fluid in the collector loop. This brings the benefit of having no need for recharging, as in the case of systems using antifreeze. A heat exchanger is needed to transfer heat from the water in the collector loop to the domestic water in the open loop. The system features a reservoir tank for the water in the collector loop. When the pump is off, the water in the collector loop drains into the reservoir tank by gravity, thereby avoiding freezing. With temperature monitoring, the system can be turned off when the water in the storage tank is too hot.

A key design consideration of a drain-back system is the appropriate sizing of the reservoir tank, which needs to be large enough to take all the water in the collector and lines when the pump is off. In a pressurized system, the reservoir also serves as an expansion tank, which means it must have a temperature- and pressure-relief valve to protect against excessive pressure. A disadvantage of this system is the heat loss when the pump is off. Also, due to the size of the pump and piping and the requirement for insulation, drain-back systems are normally more expensive than direct open-loop systems.

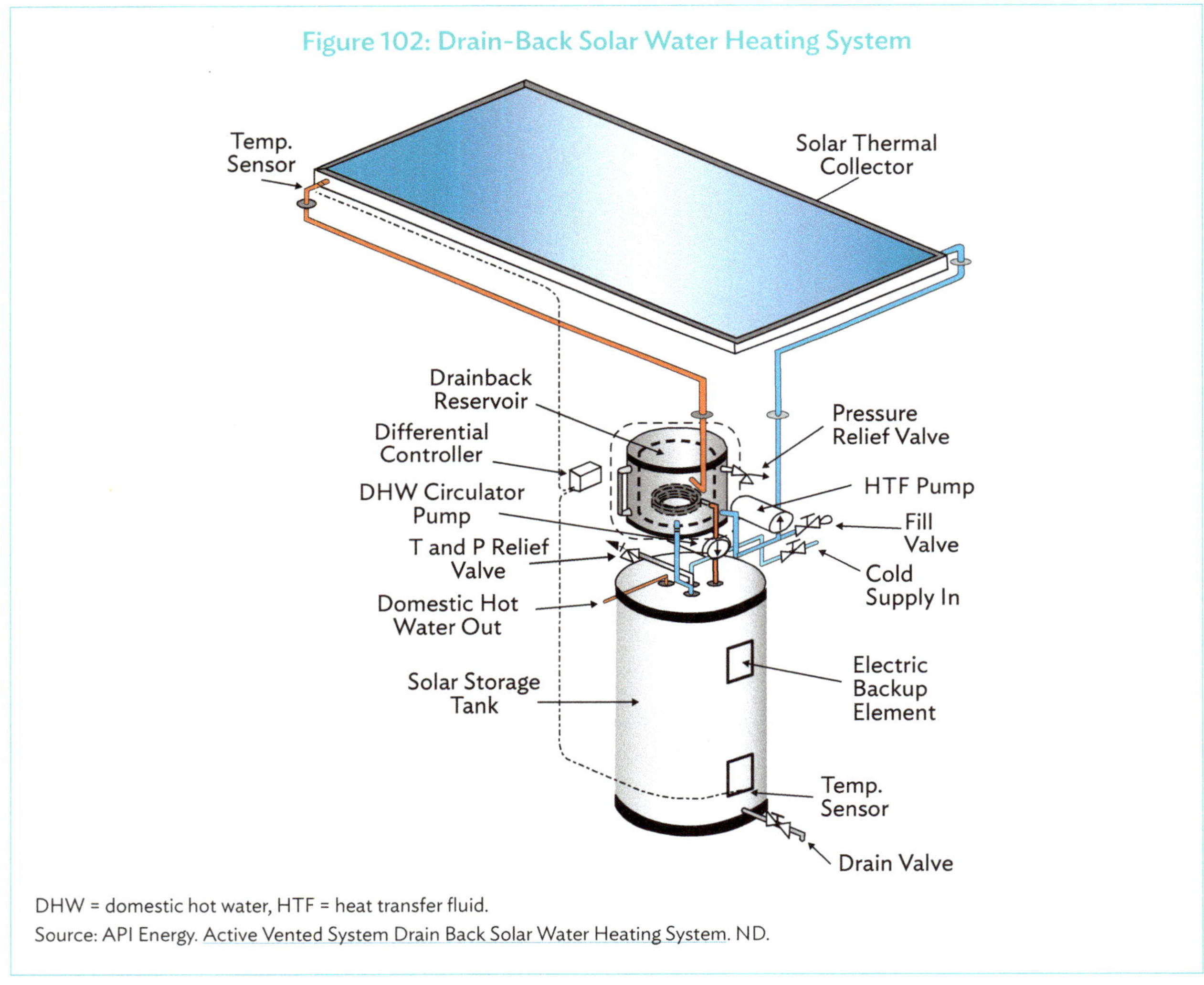

Figure 102: Drain-Back Solar Water Heating System

DHW = domestic hot water, HTF = heat transfer fluid.
Source: API Energy. Active Vented System Drain Back Solar Water Heating System. ND.

Air System

Another variant of indirect systems for water heating is the air system, which uses air rather than water or a water-antifreeze mixture as the heat transfer fluid (Figure 103). This helps the system work properly under extreme conditions such as freezing in winter or overheating in summer. As shown in Figure 104, a fan is used to circulate air through the solar collector duct, where the air is heated up. The hot air then transfers heat, through a concentric air-to-water heat exchanger, to the water in the horizontal storage tank. Fans and dampers are integral components of the system that modulate its operation. The system can produce hot water up to 80°C. The system is noncorrosive, requires less maintenance than other systems, and can be used under low temperature conditions. The system, however, has a relatively lower heating capacity and requires a larger area as a result of the use of air as the heat transfer fluid. Also, gradual air leaks will adversely affect system performance.

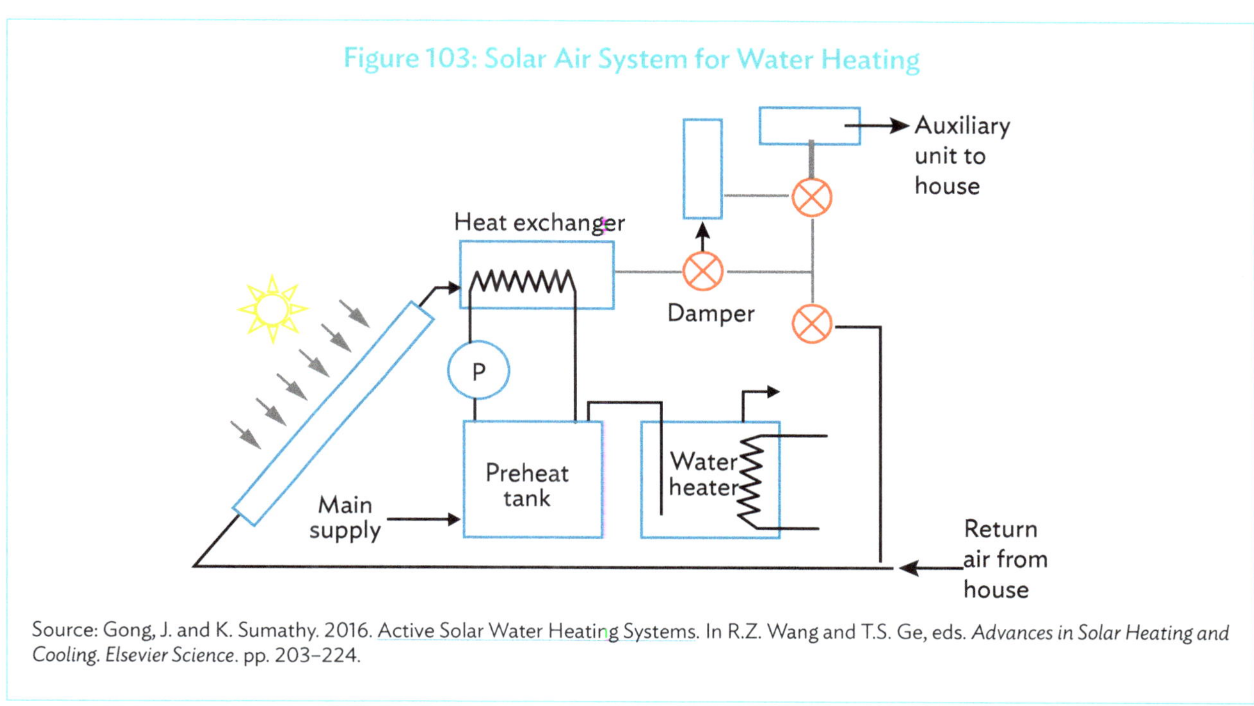

Figure 103: Solar Air System for Water Heating

Source: Gong, J. and K. Sumathy. 2016. Active Solar Water Heating Systems. In R.Z. Wang and T.S. Ge, eds. *Advances in Solar Heating and Cooling. Elsevier Science.* pp. 203–224.

Current Situation of Energy Efficiency Investment in Buildings

IFC estimates that investment opportunities in the green buildings and building construction sector in emerging markets will amount to $24.7 trillion by 2030, of which $17.8 trillion would be in East, Southeast Asia and the Pacific and $15.7 trillion would be in the residential sector (IFC 2019c). This indicates the magnitude of necessary funding and the need for sustainable business models to be able to guide financial means where they are needed. The current situation in Asia is certainly improving, but annual investment in BEE is still low.

Energy efficiency is key to solving the rising need for energy in the DMCs. With growing GDP, which will nearly triple from its base in 2017 to reach $20 trillion by 2040 in, for example, ASEAN countries, the forecasted demand for energy is expected to rise 2.8% annually from 375 million tons of oil equivalent (Mtoe) in 2017 to 714 Mtoe in 2040 (ACE, GIZ 2020). Judging from the past, it is likely that more power plants are built to meet demand than investments in energy efficiency measures are taken.

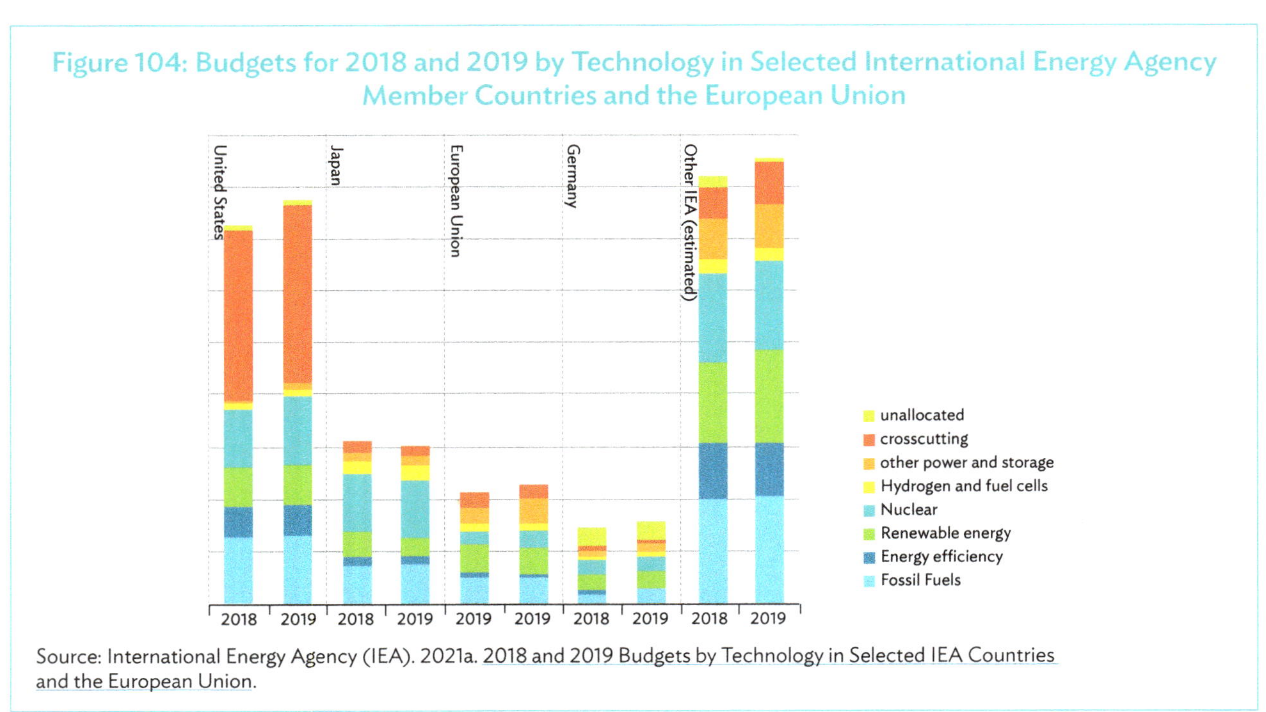

Figure 104: Budgets for 2018 and 2019 by Technology in Selected International Energy Agency Member Countries and the European Union

Source: International Energy Agency (IEA). 2021a. 2018 and 2019 Budgets by Technology in Selected IEA Countries and the European Union.

The following subchapters set the frame for understanding the characteristics of financing BEE. It provides the way for individual case diagnosis, which leads to the design of appropriate business models and financial instruments. Financing energy efficiency differs in a couple of ways from traditional financial models. The following part shows the major differences and their impacts.

Nature of Building Energy Efficiency Finance

Investing in savings. Investing in energy efficiency measures anticipates a reduction of future expenses in the form of projected energy costs. Conventionally, investors and banks seek the positive cash flow of an investment to repay a loan or pay a dividend, e.g., by growth in the output of rental space. The concept of financing savings while achieving the same output, although with fewer energy costs, is new. Traditional business models focus on maximizing usable output. For energy efficiency investments, the output occurs in the form of "negawatts," an asset that is not tangible and cannot be taken as collateral. Therefore, the financing of energy savings requires a rethinking process.

Fragmented market. The technical solutions to reduce energy consumption can be different from building to building. The size of the building, its location, the current physical condition, usage, and a number of other factors determine the technical measures that must be taken, their costs, and the amount of possible savings. While there are conventional design methods, practices, and technologies, they vary from a simple switch to an LED light, to building envelopes to space cooling and ventilation. Subsequently, the market from an investor's and banker's point of view is fragmented, consisting of multiple, relatively small modifications, and each individual measure may not be—financially speaking—significant. Transaction costs for banks and investors, however, remain the same regardless of the size of the investment. Thus, multiple small investments are not attractive to most financial providers. It is not only time-consuming, but decision-makers may not have the technical expertise to evaluate each measure's risk level.

Split incentives. An energy efficiency investment is made with the prospect of later savings, i.e., profits. Owners and landlords, however, face the problem that under net lease contracts, energy costs are paid by tenants to the utility, and any investment in energy efficiency would reduce the bill of the tenant but have often no effect on the profit margin of the owner. If the contractual arrangement is based on a gross lease, the owners pay energy expenses, and the tenant has little incentive to correct the energy use, as one would not profit from it. The dilemma[24] that owners do not benefit from their investment is referred to as "split incentives."

Mixed purpose investments. Energy efficiency investments in existing buildings are seldom done solely for the purpose of achieving less energy consumption. Often, it is a replacement investment that had to be done anyway. Buildings will see a refurbishment after 30–40 years due to a change in lifestyle or because equipment is worn out. In most cases, a change in architectural structures, layout, and usage of the building go along with it, and the energy efficiency upgrade is part of the package. In addition, a number of refurbishments may include renewable energy options like solar. The payback period for the part of the investment leading to less energy usage is thus difficult to determine.

Subsidized energy costs. Last but not least important is the fact that in many countries, energy is still subsidized.[25] Energy prices are the most sensitive individual factor to energy efficiency investments. If the true price of energy were paid by consumers, payback periods would be shorter, profits would be higher, and society would accept energy efficiency measures as a useful instrument. It is an often-heard argument, that low-income groups suffer most, if subsidies are canceled. To achieve social acceptance and the right incentives to save energy, subsidies that

---

24 Further reading: H. Nie, R. Kemp, Jin-Hua Xu, V. Vasseur, Y. Fan. 2020. Split incentive effects on the adoption of technical and behavioral energy-saving measures in the household sector in Western Europe. *Energy Policy*. 140. p.111424; I. Hamilton. 2021. *Asia launch of online course for Energy Efficiency in Buildings*. International Energy Agency.

25 Further reading: ADB. 2016. *Fossil Fuel Subsidies in Asia. Trends, Impacts, and Reforms*. Manila.

are harmful to the environment must be redirected and modified. The savings from canceled subsidies can be used to support, e.g., the urban poor to live in more energy-efficient housing estates. Dynamic tariff models can provide allowances for further direct subsidies for selected groups to allow for the affordability of energy.

Rebound effect. Critics of investments in energy efficiency argue that the "rebound effect," meaning that energy efficiency savings are offset by increased use of energy, would render energy efficiency investments impractical. Amplified consumption may stem from

- using energy-consuming services more than before (take-back effect),
- buying additional energy using devices from savings (spending effect), and
- available income is used (also in the case of government-led incentive programs) to increase economic activities and thereby indirectly increase energy demand (the investment effect).

Scientists estimate that in developing countries, a 10% improvement in efficiency might provide "only" a 9% reduction in energy use.[26] Rebound effects are often based on backlog demand. Using more energy at lower costs and securing access to energy is an improvement in the living standards of vulnerable groups in an energy-efficient way. The effect of the energy measure is therefore not nullified but put to proper use.

Box 17: Nature of Energy-Efficiency Finance—Summary

- Unfamiliarity with the concept of investing in savings while achieving the same output
- Fragmented market increase risk perception
- Dilemma of split incentives
- High transaction costs compared to small investment amount
- Very little "pure" energy efficiency investments
- Energy subsidies are obscuring energy efficiency investment profits
- Rebound effects

Source: Author.

Low-Hanging Fruit Versus Deep Renovation

A number of energy efficiency investments have been made in the last few years that have become popular, such as changing light bulbs to LEDs, which typically use about 25%–80% less energy than traditional incandescent lights. Though more expensive, the payback period is short, and given that LEDs last longer, it is very lucrative.

Similarly, the installation of a programmable thermostat that can be adjusted to temperatures according to schedule, the assessment of heating and cooling systems, the insulation of hot water pipes to prevent heat loss, and the replacement of aging and inefficient appliances are reasonable and cheap ways to save energy and achieve short payback periods. Many programs focus on these low-hanging fruits as an incentive and educational instrument to gain broader acceptance of energy efficiency measures.

---

[26] P. O'Connor. 2015. What is the Rebound Effect? — Energy Efficiency, Part 2. Union of Concerned Scientists.

Box 18: Example of Energy-Efficient Lighting

In a parking lot, 100 400-watt metal halide lamps are installed. They operate 365 days a year for 12 hours at an electricity rate of $0.14 per kilowatt-hour.

Item	Old Lamp	LED
Wattage	400 W	135 W
Energy use, kWh p.a.	175,200	59,130
Energy costs p.a.	$24,528	$8,278
Savings p.a.	$16,250	$17,967
Payback period		2.71 years

kWh = kilowatt-hour, p.a. = per annum, W = watts.

This example covers only electricity savings, additional savings are possible due to build in energy saver programs, which dim the light during the day and brighten up during the night.

Source: Author.

It is suggested specifically that renovation rate and depth should be reinforced. Deep renovation can improve buildings' performance by 30%–50%, which is needed for buildings over 25 years. Deep renovations, including investments in the building envelope, replacement of windows, and changes in space heating and cooling, require larger investment amounts with repayment periods of approximately 15–30 years. These are all useful measures over the life of a building. However, investors expect much faster profits. Therefore, they are hesitant and forgo the opportunity of substantial energy efficiency gains.

An attempt to cut down on long payback periods is to package an energy investment deal with measures that have different payback periods. Profits from quick energy-saving measures are reinvested and shorten the longer payback periods of structural investments. A sculpted repayment schedule requires a strategic and long-term view of investors. The combination of both long- and short-term payback periods is a trade-off, inviting longer-term (and useful) investments into energy efficiency.

Likewise, combining different payback periods makes it easier for the business models of companies specializing in energy efficiency, such as energy service companies (ESCOs). They act as project developers when integrating a project's design, financing, installation, and operational elements. Working with a different set of repayment schedules stabilizes their business model and makes it easier to guarantee a certain output after gaining control over all necessary data.

Box 20: Low-Hanging Fruit Versus Deep Renovation—Summary

Investments with quick payback periods are used as incentives to achieve the popularity of energy efficiency measures. Deep renovation measures require higher investments and thus longer payback periods, which, when mixed with low-hanging fruit can be shortened.

Source: Author.

Financial Instruments

The following is meant to be a practical section addressing the question of how best to finance energy efficiency in buildings, what the challenges are, and how best to overcome them. It is based on the understanding that energy efficiency investments in buildings must be economically viable and financially feasible to make sense to tenants, homeowners, developers, and investors alike. The objective of the report is to contribute to opening a sensitive avenue for the transition to a zero-carbon, efficient, and resilient building stock.

To finance energy efficiency in buildings, all commonly known financial instruments are used. They are applied by banks and investors under the same risk assessment scenarios as any other business opportunity. Given the previously discussed lack of information in terms of uncertain cost savings and data to measure performance, financing has been restrained. In response, specialized instruments have evolved that circumvent or mitigate these impediments, such as ESCO models. They also combine traditional financial instruments. The differentiation between specialized financial instruments and business models is not clearly defined. For this handbook, we loosely classify business models as more complex structures that involve financial and nonfinancial instruments.

In the following first traditional instruments (Figure 105) and then specialized instruments (Figure 110) are briefly described, and international best practice examples for their usage are given. The correct use of financial instruments will reinforce political measures.

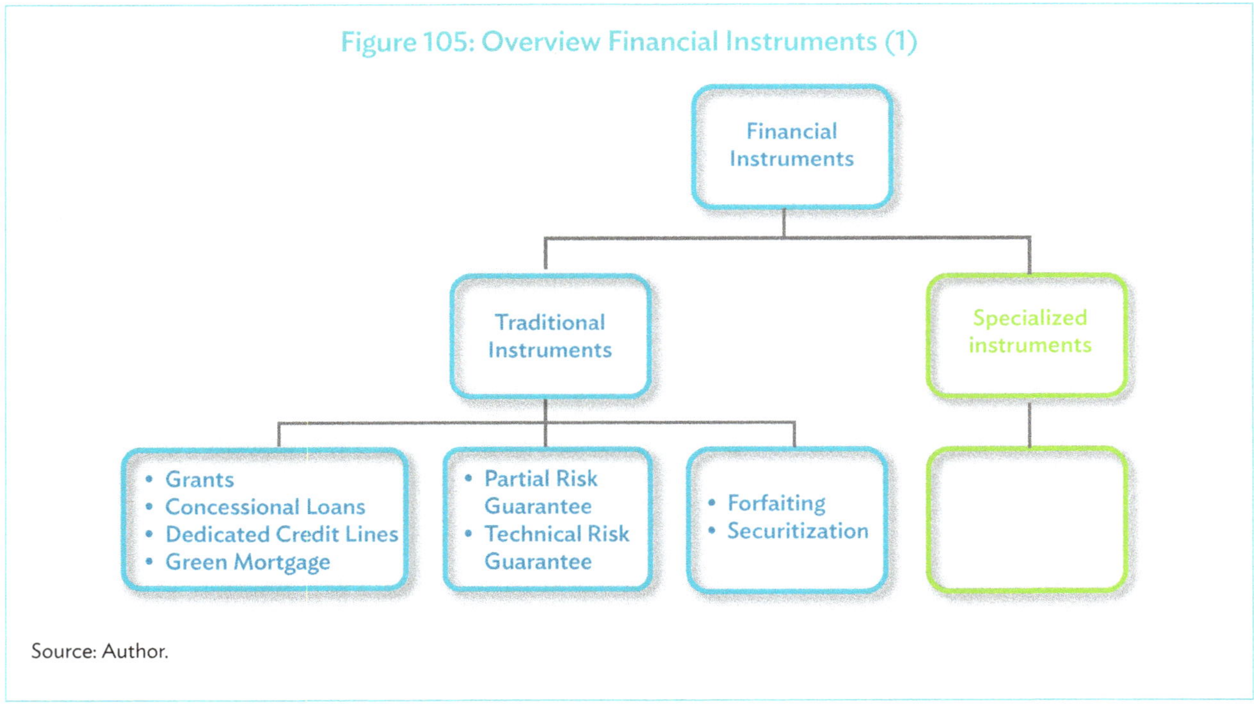

Figure 105: Overview Financial Instruments (1)

Source: Author.

Traditional Instruments

Grants and Technical Assistance

Grants are funds given for a specific purpose by governments and development banks and are not paid back. They can pay full or part of the costs of predefined investments and are seen as incentives to open markets.[27] Well-known in technical assistance (TA), they are often used to finance capacity development components to cover any additional design costs and demonstration projects involving untested and novel technologies, which is often the case in energy efficiency in buildings. An example is Germany's KfW Development Bank, which provides grants for engineering and construction supervision of energy efficiency measures to cover any additional design costs in existing buildings of up to 50% of costs, up to €4,000 (Dorendorf 2018). Grant finance reduces the costs for building owners. They are often used for a limited period to overcome clearly defined obstacles and to create awareness.

Concessional Loans

On behalf of governments, specialized banks provide loans at better than the prevailing market terms, i.e., with lower interest rates and/or longer maturities for investments that fall within specific guidelines—for example, passive houses. These "soft" loans are often combined with grant finance. For example, investors who follow the regulations of the "KfW energy efficient house" receive concessional loans with interest rates of 1%–2% for 10 to 30 years. The interest rate is among others, determined by the level of efficiency achieved; the higher the level, the lower the interest rate (Kolb 2020). Though the risk stays with the investor, soft loans reduce costs substantially. Good results have been achieved when financing the retrofitting of buildings, especially in the (private) housing sector. Borrowers, however, must be solvent to obtain a loan. If not, they need external finance (see ESCOs). There is no direct link between energy or costs saved and the repayment rate. Loans require collateral, which may be the equipment financed by the loan or other assets such as the customer's mortgage. Concessional lending does not avoid recourse to the lender.

Dedicated Credit Lines

Governments and development banks may decide on a bundle of loan alternatives to support the fragmented buildings and building construction sector. Each of the lines would target a certain market imbalance, and together they would support a larger policy measure. As a bundle, they form a selection of opportunities and have lower transaction costs. Dedicated credit lines are often used to stabilize specific market segments. They may consist of a mixture of concessional loans and grants. Dedicated credit lines are used in on-lending schemes in which commercial banks can build up a steady project pipeline with existing clients. For example, the European Bank for Reconstruction and Development (EBRD) has provided, via their funding vehicle, Green Economy Finance Facilities credit lines to local financial institutions in Bulgaria for on-lending to homeowners. Through Green Economy Finance Facilities, the EBRD offers credit lines to local partner financial institutions for on-lending to small and mid-sized green projects with loans or leasing alternatives. Credit lines are complemented with TA for capacity-raising and project assessments, and occasionally with low-intensity grants to reward end-beneficiaries who opt for advanced technologies.[28] Dedicated credit lines are a cost-efficient instrument; when designed properly, they follow a standardized pattern, may combine several efficiency objects, train the local financial market, and reduce funding shortages. Dedicated credit lines are used in on-lending schemes in which commercial banks can build up a steady pipeline with existing clients. However, on-lending schemes take time to become successful as participating banks need to train their staff and look for suitable clients in an often highly competitive market, which makes pipeline-building a medium-term process.

---

[27] The report does not include consideration of the impact of current coronavirus disease pandemic as it is still too early to make useful assumptions.

[28] EBRD. *Innovative Financing for Green Building*; further reading: Green Economy Financing Facility. Bulgarian Residential Energy Efficiency Credit Line.

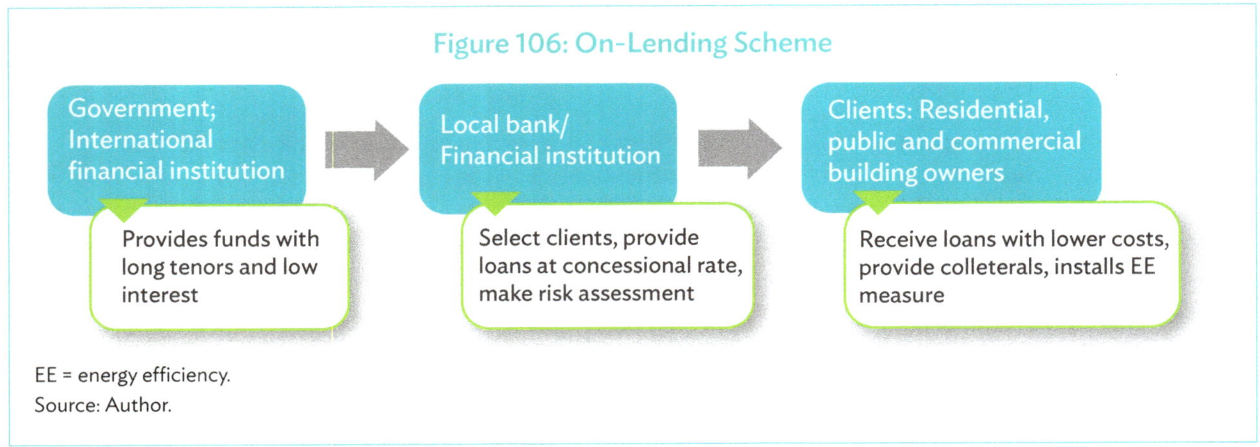

EE = energy efficiency.
Source: Author.

Green Mortgage

Homeowners or potential home buyers are offered preferential mortgage terms if they can demonstrate that the property for which they are borrowing meets certain environmental standards or if they commit to investing to improve its environmental performance. Preferential terms can be in the form of extended loan maturities, interest rate deductions or in larger loan amounts (Richardson 2020). For example, the Energy Efficient Mortgage Action of the European Union has issued two complementary programs that offer a standardized energy-efficient mortgage using commercial banks to channel private capital to homeowners. The Energy Efficient Mortgages Action Plan. assumes that energy efficiency has a risk mitigation effect for banks as it lowers energy costs for the property and increases the value of the buildings. As a result, banks take on lower risk as usual, thus reducing the cost of capital. The Energy Efficiency Data Protocol and Portal is recording data relating to energy-efficient mortgage assets and allowing connected banks to make reliable risk assessments.[29] The initiative is financially supported by the EU's Horizon 2020 Program.

The combination of market-led data analysis, to which participating stakeholders report, has provided the maximum amount of transparent and relevant information. As the green mortgage program follows a clearly defined standard, the learning curve, monitoring, and valuation of loans reduce transaction costs for financial institutions and, at the same time, become comparable. This allows the bundling of loans, their securitization (see following discussion), the mobilization of an investor base, and the building of an energy-efficient asset pool. Since the administrative effort is substantial, these platforms are best led by independent organizations.

Partial Credit Guarantees

Credit guarantees are designed as a strategy of risk-sharing and mitigation. Partial credit guarantees (PCG) remove the risk from commercial lenders such as banks and financial institutions they do not feel comfortable with. Guarantees may absorb parts or the full debt service, regardless of the background of the default. If no PCG is provided, lenders may abstain from providing loans, or the risk costs would drive interest rates up. For example, ADB offered a credit guarantee to participating partner banks in the PRC in its Energy Efficiency Multiproject Financing Program. The PCG covered the credit risks associated with energy efficiency projects, thereby mobilizing their funds for energy efficiency projects (ADB 2022). A credit risk guarantee is another instrument that can be used in a not-fully matured energy efficiency lending market. It provides confidence to commercial lenders and will, in some cases reduce the interest rate of the borrower. However, a fee for the guarantee will have to be borne by the borrower. It can be used for individual owners as well as for ESCOs.

---

[29]   Energy Efficient Mortgages Action Plan. *Energy-Efficient Mortgages Initiative*.

Energy Savings Insurance

The lower the perceived technical risk in energy efficiency, the lower the risk premium the investor must pay. Equipment insurance programs started offering specific insurance coverage for the difference in projected and actual energy savings due to technical defaults. The Energy Savings Insurance ensures guaranteed energy performance. Because of the strong linkage to possible malfunctioning of equipment, it is similar to technical risk insurance. Potential clients are ESCOs, building owners, and even dedicated programs. Eligible equipment is usually limited, such as HVAC, lighting, or building control systems. Programs are developed by governments and development banks to generate confidence in the potential for efficiency savings. When markets mature, commercial insurers also offer such schemes. For example, the Inter-American Development Bank (IDB) facilitates the development of the insurance program in alliance with national development banks. The model offers insurance to SMEs who, under the same program, sign a contract for the supply, installation, and maintenance of equipment and validation by an independent body (IDB 2020). Though the program is designed for SMEs, it also caters to hospitals, commercial buildings, and schools as the equipment eligible under the program is typical for saving energy in buildings, such as lighting and air-conditioning.[30] Those who seek insurance will have to follow stringent standardized procedures as to the equipment covered, the installation process, and the validation of equipment. The interest of the insurer is to control every single step along the way to minimize risks. Deductibles range between 5% and 20% depending on the potential energy savings.

Forfaiting

Known from export finance, the instrument can be used for energy efficiency in buildings as well. It is a means of selling medium- and long-term receivables at a discount to banks or specialized institutions to raise liquidity. ESCOs, who, for example, purchase energy-efficient equipment for their clients through a loan, could sell medium- to long-term debt to a specialized financial institution. By doing so, the ESCO frees its balance sheet of debt and can continue to grow, at the same time, it is still responsible for providing the energy service. The right to claim payments from the client is transferred to the financial institution.

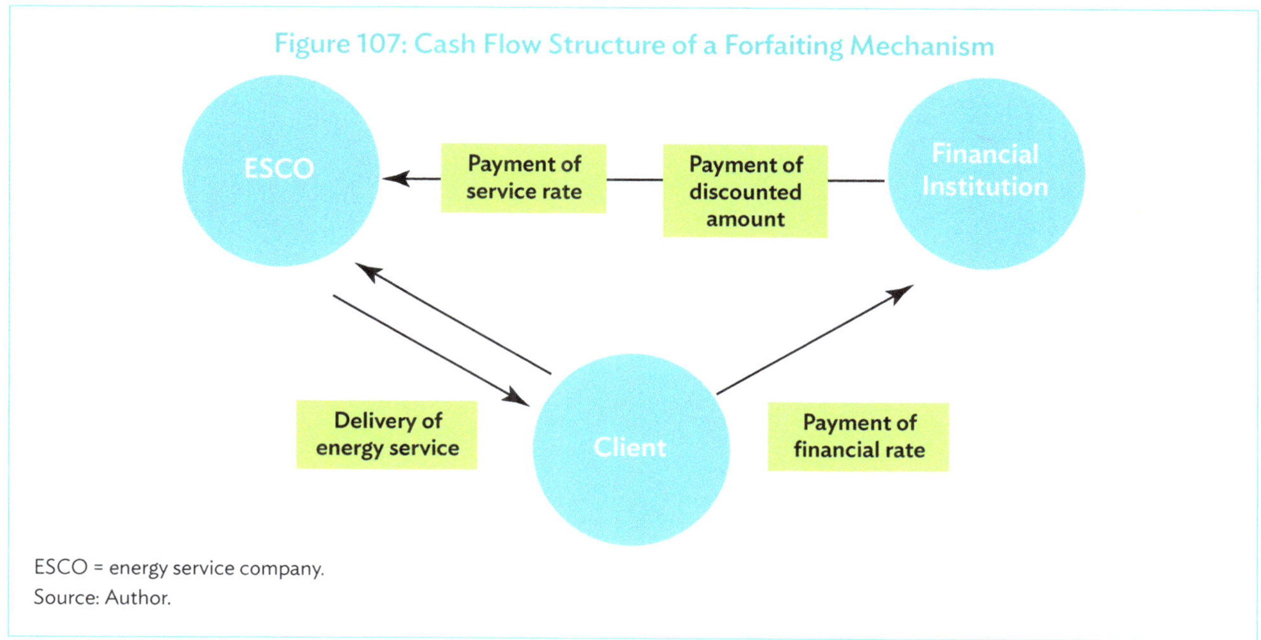

Figure 107: Cash Flow Structure of a Forfaiting Mechanism

ESCO = energy service company.
Source: Author.

---

[30]   Further reading on insurance: Base. Unlocking Investments in Energy Efficiency in Europe.

For example, the Save your bUildiNg by SavINg Energy or SUNShINE project in Latvia set up a forfaiting fund that buys receivables after 1–2 years of good performance from an ESCO, whose balance sheet has grown, allowing the ESCO to take on new loans (SUNShINE). Forfaiting comes at a price as it is a nonrecourse purchase of the financial institution, which carries the full risk of non-payment. Therefore, the discount rate the financial institution buys the receivables for depends on the risk the client of the ESCO poses. The better the client, the lower the discount rate.

Securitization

In securitization, the financial institution pools, for example, the receivables and debt from ESCOs and possibly other debt in a special purpose vehicle (SPV) and sells the vehicle to interested investors. The SPV can be a bond or another security (asset-backed security),[31] and investors receive dividends from the principal and interest cash flows collected from the underlying debt (energy savings from energy efficiency measures in the form of the fee the client must pay). Energy savings are thus a sellable asset.[32] For example, The Warehouse for Energy Efficiency Loans (WHEEL), is a public–private partnership, sponsored by the National Association of State Energy Officials of the US, that buys unsecured residential energy efficiency loans that participate in local programs (Clouse 2014). These loans are collected until the portfolio is large enough to bundle it into a bond for sale to institutional investors. Similarly, Deutsche Bank has issued a residential energy efficiency retrofit bond in which the underlying assets to be securitized were receivables from the Property Assessed Clean Energy (PACE) program.

The underlying asset, be it the EPC contract of an ESCO or the receivables from PACE, has a significant influence on the dividend payments of the bond or security. While individual defaults can be better buffered, reducing the risk, the nonstandard contracts of ESCOs make comparability difficult. When all works well, it is a triple-win situation as energy savings at the clients' end are achieved, the ESCO frees its balance sheet of debt and can expand, and investors are able to participate in a truly green instrument. Securitization needs a mature capital market.

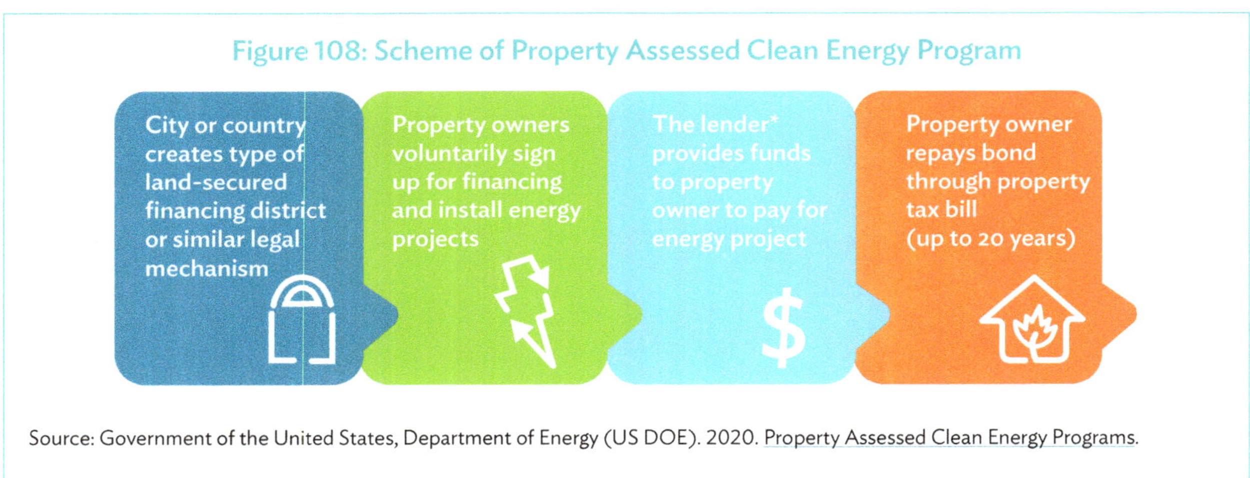

Figure 108: Scheme of Property Assessed Clean Energy Program

Source: Government of the United States, Department of Energy (US DOE). 2020. Property Assessed Clean Energy Programs.

---

[31] Securitization belongs to a group called asset backed securities (ABS) meaning a security derived from a pool of underlying assets.
[32] Further reading: Aggregation and Securitization as Key Driver for Energy Efficiency Finance; NRDC. Green Bank Network.

Specialized Instruments

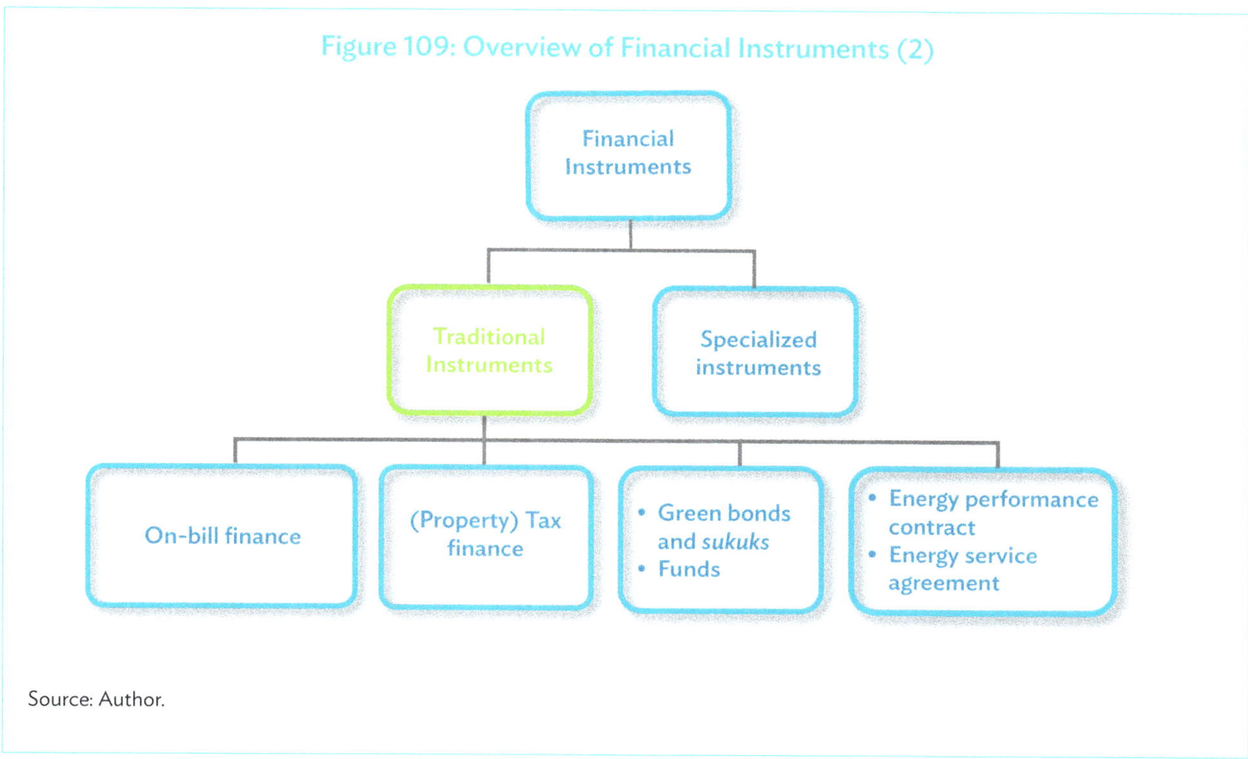

On-Bill Finance

On-bill finance is another instrument to combat the high up-front costs of energy efficiency measures. Instead of banks, the utility provides funding for the refurbishment, which is paid back via the energy bill. If calculated correctly, the bill will be lower through the energy savings than before, despite the extra payment. There are two different methodologies:

> An on-bill finance involves the utility, which acts as the lender. The customer of the utility receives financing from the utility for an energy efficiency measure and repays the loan through regular payments on an existing utility bill. Usually, this form of finance has zero or very low interest. However, it is legally not linked to the property or the utility meter.

> An on-bill tariff, where the building owner also repays the loan via the utility bill, is considered a tariff and, as such, an "essential service," which is linked to the property and part of the tariff. The obligation for repayment stays with the property and is transferred to the next owner if the property is sold.

For example, Rwanda in East Africa is supported by United Nations Environment Programme (UNEP) to replace used cooling equipment in residential homes. The program named R-COOLFI includes not only the replacement of the equipment but also the proper disposal of the used equipment. The loan is repaid through utility bills.[33] On-bill financing is suitable for building owners with little capital at hand. Because of the close relationship between the utility and the contractor, the owner often does not carry the technical risk, because the utility guarantees the

---

[33] UNEP. Rwanda.

performance. The program can be open to renewable energy measures. A great plus of on-bill finance is that it can be structured in such a manner that the debt is linked to the utility meter and not to the individual owner. This decoupling allows the debt to be transferred to the next owner. The program is driven by the utility, which in turn can be supported by the government.

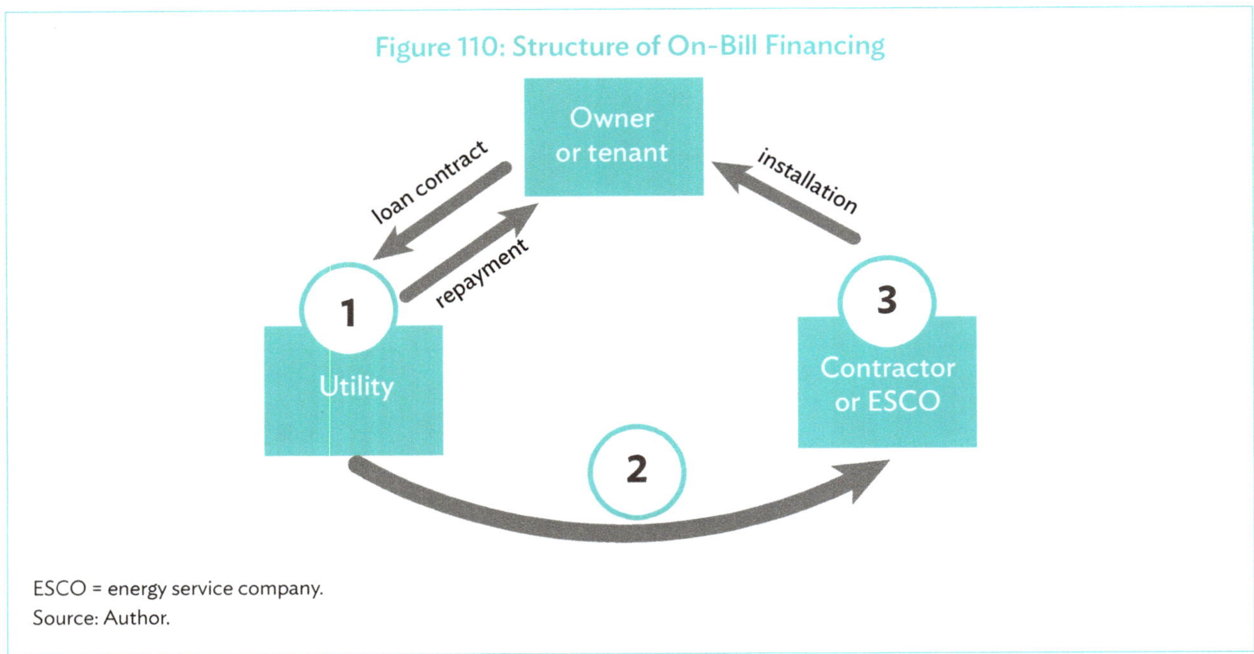

Figure 110: Structure of On-Bill Financing

ESCO = energy service company.
Source: Author.

Tax Incentives

PACE financing is a system used in the US whereby the building owner borrows money for energy efficiency measures and repays the loan via a charge on the property tax. Property tax is linked to the property and not a person or company, and financial arrangements stay with the property even if it is sold. Thus, it allows long-term investments, usually in the range of 10 years or more. The objective is to keep energy savings larger than the charge on the property tax.[34]

For example, the city of Milwaukee wanted to improve the energy efficiency of commercial and industrial buildings by 20%, but lacked capital and had to find a solution for owners who wanted the freedom to sell their property at any time without having to repay a loan. Milwaukee developed a voluntary municipal special charge that is attached to the property, not the owner (US DOE n.d. 2). C-PACE is good for larger projects that require repayment periods of 10 years or more (which do not exceed the lifetime of the equipment in question). It can finance 100%, is transferable to the next owner and overcomes the split incentive situation between landlord and tenant, because the cost savings from the utility bill as well as the property tax surcharge can be shared with the tenant. Most properties are mortgaged, and mortgage providers must usually consent to charges that relate to the asset they cover.

Tax incentives can be more direct and include reductions in import tax duties for energy efficiency equipment and credits for using energy-efficient appliances and equipment.

---

[34]	Energy.gov. Property Assessed Clean Energy Programs. A differentiation is made between C-PACE and R-PACE, depending on who the owner is. C-PACE is for properties that generate income from lease payments of that building, whereas R-PACE is for residential homeowners.

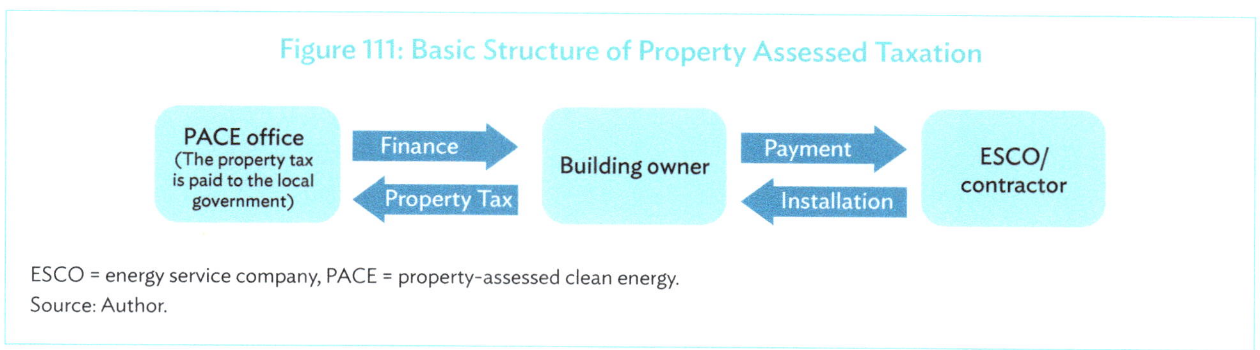

Figure 111: Basic Structure of Property Assessed Taxation

ESCO = energy service company, PACE = property-assessed clean energy.
Source: Author.

Green Bonds and *Sukuks*

To collect funding from private and public sources green bonds and *sukuks* play a growing role in energy efficiency finance. It is a fixed-income debt instrument in which the government or a financial institution (the issuer) borrows money from public and private investors to finance energy efficient projects. The structure does not differ from ordinary bonds and *sukuks* only in that the proceeds of the bond or *sukuk* must be used according to stringent guidelines for energy efficiency investments in buildings. The issuer is responsible for managing the proceeds after selling the bond to investors. They can be used to pay for the energy efficiency services of an ESCO as well as to directly finance energy efficiency equipment.[35]

For example, in 2017, DBS Bank of Singapore launched its first green bond at an amount of $500 million, which was oversubscribed on issuance day. The bond adheres to the Green Bond Principles[36] and may invest in green buildings, sustainable transportation, renewable energy, energy efficiency, waste management, and climate change adaptation projects (Sustainabbonds.com 2017).

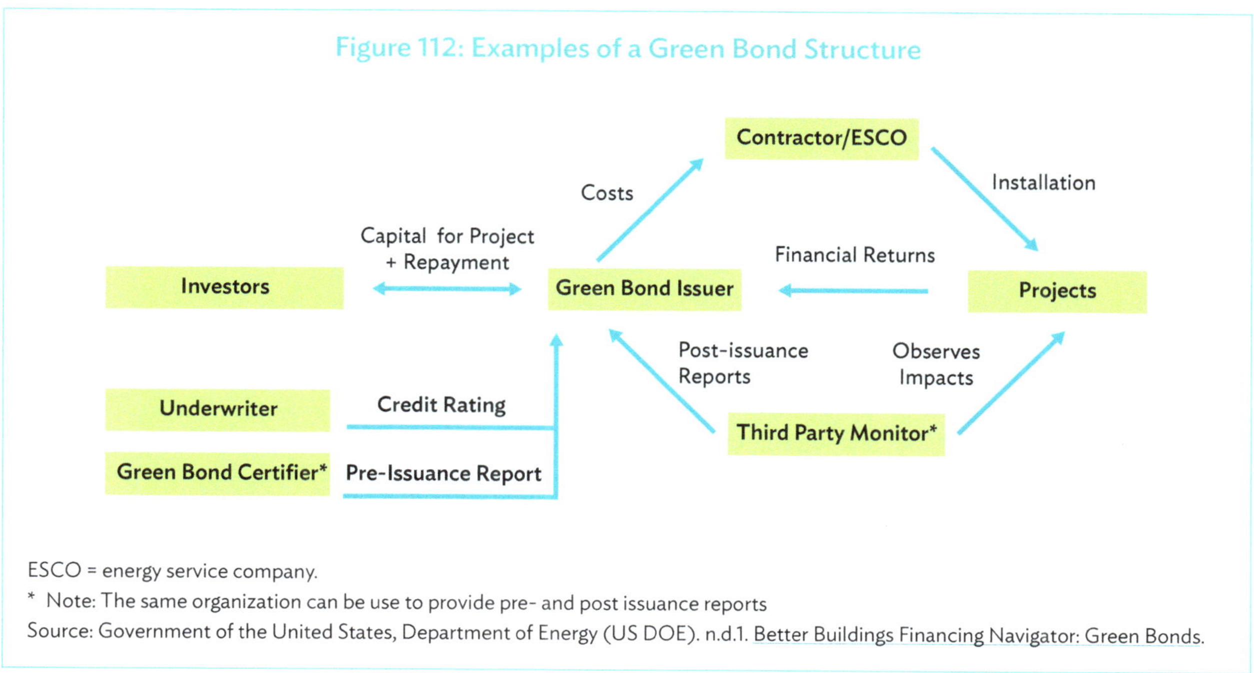

Figure 112: Examples of a Green Bond Structure

ESCO = energy service company.
* Note: The same organization can be use to provide pre- and post issuance reports
Source: Government of the United States, Department of Energy (US DOE). n.d.1. Better Buildings Financing Navigator: Green Bonds.

---

35   Further reading including other fixed income debt instruments Climate Bonds Initiative. *ASEAN Green Financial Instruments Guide.*
36   The Green Bond Principles (GBP), updated as of June 2018, is a set of voluntary process guidelines that recommend transparency and disclosure and promote integrity in the development of the green bond market by clarifying the approach for issuance of a green bond. See: ICMA. Green Bond Principles. In addition the not-for-profit organization Climate Bond Initiative-issued Climate Bonds Standard and Certification Scheme. Climate Bonds Standard.

Indonesia, Malaysia, and Singapore are the largest green bond issuers among ASEAN countries. Forty-three percent of green bond proceeds are used for green buildings. However, a few green bonds are solely used for energy efficiency (Azhgaliyeva, Kapoor, and Liu 2020).

The security Commission of Malaysia issued the first green *Sukuk* in 2017. It can be used in support of renewable energy, energy efficiency, natural resources, and community and economic development (ACE and GIZ 2019).

Green bonds are a cost-efficient instrument to finance also smaller energy efficiency projects, though the offering amount needs to be sizable—at least $10 million. It allows the issuer to structure loans with different maturities, repayment periods, and purposes along the predefined guidelines of the bond, or *sukuk*. The fundamental issue with green bonds or *sukuks* is that there is no internationally binding taxonomy, and therefore, at a minimum, a national body needs to be in place to enforce standards across green bond offerings. Possibly, issuers are faced with higher management costs for the bond due to frequent reporting requirements if the bond or *sukuk* follows voluntary standards. These are required by clients to control the proper use of funds.

Funds

The mechanism of an energy efficiency fund is to collect and allocate a pool of money for the purpose of financing energy efficiency in buildings. Sources of funding may come from various parties, the government, development banks, or the private sector. The better the fund is structured and the higher the possible dividends the more money can be leveraged by private investors.[37] Key are clear and plausible investment guidelines, and where to direct the funds. Internationally, there are several examples:

- investing in credit lines (the Thai Energy Efficiency Revolving Fund [EERF], see example below) to overcome funding shortages in the market;
- issuance of guarantees (e.g., Hungary Energy Efficiency Co-Financing Program) (Taylor et al. 2008) to bring confidence in the market and lower investment costs;
- setting up dedicated loan facilities (e.g., Bulgaria Energy Efficiency and Renewable Energy Credit Line [BEERECL]), to cover a whole range of objectives;[38]
- developing private equity funds to invest in, e.g., ESCOs and manufacturers (e.g., ADB Clean Energy Private Equity Investment fund);[39]
- developing early-stage financing funds (e.g., SCAF) to support new technologies and early-stage pipeline building for new ventures (UNEP 2020).

Funds can be fixed-term, meaning once capital is fully utilized, it will be closed, or revolving, meaning they need to be replenished, often from their own revenues. The termination date is then not set by capital extinction, but by any set date in the future.

For example, a "Green Revolving Fund" can accumulate public and private funds that are dedicated to funding energy efficiency in buildings that generate cost savings. Part of those savings is then used to replenish the fund, opening additional funding opportunities for investments in new projects with similar features. The revolving nature creates a continuous funding vehicle, which supports long-term government policies due to the availability of funds and creates confidence in the market as an established instrument. For example, the Thai Energy Efficiency Revolving

---

[37] Given that the lion's share of funding to combat climate change will need to come from the private sector UNEP made an assessment of leverage ratios and came to the conclusion that typical leverage ratios range from 3 to 15:1. This means an investment from the public side of $10 million could lead to a total of $50 million–$150 million. UNEP. 2008. Public Finance Mechanisms to Mobilise Investment in Climate Change Mitigation. Paris.

[38] Bulgaria Energy Efficiency and Renewable Energy Credit Line (BEERECL).

[39] Case Study 18: Asia – Clean Energy Private Equity Investment Funds.

Fund (EERF) started in 2003 as part of the framework of the Energy Conservation Program. It was set up to stimulate interest in energy efficiency, encourage local banks to lend to energy efficiency projects, develop a commercially stable market, and help with the deployment of other climate-related programs. The fund's initial capital was provided from the Government of Thailand's budget via revenues from a petroleum tax. As a revolving fund, it is replenishing itself, meaning there is a constant inflow and outflow of funds. The source of the replenishment stems from the interest rate differential banks charge to their clients and the charges banks can lend for.[40]

Funds require a stable institutional background as a general base. To address specific market impediments a clear analysis of the barriers must be made, and a feasible business case must be developed for private capital to come in. Administration costs of funds can be high; however, the cost of capital on an individual basis can be low. Funds need to be managed with transparent guidelines and organized in a lean manner.

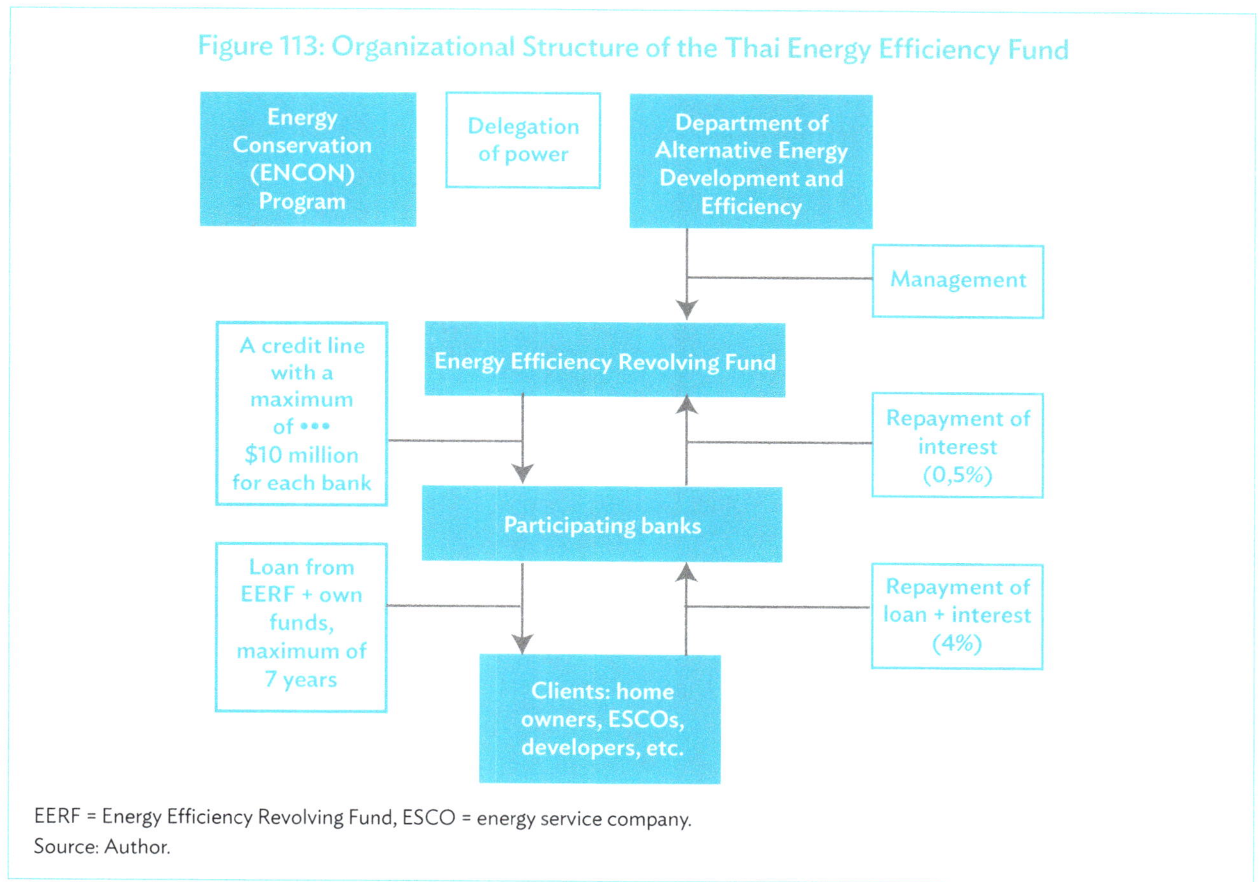

EERF = Energy Efficiency Revolving Fund, ESCO = energy service company.
Source: Author.

Energy Performance Contract

The contractual arrangements between an ESCO and a client are called energy performance contracts.[41] They are all built around ESCOs, which, in addition to energy consulting services and equipment supply, can also finance or arrange financing for the operation of energy efficiency measures. Their revenue stream is directly linked to the energy savings achieved, which incentivizes good results. There are three key contract types:

---

40   Frankfurt School UNEP Collaborating Center. 2012. *Case Study: The Thai Energy Efficiency Revolving Fund.*
41   In the US, the Energy Savings Performance Contract (ESPC).

Shared savings. The ESCO assumes the capital costs of installing energy-efficient equipment and shares with the client the energy savings throughout the life of the contract. The ESCO is paid for its services from the cost savings and the client retains the balance. After the contract expires, the energy savings will stay completely with the client. Depending on the contract specifics, the equipment can already be paid off (build–operate–transfer case) or can be bought by the client at market price.

Guaranteed savings. The ESCO guarantees the amount of energy saved after installation of equipment based on a pre-agreed floor price for the energy consumed. This model works well for clients who are skeptical as to the technical performance of the ESCO; however, if energy prices increase, they do not benefit.

Vendor finance. The manufacturer of energy-efficient equipment assumes the role of the ESCO, installs the equipment and is paid from saved energy costs. The vendor assumes the risk of nonperformance. This model works well, when a manufacturer enters a new market, when there is little choice in energy-efficient equipment and when certain efficiency measures (e.g., cooling) should be pushed forward.[42] Many vendors of energy efficiency equipment also offer lease agreements that allow the client to use the equipment without needing to buy it. It's typically part of the EPC. Two types are most common, on-balance sheet capital leases and off-balance sheet operating leases. At the end of the agreement, customers have the option to purchase the equipment at market price, return the equipment, or extend the contract.

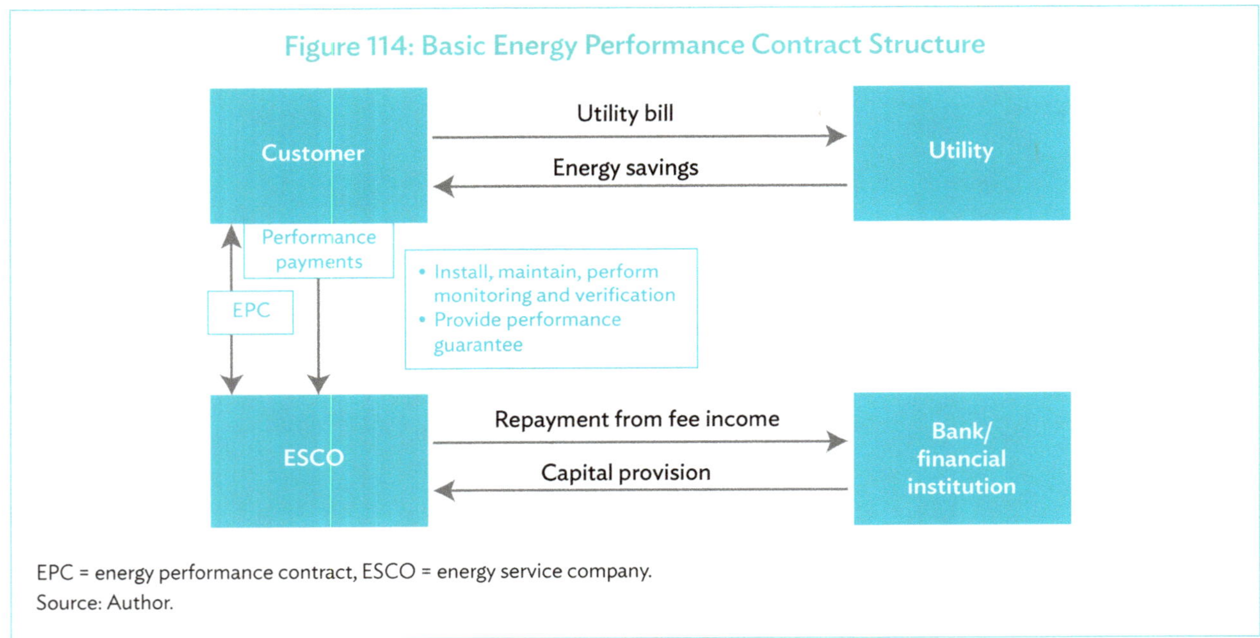

Figure 114: Basic Energy Performance Contract Structure

EPC = energy performance contract, ESCO = energy service company.
Source: Author.

The basic design of an EPC[43] is that the owner agrees to an energy efficiency measure; costs, and risks are, however, as much as possible transferred to the ESCO. This puts ESCO financially in a difficult position as their balance sheet is quickly overloaded with several contracts, and it will be difficult to continue to borrow from banks for new investments. An aggregating factor is that the EPC contract is classified as "debt" by banks because it is money owed to the client. It is not seen as a source of income based on its revenue-receiving nature. Lately, new features of EPCs have emerged, emanating from impediments in the market. The way EPCs are assessed has changed, leading to modifications in the regulatory framework. Some examples of adjustments that appear small but may have large implications are as follows:

---

42   ADB supports this type in: Southeast Asia Energy Efficiency Project.

43   Further reading: IEA. 2017. Energy Performance Contracts and SGBC. Green Procurement for Green Buildings.

- Overcome the barrier in the rental housing sector called split incentives
    - ° If owners are allowed to make EPC fees or a share thereof part of the recoverable housing charges, they may opt for consumption-based cost allocation and reduce their overall cost by implementing more and better energy efficiency equipment. This is legally very often not possible to avoid the shifting of the complete energy efficiency cost to the tenant.

- Broaden the spectrum of energy efficiency measures and lower the financial burden on ESCOs
    - ° In countries where the value-added tax (VAT) on energy-efficiency appliance and equipment is lowered, ESCOs may be excluded from the benefit, because their overall service is charged. This is why owners often finance the purchase, even though they are limited in their financial capability which in consequence reduces the number of useful energy efficiency measures. The regulations must be amended to allow the purchase of VAT-reduced equipment.[44]

- Make deep renovations more feasible
    - ° When up-front costs and finance costs are too high to be repaid from energy savings, governments often provide grants or subsidized loans to the owner to pay part of their investment. When an ESCO makes the investment, this support instrument sometimes fails, because the ESCO operates on behalf of the client and only the client is eligible. The claim for subsidies could be made transferable. Attention should be given to the fact that ESCO is a private company, and a transfer can potentially cause market distortion. Transparent financial reporting could mitigate this.

- Facilitating the market for EPCs
    - ° To build confidence in the ESCO market, public entities are often called on to go ahead. Their motives for not moving ahead are like the ones of the private sector, which are a mixture of ignorance, risk avoidance, and lack of market knowledge. By creating an honest broker as an entity between ESCO and the public entity that facilitates the process, sets standard procedures, and contracts, a service is provided that has upscaling potential and can be replicated, e.g., in other cities.[45]

- Deepening the funding base for ESCOs
    - ° The pooling of several buildings under one contract has the advantage for the ESCO that if one building fails, the loss can be mitigated by the others, and the overall contract target remains intact. Also, pooling offers the opportunity that different types of interventions can be realized, both low-hanging fruit and deep renovation, as they can be engineered to level off. Houses of the same building owner (e.g., the public sector) could be pooled as well as from different owners.
    - ° The model can be developed further in the sense that the ESCSO creates an SPV for the pool, which is financed by the savings cash flow. The ESCO provides the initial equity and may attract other equity investors. Because of the diversified risks, banks and other financial institutions are more inclined to lend, giving the entire structure credibility.

The advantage of an EPC contract for any building owner is that two main risks can be transferred to the ESCO. First, there is the risk of borrowing money for the investment, and second, there is the technical performance risk. The assumption of risks by the ESCO opens financing options for the building owner while at the same time improving the technical facilities and value of the building. In family-run businesses, however, the contract duration of an ESCO is deemed to be too long and thus too binding.

---

[44] In the Republic of Korea, the Building Retrofit Program (BRP) of the Seoul Metropolitan Government allows building owners, tenants as well as ESCOs and energy saving equipment suppliers to apply for loans. Building Retrofit Program.

[45] The Berlin Energy Agency (BEA) was mandated by the Berlin government to support the energy savings partnership program. BEA manages the entire process till the EPC is signed. Meanwhile BEA bundles several public buildings and manages the procurement to offer a larger package to ESCOs. The model has been replicated in other cities and receives competition from the private sector. See also chapter business models at BEA.

Energy Service Agreement

Widely seen as the sister to a project finance-type power purchase agreement (PPA) in the power sector, energy service agreements (ESAs) are an off-balance sheet financing solution that allows the building owner to undertake energy efficiency projects with zero up-front capital expenditures. What makes it similar to a PPA is that once the project is operational, the customer pays a service charge based on the actual savings achieved ("negawatts"). It can be a percentage of the customer's utility rate or a fixed amount per kilowatt-hour saved. The latter can be a hedge against an increase in utility tariffs.

The ESA provider develops, finances, owns, operates, and maintains the energy efficiency measures during the term of the ESA, employing an ESCO to do the installation, measurement, verification, etc. Depending on the contract structure, the customer may purchase the equipment at the end of the contract at the market price or renew the contract and receive yet again new equipment. For each new contract, the ESA provider establishes an SPV for which funding sources, like in a typical project finance structure, come from equity and debt providers. They are repaid from the revenue streams the customer pays for energy savings, possible tax rebates, and other supporting measures. The cost of funding depends strongly on the credibility of the ESA provider and the ESCO.[46]

For example, Metrus Energy, a private company that develops and finances sustainable energy projects at commercial, industrial, and institutional facilities, signed an energy service agreement with Kuakini Medical Center in Hawaii for a new central chiller plant, an LED lighting retrofit, boiler plant upgrades, variable-frequency drives, and an energy management system. The investment amount was fully financed by Metrus and repaid by energy savings, with zero capital expenditure for the client. The contract tenure was 10 years; annually, \$1.18 million and 3,234 $MtCO_2$ were saved with initial investments of \$5.96 million.[47]

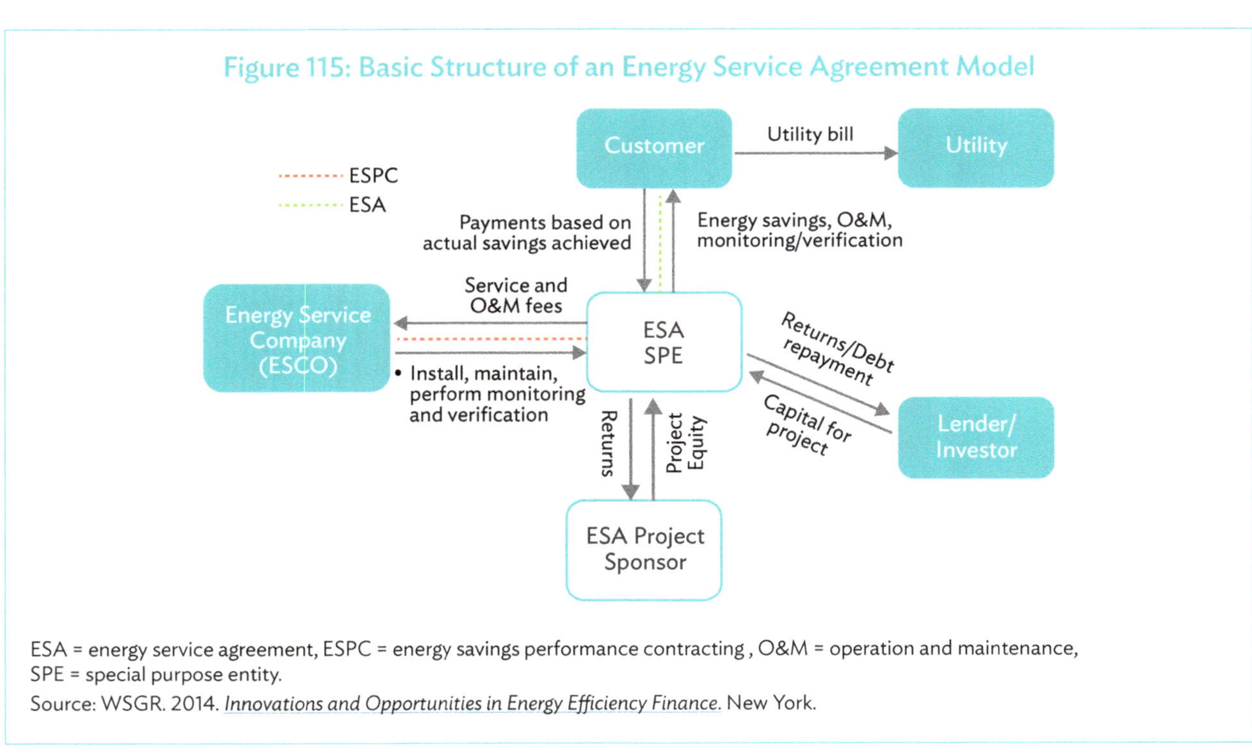

Figure 115: Basic Structure of an Energy Service Agreement Model

ESA = energy service agreement, ESPC = energy savings performance contracting , O&M = operation and maintenance, SPE = special purpose entity.
Source: WSGR. 2014. *Innovations and Opportunities in Energy Efficiency Finance*. New York.

---

[46]   Further reading: U.S. Department of Energy. Better Buildings.
[47]   METRUS.

ESAs build on the successful PPA model in project finance, where third-party project developers and investors provide the up-front capital for energy efficiency improvements, which is repaid over time by the customer through energy savings. Because the ESA provider is paid based on the amount saved, there is an incentive to implement newer technology to maximize revenues. As in project finance, smaller energy efficiency projects (below the $1 million annual energy expenditure threshold) have difficulties using private sector funding, as the amount is deemed too small for the structural costs of an ESA. Because ESAs deliver stabilized utility costs without capital expenditure at no performance risk for the client, they are a practical instrument for multi-use commercial buildings such as shopping centers and warehouses, as well as for office buildings. The ESA fee is for operational costs, which are usually tax-deductible.

Carbon Finance

According to the UNFCCC, "carbon pricing curbs greenhouse gas emissions by placing a fee on emitting and/or offering an incentive for emitting less. The price signal created shifts in consumption and investment patterns, making economic development compatible with climate protection." (UNFCCC 2022). Carbon pricing instruments come in three forms, whereby, worldwide, mostly hybrids are used:[48]

- **Emission trading systems**. A limit is set for participating entities, such as buildings on the emissions that can be emitted. If the building emits more than the allowance, owners must buy additional emission units at the market to offset the exceeding greenhouse gas emissions. The owner can also sell if the building emission are less. The idea is that refurbishment is cheaper than buying credits at market rates. The price is set based on supply and demand.[49]
- **Emission reduction funds.** The government sets up a fund, usually financed by the taxpayer, that buys or sells emission units for eligible projects.[50]
- **Carbon tax**. A tax is applied to the use of fossil fuels making their use more expensive. The price signal would be felt not only in the buildings and building construction sector but throughout the entire economy. No minimum greenhouse gas emissions can be guaranteed for the ETS and emission reduction fund, but it allows for additional tax income to be used for energy efficiency projects.[51]

For example, in Germany, a new price on greenhouse gas emissions in the transport and buildings and building construction sectors from 2021 is applied. It exists in parallel to the EU–ETS trading system and covers a number of those emissions currently not included there. The tax is on heating oil, natural gas, gasoline, and diesel, which both sectors use heavily, contributing more than one-third of greenhouse gases in 2020 in Germany.

The government sets an annual total emissions limit for transport and heating fuels ("cap and trade" system). For the first 5 years, the price is fixed, starting at €25 per allowance (ton of CO_2 equivalent) and going up to €55 in 2025. For 2026, allowances will be auctioned within a price corridor of €55–€65. From 2027 onward, a market price is foreseen. Emission allowances then become fully transferable and can be traded. The government expects revenues in the amount of €40 billion in the first 4 years, which will be used for relief measures for citizens and industry, and for climate action support programs. The government expects the price to save 3.1 $MtCO_2$ in 2025, 7.7 million tons in 2030, and 12.4 million tons in 2035. To avoid carbon leakage, the government also issued a regulation that provides compensation for certain industries. It is supposed to prevent harmful industries from moving production abroad (Wettengel 2021).

---

[48] Further reading: World Bank. State and Trends of Carbon Pricing 2021.
[49] The PRC has set up a national ETS system. Further information: IEA. China's Emissions Trading Scheme.
[50] Australia has introduced an emission reduction fund. Further reading: Government of Australia. ACCU Scheme.
[51] For examples on carbon tax in use: Center for Climate and Energy Solutions. Carbon Tax Basics.

Considering its long-term investment cycles, fragmented ownership structure, and persisting non-economic barriers, a carbon price is part of the solution to reduce greenhouse gas emissions in the buildings and building construction sector. It must, however, be accompanied by other measures. The long-term nature of a carbon tax system and the reliability of price increases can, in addition, make way for urgently needed deep renovation, as pay-off periods will drop. The price volatility of a market-based ETC. system may not offer a reliable price forecast for investors; however,
it incentivizes the introduction of greenhouse gas reduction measures.

Any sort of carbon finance system needs to be implemented at the end of the user, be it building owners or those who supervise the energy use. The implementation needs technical and financial run-up time. While neither system is perfect, a mixture of both can address a larger variety of impediments.

Latvian Baltic Energy Efficiency Facility

Independent of the type of building, the following features are important to determine methodologies for business models:

- Age: newly built or existing buildings that need retrofitting
- Ownership: who pays for the energy efficiency measure?
- Beneficiaries: whose energy bill will be reduced?
- Risk takers: who takes what risk?
- Usage: some building types have limited energy usage, some 24/7.

Background

Latvia is one of the countries that performed well after the fall of the Soviet Union and became an EU member only five years after gaining independence. The Ministry of Economics is responsible for energy and energy efficiency. Four main policies have been issued in the past years based on the Latvian Law on Energy Performance of Buildings of 2013, of which the Latvian Energy Strategy 2030 is most important for energy efficiency in buildings. It proposes an average energy intensity of 100 kWh/m^2.

Latvian building stock in the multi-family sector is old and constructed mainly in the period from 1941 to 1990. Many of these Soviet-style buildings have been privatized, so 83% are in the hands of individual families. All of them need to agree to energy efficiency measures. Many of the buildings slowly fall apart because the owners do not have enough capital for renovations. There are about 39,000 multi-family buildings covering a total floor space of about 55,000 m^2. Average energy consumption for heating is around 160–180 kWh/m^2.

Prior to the Latvian Baltic Energy Efficiency Facility (LABEEF),[52] a private ESCO called RenEsco[53] started out renovating 15 multi-family homes based on EPC contracts. Buildings are typically connected to district heating.

---

[52] Key words: deep renovation of multi-family buildings, support of long-term finance of ESCOs through a combination of financial instruments (forfaiting and on-bill finance), model possibly replicable in former Soviet Union countries, finance platform transferable to other type of buildings. Source and further reading:

- European Commission. *Energy.*
- Federal Ministry for the Environment, Nature Conservation and Nuclear Safety (BMU) *LABEEF in Latvia.*
- K. Jörling, M. Schäfer. 2018. LABEEF in Latvia.
- SUNSHINE: EPC for deep renovation of condominiums, forfeiting fund for ESCOs Latvia. (Youtube)
- Funding for Future. Building a Green Future: Investment that Will Make an Impact.

LABEEF has been replicated in Bulgaria (D. Dukov. 2010. *Accessible Financing Instruments for Energy Efficiency and Renewable Energy Projects.* International Energy Efficiency Forum. Astana, Kazakhstan. 28–30 September.) and is planned in other countries in Europe.

[53] RenEsco. RenEsco Ltd. was founded in 2008 to develop and implement energy efficiency projects for residential housing.

Therefore, the renovation targeted the building envelope, heat distribution pipes, heat control, and energy management. The measures were eligible to a 40% grant from the European Regional Development Fund.

EPC contracts were signed for 20 years. The average payback period was 9–10 years (including the grant), whereby an on-bill finance model was used. Homeowners paid the same amount throughout the contract period and accepted an increase of 15% on their energy bill. RenEsco received a fee based on the amount of saved energy from a house maintenance company. The benefit to homeowners was not only energy savings of 45%–65%, but also an increase in value of the apartment, estimated to be 20%–30% in a building, with an extended lifetime of about 30 years or more.

RenEsco received funding from a local commercial bank of up to 60%, which was backed by a third-party guarantee from the Dutch Housing Institute of 40%. No further collaterals was provided. It soon became clear that the balance sheet of RenEsco filled up and no further EPC contracts could be signed, even though the payment morale of homeowners was very high. This led to the creation of SUNShINE (Save your bUildiNg by SavINg Energy), a project funded by the EU-Horizon2020 platform, which led to LABEEF.

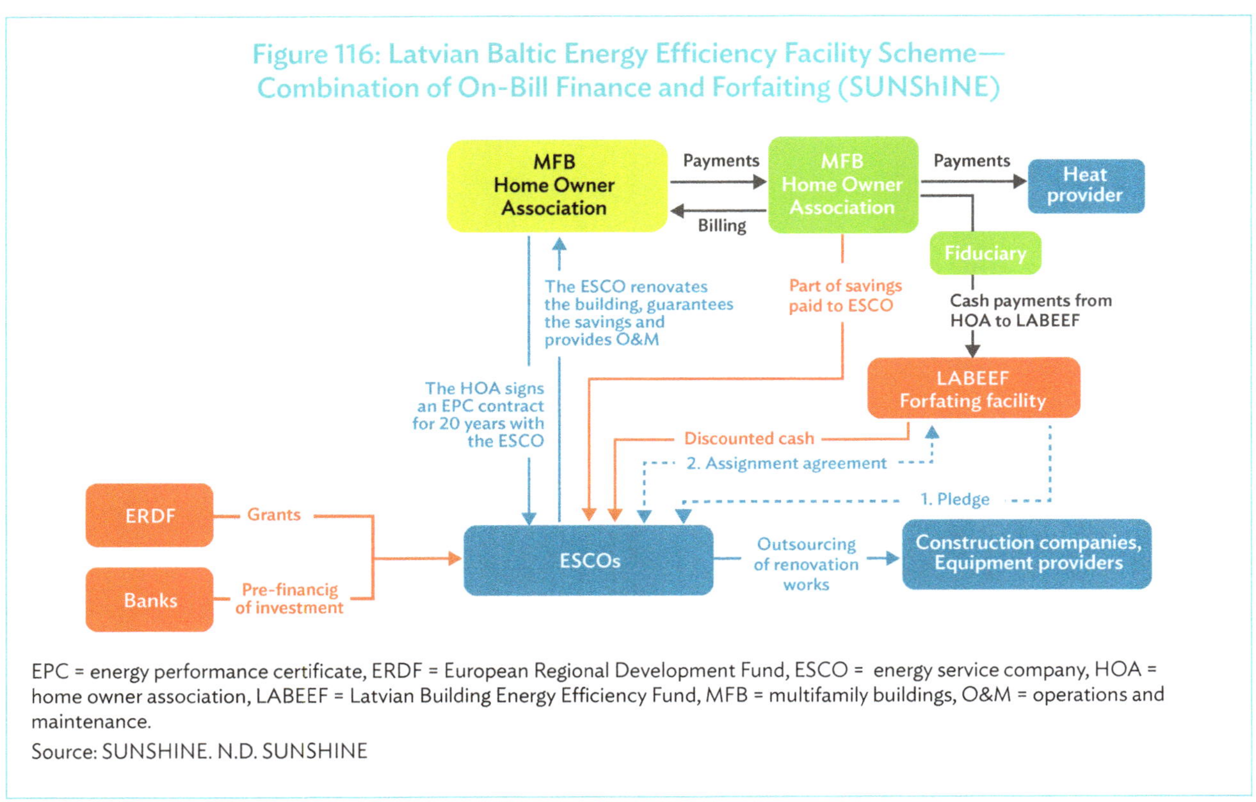

Figure 116: Latvian Baltic Energy Efficiency Facility Scheme— Combination of On-Bill Finance and Forfaiting (SUNShINE)

EPC = energy performance certificate, ERDF = European Regional Development Fund, ESCO = energy service company, HOA = home owner association, LABEEF = Latvian Building Energy Efficiency Fund, MFB = multifamily buildings, O&M = operations and maintenance.
Source: SUNSHINE. N.D. SUNSHINE

Basic Model

LABEEF was founded in 2016 in support of the Latvian ESCO business. It aims at reducing energy use in prefabricated Soviet-style multi-family buildings with diverse ownership based on EPCs.

The facility has a fund size of €30,000,000. It receives capital in the form of a loan from the EBRD and technical support from the EU Horizon project. Funding for Future B.V., a management company, is a shareholder in LABEEF.

LABEEF's objective is to modernize 20% of multi-family buildings by 2022, which is currently projected to result in the saving of 21 kg CO_2e/m^2. It is estimated that €7,600 is needed to avoid 1 ton of CO_2 emissions in the first year after renovation. This may go down to €250 per ton when considering the extended lifetime of the building due to renovation. In addition, an information technology (IT) platform is created to simplify administrative work such as applications, templates, and reporting, as well as provide financial, technical, and economic tools and training. The structure of the facility is lean and works without a delivery unit.

ESCOs sign 20-year EPCs with the building owners[54] and finance the refurbishment measures with bank loans. Once measures are implemented and their guaranteed savings verified—usually after 1.5–2 years—LABEEF forfeits the EPC contracts and receives EPC fees from building owners until the contract terminates. Please also refer to Figure 119. ESCOs receive discounted cash for their future receivables, minus an amount for operation and maintenance and a guarantee. The mechanism allows the ESCO to take on new EPCs as their balance sheet is cleared of debt. On the other side, ESCOs are incentivized to continue delivering good-quality work, as they will otherwise not receive any payments for their efforts. For the first 1.5-2 years 100% of the risk stays with the ESCO; for the remaining 18 years, the risk is shared between the two of them, with LABEEF taking 80% of the net capital expenditure receivables and the ESCO 20%.

Success Factors

The success of the platform LABEEF is first, energy savings, which are achieved through the renovation of multi-family homes. Second, LABEEF frees up long-term finance by taking on long-term risks. Third, the facility provides standardized and internationally accepted procedures. Fourth, the project learned from previous experience and built on it. Below are some of the success factors in more detail:

- A challenge for multi-family-owned buildings is to receive the commitment of all owners. Even though the need to act may have been obvious, owners were severely short of capital. The acceptance of homeowners has been achieved, probably due to a mixture of factors:
  - They were approached by a professional company and did not need to organize anything on their own.
  - The additional amount of 15% on their energy bill did not touch any savings and was manageable and stable for the entire period.
  - Homeowners did not directly sign any contracts. In case of difficulties, they could approach the house management company, which they knew and trusted.
  - Because no individual contract was signed, homeowners could sell their apartments at any time. The increased energy fee is attached to the apartment and transferred to the new owner.
  - The value of their apartments increased as did the quality of their housing conditions.

- The structure of the facility is based on earlier experience, involves long-term stakeholders, and is lean.
  - RenEsco has worked before in the sector, and the creation of LABEEF is a back-to-back consequence of direct experience that helped to gain confidence at every angle and with all stakeholders.
  - Most stakeholders were approached at a very early stage. It helped to shape the structure and individual mechanisms to match each party's requirements. Therefore, processing was carried out with few adjustments contributing to a stable working environment and trust.
  - EPCs, contracts, procedures, and standards thus follow international standards, which helped to gain investors' confidence.

---

[54] Or by proxy with the homeowners' association.

- ° The IT platform is also open to businesses that do not operate under the LABEEF umbrella, carrying standardized procedures even beyond the LABEEF border, meaning that the EPCs are replicable.
- ° The structure works without a project delivery unit. Each organizational division is directly responsible for a certain part of the internal value chain of the project. Thus, all participating organizations are directly linked and part of the success. It also keeps admin and transaction costs down.

- Even though Latvia has a history of corruption in the buildings and building construction sector, the use of a fiduciary system supports transparency and good governance.
- LABEEF is a market-based instrument. Eventually, it will work without public support. This may have increased investors' confidence.

The long-term development and experience in the market have created a new sector in building construction as well as in maintenance and most likely led to new jobs. The more people who work in the sector, the more accepted renovations of buildings get.

Building Sector Energy Efficiency Project, Malaysia

Background

Malaysia wanted to reduce energy consumption by 40% in 2020 based on consumption patterns from 2003. One of the main consumers is the buildings and building construction sector, with commercial and government buildings ahead of residential buildings.

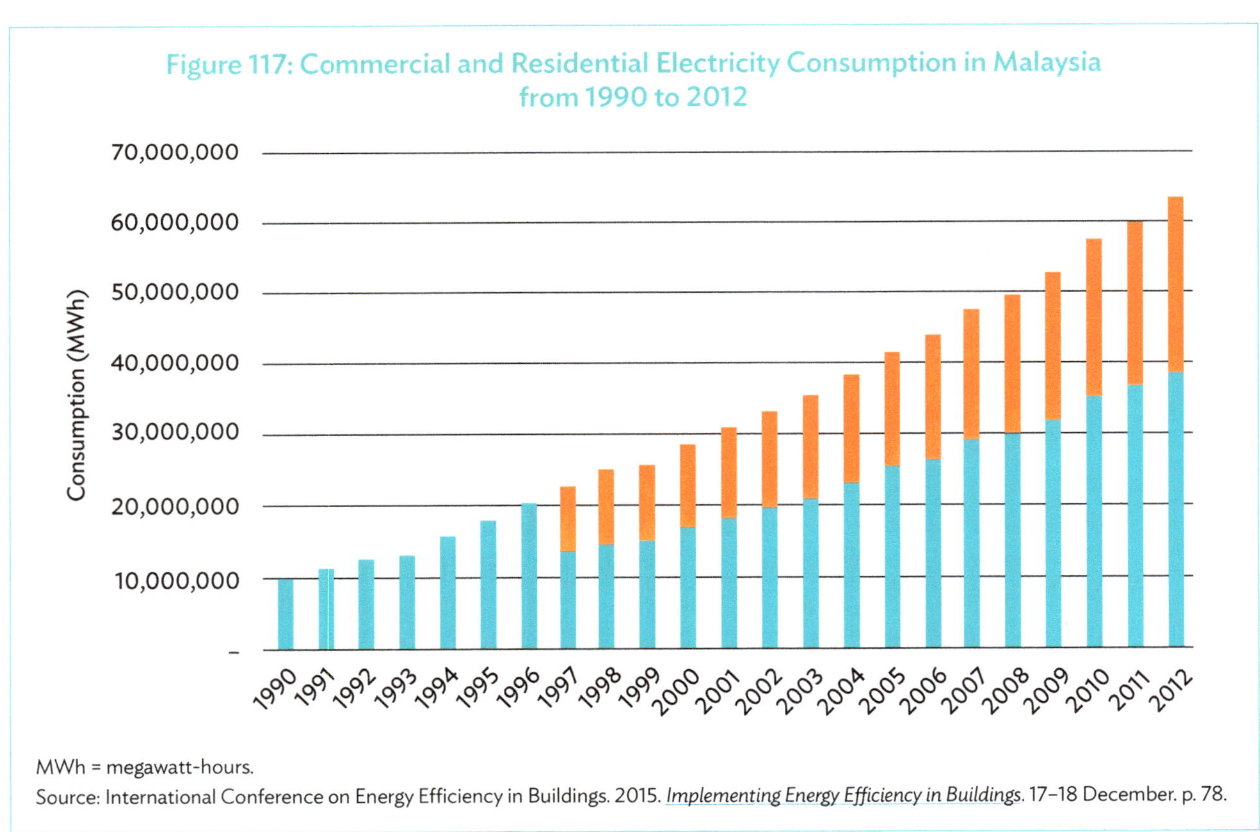

MWh = megawatt-hours.

Source: International Conference on Energy Efficiency in Buildings. 2015. *Implementing Energy Efficiency in Buildings*. 17–18 December. p. 78.

During the preparation period of the Building Sector Energy Efficiency Project (BSEEP),[55] the following main barriers were identified:

- **Fragmented institutional and regulatory environment.** At the time of the establishment of the fund, the responsibility for energy efficiency lay with the Ministry for Energy, Green Technology and Water (KeTTHA), which had a number of implementing agencies, whose scope of work was not aligned. This led to gaps in policy execution and interference. At the same time, the Ministry of Housing and Local Government had the mandate to regulate building energy design. As is the situation in Malaysia, the bylaws were never fully implemented by the federal and lower levels, as they are not mandatory. Also, the execution was not supervised.

- **No adoption of action plan.** When KeTTHA drafted an energy efficiency action plan in 2013, it was only for 10 years. This provided insufficient planning security for the industry and investors. The plan was based on voluntary reduction based on the thought that if people know how energy can be saved they will do it. This assumption proved to be unsustainable. At the same time, the action plan was issued, and the Sustainable Consumption and Production Plan for Malaysia's 11th Five-Year Plan came into light, supported by the Switch Asia program, creating an overlap of activities.

- **No database.** Despite Malaysia being able and permitted to collect individual consumption data of buildings, it has done so only irregularly. Therefore, the effect of policy and support activities cannot be measured.

- **Limited access to finance.** While there are some successful measures by the government to support energy efficiency in buildings, the financial sector fails to provide finance to the private sector. The reasons are not unique to Malaysia. Banks and financial institutions lack experience in evaluating energy efficiency projects. The concept of financing future savings translated into banking language means that decision-makers must accept future cash flows instead of collateral. This, as can be seen in the much more popular field of project finance, has always been a stumbling block. Third, and not less well-known, is the high transaction costs for relatively small-sized loans compounded by a much more complex financial transaction.

Basic Model

Under the responsibility of Jabatan Kerja Raya (Public Works Department), the agency, under the Ministry of Works and financed by GEF BSEEP, aimed at enhancing the energy conservation design of new buildings and reducing the energy use in existing ones. The project closed in December 2015. It had five components that relate to the barriers identified:

(i)   institutional capacity development,

(ii)   policy development and regulatory framework,

(iii)   energy efficiency finance capacity improvement,

(iv)   information and awareness enhancement, and

(v)   BEE demonstrations.

---

[55] Key words: transfer of knowledge from Europe (Croatia) to Asia (Malaysia) and lessons learned; commercial buildings and government sector; new and existing buildings. Sources and further reading:

- UNDP. 2018. Terminal Evaluation for the Building Sector Energy Efficiency Project.
- M. Ahmad. 2011. Introduction to Building Sector Energy Efficiency Project (BSEEP). 14 December. Renaissance Hotel, Malaysia.
- United Nations Development Programme (UNDP). Project Document: Malaysia.

Connecting policy, regulatory, and financial objectives were to be linked to specific outcomes, which for the financial part was the availability of financial and institutional support for initiatives on energy efficiency building technology applications.

The outcomes were formulated to be:

- For 1: Clear and effective monitoring system
- For 2: Implementation of, and compliance to, favorable policies that encourage the application of energy efficiency technologies and practices in the country's buildings and building construction sector
- For 3: Availability of financial and Institutional support for initiatives on energy efficiency applications
- For 4: Enhanced awareness of the government, public, and the buildings and building construction sector on energy efficiency building technology applications
- For 5: Improved confidence in the feasibility, performance, energy, environmental and economic benefits of energy-efficient building technology applications.

The outcomes were then broken down into specific outputs. The situation in Malaysia is not untypical for many countries: a number of laws, regulations, financial support instruments, and fiscal incentives exist; they are not linked to a coherent strategy. Malaysia had a number of governmental financial support programs, which, however, were not sufficiently effective due to an uncoordinated approach between policy, regulation, and financial measures. They have since been combined and renewed.[56]

Under the project it was also proposed to

- Establish a dedicated credit line for EPC contracts
- Establish an on-bill finance scheme for commercial building retrofits
- Establish an on-bill finance scheme for household appliances

As a result, an Energy Performance Contracting Fund was provided by Malaysian Debt Ventures (MDV), a corporation under the Minister of Finance, in 2016 with a fund size of RM200 million. In addition, a credit guarantee fund of RM15.8 million from the Ministry of Energy, Green Technology and Water and a RM2 million contribution from the JKR BSEEP, funded by the United Nations Development Programme-Global Environment Facility. Altogether, $52 million was supplied in support of ESCOs implementing energy-efficient projects in the buildings and building construction sector. It combines typical ESCO business with on-bill finance and opens up a spectrum of financing alternatives.

Figure 118 describes the simple mechanism of an ESCO model funded by MDV. Figure 119 shows how, in the next step, MDV as the financing vehicle is secured by a guarantee facility funded by KeTTHA and BSEEP. MDV does the management of the guarantee facility (CGC) in-house. In addition, KeTTHA provides another facility to reduce interest rates for ESCOs from 8% to 7%, thus profiling the repayment scheme. Also, this facility is managed by MDV.

---

56   The Green Investment Tax allowance (GITA) and the Green Investment Tax Exemption (GITE).

Figure 118: Malaysian Debt Ventures Fund Scheme

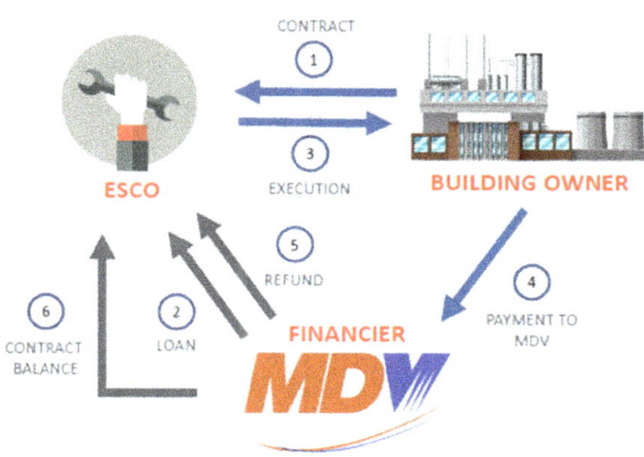

ESCO = energy saving company, MDV = Malaysian Debt Ventures.
Source: Malaysia Debt Ventures Berhad. 2017. The Energy Performance Contacting (EPC) Fund by Malaysia Debt Ventures.

Figure 119: Basic Malaysian Debt Ventures Fund Financial Incentive Scheme

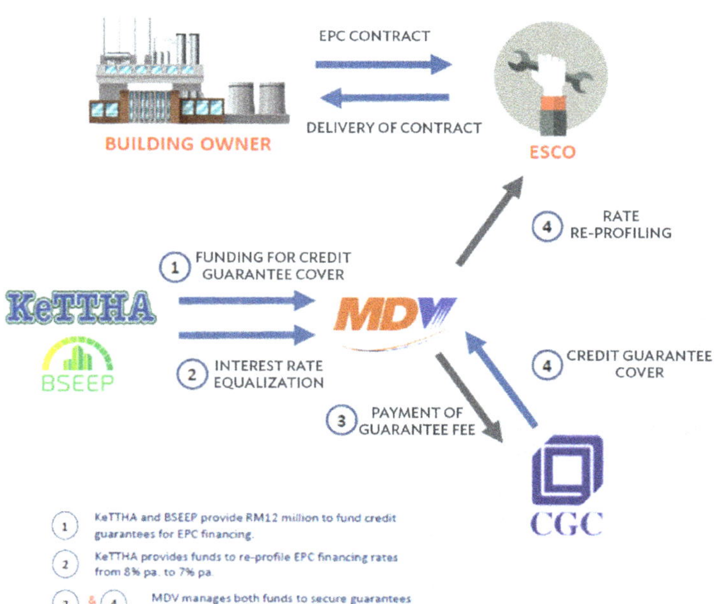

BSEEP = Building Sector Energy Efficiency Project, CGC = Credit Guarantee Corporation, EPC = Energy Performance Contacting,
ESCO = energy saving company, KeTTHA = Ministry for Energy, Green Technology and Water (Malaysia), MDV = Malaysian Debt Ventures.
Source: Malaysia Debt Ventures Berhad. 2017. The Energy Performance Contacting (EPC) Fund by Malaysia Debt Ventures.

Success Factors

- Review of tariff policy: removal of primary fuel subsidies to reflect true economic and market costs allowed gradually increasing electricity tariffs.

- BSEEP developed better ties and communication channels between various actors, such as Jabatan Kerja Raya (Public Works Department) and KeTTHA, which helped mending the fragmented sector and apply an integrated approach to energy efficiency in buildings.

- The work of ESCOs is based on newly established and supervised policy and regulatory measures developed under BSEEP, such as BEE codes (MS1525), the application of MEPs, and the disclosure of building energy intensity (BEI). The development of technical measures created synergies between policy and economic instruments. It also gave banks and financial institutions a standard they could tick off when assessing loan proposals.

- The credit guarantee facility is set up as a market-based instrument that covers and leverages private sector funding, while directly supporting public sector funding.

- Balancing policy, regulatory, fiscal, and financial instruments.

- Saving costs by building on the experience of the Croatian Environment Protection and Energy Efficiency Fund. Due to the involvement of the United Nations Development Programme (UNDP) in the project, which supported a similar project in Croatia, parallels were drawn. The energy management system developed by UNDP Croatia was transferred and customized to meet the needs of Malaysia. It facilitates the linkages with existing databases and information systems developed by different ministries and thus can be used to enforce the management and supervision of energy efficiency regulations.

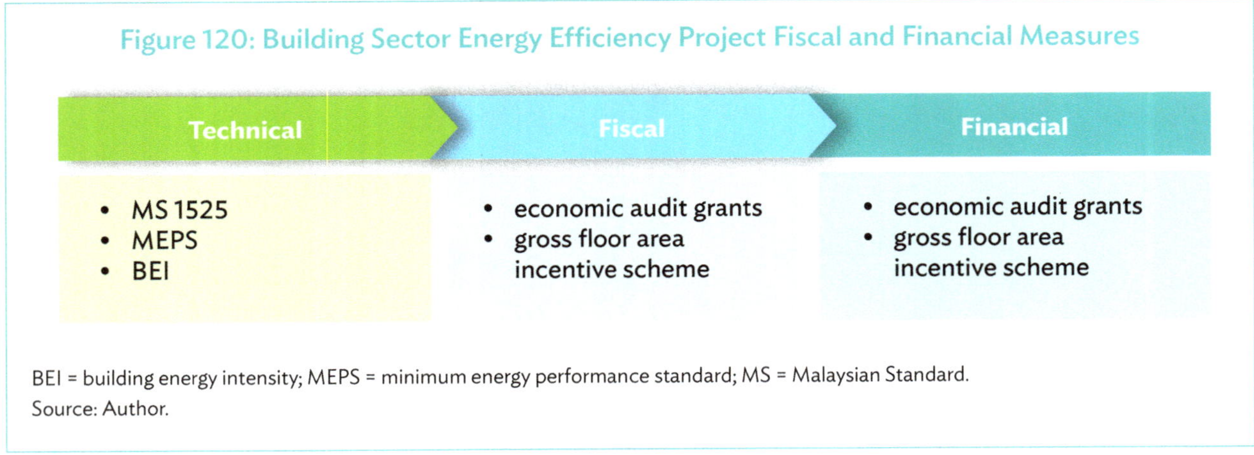

Figure 120: Building Sector Energy Efficiency Project Fiscal and Financial Measures

BEI = building energy intensity; MEPS = minimum energy performance standard; MS = Malaysian Standard.
Source: Author.

Super-Energy Saving Company: Berlin Energy Agency

Background

In the early 1990s, the city of Berlin faced high energy bills for its public buildings, but public budgets were reduced also due to the reunification of Germany. Energy costs were created mainly by the cost of heating and lighting adding to the disadvantages of an aging building stock. Berlin Energy Agency (BEA) was established in 1992 with equal shareholdings by four partners, two of which are utilities (Vattenfall Wärme Berlin and private sector GASAG AG), KfW Development Bank and the Federal State of Berlin. Together with the Berlin Senate, BEA developed the Energy Savings Partnership (ESP) and acted as the manager. The objective was to reduce energy consumption by 26%, reduce CO_2 emissions and develop EPCs for large public buildings such as schools, office buildings, and prisons.

Basic Model

Under the ESP, BEA[57] facilitates as an independent advisor retrofit by acting as the arranger between a pool of (public) buildings and ESCOs. It thereby overcomes technical and economic knowledge gaps on the part of the public building owners, which may impact their interest in such savings.

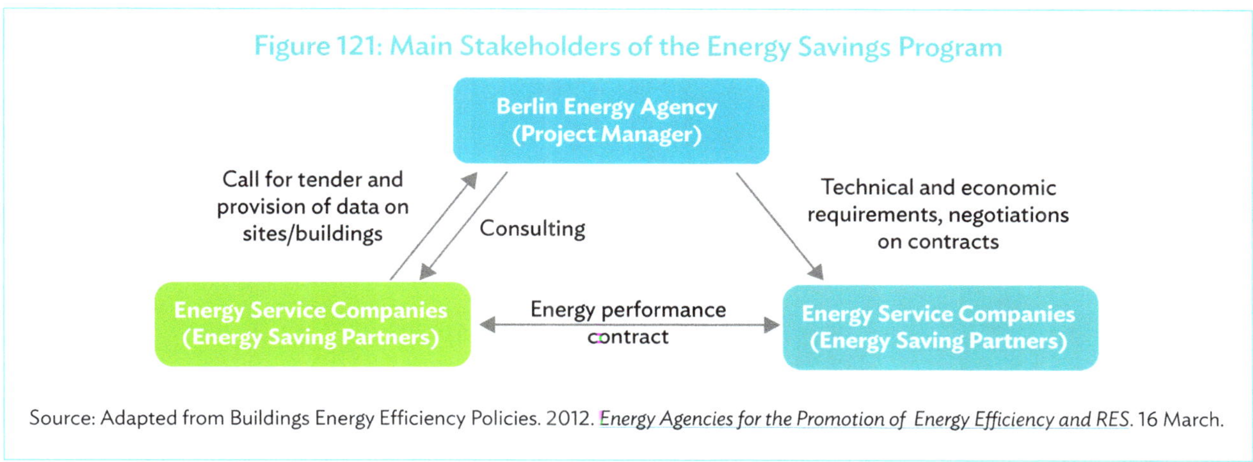

Figure 121: Main Stakeholders of the Energy Savings Program

Source: Adapted from Buildings Energy Efficiency Policies. 2012. *Energy Agencies for the Promotion of Energy Efficiency and RES*. 16 March.

The main components of the ESP are as follows:

- Building owners may pool several buildings. Required is a total energy bill of approximately $270,000 annually. It allows building owners to club together bigger and smaller projects such as universities with small daycare centers.
  - ° Buildings cannot be sold over the next 10 years after signing of the contract.
  - ° BEA must see a certain minimum savings potential.
  - ° The building must be in use with a similar occupancy rate for 3 years and for the period of the contract.
  - ° No limitation on retrofitting heating, cooling, and ventilation systems.
- BEA sets the baseline, analyzes the technical necessities on behalf of the building owner, helps to set up the bidding documents, and guides though the tendering process. Because of the market overview BEA will also supports in the negotiation of the financial terms of the contract. Later it will monitor the implementation.
- All participating ESCOs must be accredited.
- According to the EPC contract at least 25% cost savings are targeted, of which 6% go back to the city government and the remainder is used to finance the investments, i.e., the building owners do not pay for the retrofit.
- Average payback periods are between 8–12 years.
- Amounts exceeding the guaranteed cost savings go to the ESCOs.

---

[57] Key words: possible roles for energy agencies, PPP-led program, public sector develops best practice examples, pooling of buildings to save costs, replication potential in other countries.

Sources and further reading:
- New York City Global Partners. 2011. *Best Practice: Public-Private Partnership for Building Retrofits*.
- BEA. *Energy Performance Contracting*.
- Achim Neuhäuser. 2012. *Energy Agencies for the Promotion of Energy Efficiency and RES. Buildings Energy Efficiency Policies*. 16 March.

In Berlin, more than 25 pools are in operation involving more than 15 different ESCOs, which have at least 100 subcontractors. These are often regional companies that gain experience in the safe framework of an expert led program. Energy savings range between 25% and 32%.[58] The partnership has now opened to commercial buildings.

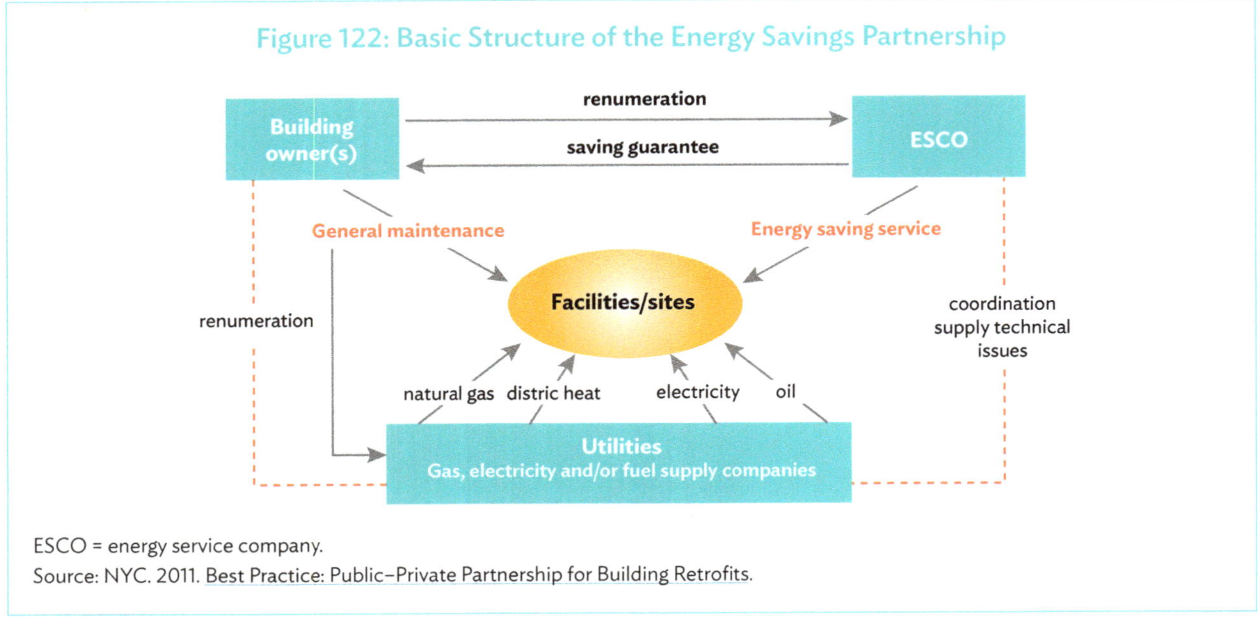

Figure 122: Basic Structure of the Energy Savings Partnership

ESCO = energy service company.
Source: NYC. 2011. Best Practice: Public–Private Partnership for Building Retrofits.

Success Factors

- The setup of BEA, with four equal shareholders, brings in political will, technical expertise, and private sector thinking.
- BEA is an expert company providing independent advice and acting as an honest broker between stakeholders.
- A stable legal and regulatory framework on which enforceable standards, EPCs, and tenders were based eases the implementation of the ESP as a program.
- Transparent and standardized procedures.
- Kick-starting ESCO business and/or EPC contracts with public support.
- Guaranteed savings are for building owners, and savings are used to finance investment costs; therefore, there is no direct financing burden for building owners.
- Ability to club buildings into one contract, allowing larger building complexes to participate.
- Success on the public buildings and building construction sector side incentivized also larger private owners to participate.

Further References

In India, ADB has provided loans to Energy Efficiency Services Limited (EESL), which is a public-owned super-ESCO, to expand energy efficiency investments in India. EESL was established in 2009 as a joint venture of four public sector undertakings of the Ministry of Power to pursue large-scale energy efficiency investments, providing a comprehensive package of project design, implementation, monitoring, and investment. ADB's loans in 2016 and 2019 were directed to investments in efficient lighting and appliances, smart meters, distributed solar PV, and electric vehicles.

---

[58] Established energy savings figures are old. As of 2012, over 1,500 buildings had been retrofitted, leading to a CO_2 reduction of 70,000 metric tons per annum. It has led to $74 million in private investments and financial savings of approximately $14.3 million, which represented 26% of the total energy bill.

8 CONCLUSION

The handbook provides a wealth of technical information for energy efficiency practitioners in the buildings and building construction sector, covering a wide range of policies, technologies, financial instruments, incentives, information and awareness mechanisms, strategies, plans, roadmaps, and business models. An energy-efficient building will reduce energy costs for the building owner, create a higher asset value, improve the health, comfort, and well-being of the residents, and support poverty reduction.

The buildings and building construction sector offers significant opportunities for improving energy efficiency and reducing greenhouse gas emissions, as well as enhancing energy security, affordability, and resilience. The energy demand and emissions from the buildings and building construction sector are expected to grow rapidly in the coming decades, especially in developing countries, driven by population growth, urbanization, economic development, and rising living standards. There are various barriers that hinder the uptake of energy-efficient technologies and practices in the buildings and building construction sector, such as lack of information, awareness, and skills, high up-front costs, split incentives, and regulatory gaps. To overcome these barriers, a comprehensive and integrated approach is needed, combining technical, policy, regulatory, financial, and behavioral measures, as well as stakeholder engagement and collaboration.

First, to accelerate the adoption of energy efficiency in the buildings and building construction sector (both new and existing buildings), it is imperative that governments reconsider the use of inefficient fossil fuel subsidies, improve availability and access to building energy use data, set national energy efficiency targets for buildings, and enable local governments to adopt and effectively enforce mandatory building energy codes. The energy performance of buildings can be improved by applying passive design principles, optimizing building envelopes, using efficient HVAC systems, lighting, appliances, and equipment, integrating renewable energy sources, and implementing smart building management systems and controls. The choice and combination of these technologies and practices depend on various factors, such as climate, building type, function, occupancy, user behavior, and cost-effectiveness. The energy savings potential of these technologies and practices can vary significantly across regions and building segments, but in general, they can achieve a 30%–80% reduction in energy consumption compared to conventional buildings.

Second, green building certification and energy labeling can help building owners identify buildings that go beyond the national standards and assist the government in providing incentives to cover the incremental costs or even expedite the permitting process of green buildings. The main types of incentives for stimulating energy efficiency investments in buildings include fiscal incentives (such as tax credits, rebates, subsidies, and grants), financial incentives (such as preferential loans, guarantees, and green bonds), and nonfinancial incentives (such as recognition, awards, and public procurement). The design and implementation of these policy and regulatory frameworks and incentives require careful consideration of the local context, stakeholder interests, institutional capacities, and monitoring and evaluation mechanisms.

Third, capacity building of government staff and the private sector will be important to ensure there is a sufficient skilled workforce to deliver and invest in sustainable and efficient buildings in the country. Energy efficiency can contribute significantly to job creation, particularly in building retrofits. The adoption of energy-efficient technologies and practices requires adequate technical capacity, quality assurance, performance monitoring, and feedback mechanisms, as well as user education and engagement. Project development and financing are key steps for realizing energy efficiency investments in the buildings and building construction sector, which involve identifying, assessing, and prioritizing opportunities, preparing bankable project proposals, and securing adequate and affordable sources of funding. The main challenges for project development and financing of energy efficiency investments in the buildings and building construction sector include lack of data and benchmarks, high transaction costs, low profitability and long payback periods, perceived risks and uncertainties, and limited access to capital and credit. To address these challenges, various tools and methodologies can be used, such as energy audits, energy performance contracting, measurement, and verification, ESCOs, and blended finance. The main sources of funding for energy efficiency investments in the buildings and building construction sector include public funds, private funds, and multilateral and bilateral funds, which can be mobilized and leveraged through different instruments and mechanisms, such as grants, loans, equity, guarantees, and carbon finance.

Fourth, it is essential to have enhanced information and awareness on energy efficiency policies and technologies at different levels of government, including building owners, developers, real estate companies, and even the banking and utility sectors. This handbook aims to raise awareness of BEE measures, provide a better understanding of the potential, proven technologies, and financing instruments applicable to this sector, and foster project opportunities within DMCs. To address this point, ADB has prepared this handbook as its contribution to opening a sensible avenue for the transition to a zero-carbon, efficient, and resilient building stock.

2030 Palette. 2021a. East/West Shading.

———.2021b. Form for Cooling.

———.2021c. Shading Devices.

ACE and GIZ. 2019. *Mapping of Energy Efficiency Financing in ASEAN*.

———.2020. The 6th ASEAN Energy Outlook.

ACEEE, Alliance to Save Energy, The Business Council for Sustainable Energy. 2020. Co-Benefits with Energy Savings.

Air Movement and Control Association. 2017. *Introducing High Performance Air Systems*.

Airtècnics. 2021. Air Handling Unit.

Akbari, H. 2005. Energy Saving Potentials and Air Quality Benefits of Urban Heat Island Mitigation.

Ali, M. et al. 2018. Performance Analysis of Solar-Assisted Desiccant Cooling System Cycles in World Climate Zones. *Journal of Solar Energy Engineering*. 140 (4).

Ampofo, F., Maidment, G.G. and Missenden, J.F. 2006. Review of Groundwater Cooling Systems in London. *Applied Thermal Engineering*. 26 (17–18). pp. 2055–2062.

APEC Energy Working Group. 2012. Cool Roofs in APEC Economies: Review of Experience, Best Practices and Potential Benefits.

APERC. 2019. *APEC Energy Demand and Supply Outlook 7th Edition 2019*. Vol. 1.

API Energy. Active Vented System Drain Back Solar Water Heating System. ND.

ARUP. 2008. *Integrated Building Environmental Design—Beijing Parkview Green*.

———.2021. Hong Kong's First Zero Carbon Building.

ARUP Associates. 2013. BskyB Sky Studios.

ASHRAE. 2016. 2016 ASHRAE Handbook – HVAC Systems and Equipment.

———.2017. 2017 ASHRAE Handbook: Fundamentals.

Asian Development Bank (ADB). 2020. Sector-Wide Evaluation: ADB Energy Policy and Program, 2009–2019. Independent Evaluation Department.

———.2021. ADB's Clean Energy Investments.

———.2022. China, People's Republic of: Energy Efficiency Multi-Project Financing Program.

Asphalt Roofing Manufacturers Association. 2017. Multi-Ply Built-Up Roofing.

Azhgaliyeva, D., A. Kapoor, and Y. Liu. 2020. Green Bonds for Financing Renewable Energy and Energy Efficiency in Southeast Asia: A Review of Policies. *ADBI Working Paper*. No. 1073. Asian Development Bank Institute

BC Housing. 2015. *Heat Recovery Ventilation Guide for Multi-Unit Residential Buildings*.

BEBuilding Energy Efficiency Project (BEEP). 2019. *Case Study on "Green" Affordable Housing: Smart GHAR III Project*.

———. 2020a. Earth Air Tunnel.

———. 2020b. Shading (EMSYS).

BECWG. 2021. *Building Energy Codes and Other Mandatory Policies Applied to Existing Buildings*. Birmingham, United Kingdom.

Bisoniya, T.S. 2015. Design of Earth–Air Heat Exchanger System. *Geothermal Energy*. 3 (18).

BP. 2020. *BP Energy Outlook 2020 edition*.

BREEAM. 2020. MEGA Silk Way, Kazakhstan.

Broadwater, M. D. 2016. Mobilizing the buildings sector for climate action. World Bank Blogs. 18 February.

BSEEP Malaysia. 2017. Building Energy Efficiency Technical Guideline for Passive Design.

Building and Construction Authority (BCA). 2019. *BCA Awards 2019 – Recognising Excellence in the Built Environment*.

Building Research Establishment Environmental Assessment Method (BREEAM). 2016. Baku White City, Azerbaijan.

Building.com.hk. 2011. *Parkview Green Beijing: Microclimatic Envelope*.

Bureau of Energy Efficiency. 2009. *Energy Conservation Building Code TIP Sheet - Building Envelope*.

———. 2013. *Passive Design Strategies*.

———. 2014. *Design Guidelines for Energy-Efficient Multi-Storey Residential Buildings (Composite and Hot-Dry Climates)*.

———. 2016. *Design Guidelines For Energy-Efficient Multi-Storey Residential Buildings (Warm-Humid Climates)*. New Delhi.

———. 2017. *Cool Roofs for Cool Delhi*.

———. 2021a. *Building Envelope Solution Sets (v 1.0) for Eco-Niwas Samhita 2018 for External Walls, Roof, Window Shading, and Glazing*.

———. 2021b. *External Movable Shading Systems (EMSyS) Manual*.

C40. 2019. How to Set Energy Efficiency Standards for New Buildings.

Cabeza, L.F. et al. 2011. Materials Used as PCM in Thermal Energy Storage in Buildings: A Review. *Renewable and Sustainable Energy Reviews*. 15 (3). pp. 1675–1695.

Calautit, J.K., D. O'Connor, and B.R. Hughes. 2014. Determining the Optimum Spacing and Arrangement for Commercial Wind Towers for Ventilation Performance. *Building and Environment*. 82. pp. 274–287.

CAREL. 2021. Evaporative Cooling.

Chen, X. et al. 2020. Recent Progress in Liquid Desiccant Dehumidification and Air-Conditioning: A Review. *Energy and Built Environment*. 1 (1). pp. 106–130.

Chen, Y., Y. Yin, and X. Zhang. 2014. Performance Analysis of a Hybrid Air-Conditioning System Dehumidified by Liquid Desiccant with Low Temperature and Low Concentration. *Energy and Buildings*. 77. pp. 91–102.

Chetia, S. et al. 2020. A Case Study on Design of Thermally Comfortable Affordable Housing in Composite Climate: Simulation Results and Monitored Performance. In *Energise 2020: Energy Innovation for a Sustainable Economy*.

Chokchai, S. et al. 2018. A Pilot Study on Geothermal Heat Pump (GHP) Use for Cooling Operations, and on GHP Site Selection in Tropical Regions Based on a Case Study in Thailand. *Energies*. Vol. 11(9).

Chwieduk, D.A. 2016. Active Solar Space Heating. In R.Z. Wang and T. S. Ge, eds. Advances in Solar Heating and Cooling. *Elsevier Science*. pp. 151–202.

CIBSE. 2012. Guide F Energy Efficiency in Buildings.

———. 2014. Module 65: Applying Chilled Beams to Reduce Building Total Carbon Footprint.

———. 2016. Guide B essentials – Air Conditioning and Refrigeration.

———. 2019. Opening the Loop – CP3 Guide.

———. 2020. Ensuring Robust Selection and Design of Surface Water Source Heat Pump Systems.

CLEAR. 2004. Building Shape – Surface Area to Volume Ratio.

Clouse, C. J. 2014. WHEEL: Aligning Energy Efficiency and Securitization.

Cool Roof Rating Council. 2021. What is a Cool Roof?

Coskun, S. 2020. Performance analysis of a solar-assisted ground-source heat pump system in climatic conditions of Turkey. *Thermal Science*. 24. pp.977–989.

Cowie, J. HPAC Engineering. 2017. The Case for High-Performance Variable-Air-Volume (VAV) Systems.

Cucchiaro, R. 2011. Beijing's Linked Hybrid: one of the largest geothermal cooling and heating systems in the world.

Cui, Y. et al. 2017. A Review on Phase Change Material Application in Building. *Advances in Mechanical Engineering*. 9 (6). Sage Journals..

Delta Pyramax. 2021. Windcatcher: Feature.

Designing Buildings. 2020. Types of Cool Roofs.

DGNB. 2020. *Building for a better world: How buildings contribute to the UN Sustainable Development Goals (SDGs)*. Stuttgart: DGNB.

Dhir, R. The Hindu. 2021. Creating a Sustainable Future with Green Buildings.

Dincer, I. and Rosen, M.A. 2007. Chapter 6 - Exergy Analysis of Psychrometric Processes. In I. Dincer and M. A. Rosen, eds. Exergy. *Elsevier Science*. pp. 76–90.

Dorendorf, B. 2018. KfW Promotional programs for energy efficiency in building: Main elements and success factors. National Roundtable on Financing Energy Efficiency in Latvia.

DRI. 2021. Indirect-Direct(2/3 Stage) Evaporative Coolers.

Dwyer, T. 2017. Module 109: Evaporative Cooling for Building Environmental Systems.

———. 2018. Module 135: Direct Evaporative Cooling for Comfort Applications.

Dynamic Air. 2012. Energy Recovery Heat Pipes | 5 Advantages of Using Heat Pipes for Air-To-Air Energy Recovery.

Economic Research Institute for ASEAN and East Asia. 2021. *Mongolia's Energy Efficiency Indicators 2019*.

Efficient Windows Collaborative. 2011. *The Efficient Windows Collaborative Tools for Schools*.

———. 2021a. Glossary.

———. 2021b. Window Fact Sheet.

———. 2021c. Window Technologies: Glazing Types - Double Low-E Glazing.

———. 2021d. Window Technologies: Glazing Types - Double Low-E Glazing with Roomside (4th surface) Low-E.

———. 2021e. Window Technologies: Glazing Types - Triple Low-E Glazing.

Elsarrag, E. 2008. Evaporation Rate of a Novel Tilted Solar Liquid Desiccant Regeneration System. *Solar Energy*. 82 (7). pp. 663–668.

Energy Star. 2021. How it Works — Heat Pump Water Heaters (HPWHs).

Enteria, N. et al., 2016. Application of Desiccant Heating, Ventilating, and Air-Conditioning System in Different Climatic Conditions of East Asia Using Silica Gel (SiO_2) and Titanium Dioxide (TiO_2) Materials. In N. Enteria, H. Awbi, and H. Yoshino, eds. *Desiccant Heating, Ventilating, and Air-Conditioning Systems*. Springer. pp. 271–299.

European Concrete Platform. 2009. *General Guidelines for Using Thermal Mass in Concrete Buildings*.

Fairconditioning. 2021a. Form and Orientation.

———. 2021b. Glazing.

———. 2021c. Radiant Cooling Case Studies.

———. 2021d. Radiant Cooling Working Principle.

———. 2021e. Shading.

———. 2021f. Window Shape and Sizing.

Falconer, J. 2017. Making The Case For Evaporative Cooling.

Feuvre, P. Le and C.S.J. Cox. 2009. *Ground Source Heating and Cooling Pumps – State of Play and Future Trends*.

Finocchiaro, L., L. Georges, and A.G. Hestnes. 2016. Passive Solar Space Heating. In R.Z. Wang and T.S. Ge, eds. *Advances in Solar Heating and Cooling*. Elsevier Science. pp. 95–116.

Fong, K.F. et al. 2011. Investigation on Solar Hybrid Desiccant Cooling System for Commercial Premises with High Latent Cooling Load in Subtropical Hong Kong. *Applied Thermal Engineering*. 31 (16). pp. 3393–3401.

Frenger Systems. 2012. *An Introduction to Chilled Beams and Ceilings*.

Gainwell. 2021. About Unnati.

GAISMA. 2021. Sunrise, Sunset, Dawn and Dusk Times Around the World.

Gawad, A.F.A. et al. 2015. Computational and Artificial Intelligence Study of the Parameters Affecting the Performance of Heat Recovery Wheels. *American Journal of Energy Engineering*. 3 (4). p. 79.

Ge, T.S. et al. 2010. Performance Comparison Between a Solar Driven Rotary Desiccant Cooling System and Conventional Vapor Compression System (Performance Study of Desiccant Cooling). *Applied Thermal Engineering*. 30 (6–7). pp. 724–731.

Ge, T.S. and J.C. Xu. 2016. Review of Solar-Powered Desiccant Cooling Systems. In R.Z. Wang and T.S. Ge, eds. *Advances in Solar Heating and Cooling. Elsevier Science*. pp. 329–379.

German Aerospace Center. 2021. Institute of Robotics and Mechatronics.

Gibson, S. 2016. Is This Ground-Source Heat Pump Plan Workable?

Glassworks. 2018. *Low-E Glass Coatings Explained*.

GlobalADB/IEA/UNEP. 2020. *GlobalABC Regional Roadmap for Buildings and Construction in Asia 2020-2050: Towards a Zero-Emission, Efficient, and Resilient Buildings and Construction Sector*.

Global Cool Cities Alliance. 2012. *A Practical Guide to Cool Roofs and Cool Pavements*.

Glumac. 2016. *Maximize Displacement Ventilation Opportunities*.

Gong, J. and K. Sumathy. 2016. Active Solar Water Heating Systems. In R.Z. Wang and T.S. Ge, eds. *Advances in Solar Heating and Cooling. Elsevier Science*. pp. 203–224.

Government of the Philippines, Department of Public Works and Highways. 2015. Philippine Green Building Code.

Government of the United States, Department of Energy (US DOE). 2020. Property Assessed Clean Energy Programs.

———. n.d.1. Better Buildings Financing Navigator: Green Bonds.

———. n.d.2. City of Milwaukee: Property Assessed Clean Energy Program to Finance Efficiency Improvements.

Government of the United States, Department of Energy, Office of Energy Saver. 2021a. Cool Roofs.

———. 2021b. Geothermal Heat Pumps.

———. 2021c. Window Types and Technologies.

Graham, C.I. 2016. High-Performance HVAC.

Green Building Alliance. 2021. Cool Roofs.

Gruner, A. and A. Zinecker. 2020. *Better Design for Cool Buildings: How Improved Building Design Can Reduce the Massive Need for Space Cooling in Hot Climates*.

Hanson, P. 2019. Vacuum Insulated Tubing for Deep Borehole Heat Exchangers.

Harrison C., Muething L. 2021. *Sustainable Global State of the Market 2020*. Climate Bonds Initiative.

Hassan, H.Z., A. Mohamad, and R. Bennacer. 2011. Simulation of an Adsorption Solar Cooling System. *Energy*. 36 (1). pp. 530–537.

Heat Pump Association. 2021. The Vapour Compression Cycle.

Heidarinejad, G. et al. 2009. Experimental Investigation of Two-Stage Indirect/Direct Evaporative Cooling System in Various Climatic Conditions. *Building and Environment*. 44 (10). pp. 2073–2079.

Henning, H.-M. et al. 2001. The Potential of Solar Energy Use in Desiccant Cooling Cycles. *International Journal of Refrigeration*. 24 (3). pp. 220–229.

Hewitt, V., E. Mackres, and K. Shickman. 2014. *Cool Policies for Cool Cities*.

HKGBC. 2021. ZCB.

HPT. 2021. DHP Dehumidification Heat Pipe Series.

Huan, C. et al. 2019. A Performance Comparison of Serial and Parallel Solar-Assisted Heat Pump Heating Systems in Xi'an, China. *Energy Science and Engineering*. 7 (4). pp. 1379–1393.

Hwang, B.-G. 2018. Green Building Rating Systems: Practices and Research Efforts. In *Performance and Improvement of Green Construction Projects: Management Strategies and Innovations*. Elsevier Science and Technology. pp. 23–44.

Inter-American Development Bank (IDB). 2020. *Energy Savings Insurance: Advances and Opportunities for Funding Small- and Medium-Sized Energy Effciency and Distributed Generation Projects in Chile.*

In Balance Energy. 2021. Solar Thermal System Design.

Integrated Design Associates. 2021. Parkview Green (FangCaoDi).

Intelligent Glass Solutions. 2019. A Sustainable Urban Pyramid of Glass and Light.

International Energy Agency (IEA). 2013. *Technology Roadmap: Energy Efficient Building Envelopes.*

———. 2017. Energy Technology Perspective 2017.

———. 2019a. Multiple Benefits of Energy Efficiency.

———. 2019b. World Energy Outlook 2019

———. 2020a. Annual Investment in Building Energy Efficiency by Region, 2014–2019.

———. 2020b. Number of Clean Energy Technology Designs and Components Analyzed in the ETP Clean Energy Technology Guide.

———. 2021a. 2018 and 2019 Budgets by Technology in Selected IEA Countries and the European Union.

———. 2021b. *Energy Efficiency 2021.*

———. 2021c. India Energy Outlook 2021.

———. 2021d. *Net Zero by 2050: A Roadmap for the Global Energy Sector.*

———. 2021e. Tracking Buildings 2021.

IEA and EU4Energy. 2020. *Clean Household Energy Consumption in Kazakhstan: A Roadmap.*

International Finance Corporation. 2019a. ArthaLand Century Pacific Tower.

———. 2019b. *EDGE Methodology Report (version 2.0).*

———. 2019c. Green Buildings: A Finance and Policy Blueprint for Emerging Markets.

———. 2021. The Business Case for the Profitability of Green Buildings.

International Monetary Fund. 2021. List of Continents by GDP.

Jensen, O.M., A.R. Hansen, and J. Kragh. 2016. Market Response to the Public Display of Energy Performance Rating at Property Sales. *Energy Policy.* 93. pp. 229–235.

Joinery Developments. 2010. Background Ventilation.

Kellert, S.R., J. Heerwagen, and M. Mador. 2008. *Biophilic Design: The Theory, Science and Practice of Bringing Buildings to Life.* Wiley.

Kenisarin, M. and K. Mahkamov. 2007. Solar Energy Storage Using Phase Change Materials. *Renewable and Sustainable Energy Reviews.* 11 (9). pp. 1913–1965.

Kevin J.L. 2007. Architectural Design of an Advanced Naturally Ventilated Building Form. *Energy and Buildings.* 39 (2). pp. 166–181.

Khalid, A. et al. 2009. Solar Assisted, Pre-Cooled Hybrid Desiccant Cooling System for Pakistan. *Renewable Energy.* 34 (1). pp. 151–157.

Khalil. A. 2012. An Experimental Study on Multi-Purpose Desiccant Integrated Vapor-Compression Air-Conditioning System. *International Journal of Energy Research.* Vol. 36(4). pp. 535–544.

Kim, D.S. and C.A. Infante Ferreira 2008. Solar Refrigeration Options – A State-of-the-Art Review. *International Journal of Refrigeration*. 31 (1). pp. 3–15.

Kolb, C. 2020. Energy Efficiency: Energy Transition 2.0.

Kuznik, F. et al. 2011. A Review on Phase Change Materials Integrated in Building Walls. *Renewable and Sustainable Energy Reviews*. 15 (1). pp. 379–391.

Kwong, Q.J. et al. 2017. Evaluation of Energy Conservation Potential and Complete Cost-Benefit Analysis of the Slab-Integrated Radiant Cooling System: A Malaysian Case Study. *Energy and Buildings*. 138. pp. 165–174.

La, D. et al. 2010. Technical Development of Rotary Desiccant Dehumidification and Air Conditioning: A Review. *Renewable and Sustainable Energy Reviews*. 14 (1). pp. 130–147.

La, D. et al. 2011. Case Study and Theoretical Analysis of a Solar Driven Two-Stage Rotary Desiccant Cooling System Assisted by Vapor Compression Air-Conditioning. *Solar Energy*. 85 (11). pp. 2997–3009.

Lawrence Berkeley National Laboratory. 2009. Cool Roof Q & A.

Li, Y., L. Lu, and H. Yang. 2010. Energy and Economic Performance Analysis of an Open Cycle Solar Desiccant Dehumidification Air-Conditioning System for Application in Hong Kong. *Solar Energy*. 84 (12). pp. 2085–2095.

Lim, H.S. and G. Kim. 2018. Analysis of Energy Performance on Envelope Ratio Exposed to the Outdoor. *Advances in Civil Engineering*.

Link, K. 2017. Example of Geothermal Cooling with Ground Source Heat Pumps in South-East Asia.

Linquip. 2020. Stack Ventilation: A Comprehensive Overview of the Principles and Applications.

Maidment, G.G. and A. Paurine. 2012. Solar Cooling and Refrigeration Systems. In A. Sayigh, ed. *Comprehensive Renewable Energy*. Elsevier. pp. 481–494.

Malaysia Debt Ventures Berhad. 2017. The Energy Performance Contacting (EPC) Fund by Malaysia Debt Ventures.

Marinakis, V. 2020. Big Data for Energy Management and Energy-Efficient Buildings. *Energies* 13, no. 7: 1555.

Maven Double Glazing. 2017. Low-E Coatings.

Mei, L. and Y.J. Dai. 2008. A Technical Review on Use of Liquid-Desiccant Dehumidification for Air-Conditioning Application. *Renewable and Sustainable Energy Reviews*. 12 (3). pp. 662–689.

Ministry of Housing and Urban-Rural Development. 2015. *Design Standard for Energy Efficiency of Public Buildings (GB 50189-2015)*.

Ministry of Housing and Urban-Rural Development. 2013. Design standard for energy efficiency of residential buildings in hot summer and warm winter zone (JGJ 75-2012).

Mohammad, Abdulrahman. et al. 2013. Historical Review of Liquid Desiccant Evaporation Cooling Technology. Energy and Buildings. 67. pp. 22–33.

Monodraught. 2010. Natural Cooling: Cool-Phase.

Natural Resources Defense Council. 2020. *Cool Roofs*.

Niu, X., F. Xiao, and Z. Ma. 2012. Investigation on Capacity Matching in Liquid Desiccant and Heat Pump Hybrid Air-Conditioning Systems. *International Journal of Refrigeration*. 35 (1). pp. 160–170.

NUS. 2020. NUS SDE4 is First in Southeast Asia to Achieve ILFI Zero Energy Certification.

———. 2021a. Net Zero Energy Building.

———. 2021b. NUS SDE is First in Singapore to Achieve the IWBI WELLTM Health-Safety Rating, and First Worldwide in the Educational Sector.

NYC. 2011. *Best Practice: Public–Private Partnership for Building Retrofits*.

NZEB Alliance. 2019. Green Conversations: Unnati.

———. 2021a. Bayalpata Hospital Achham, Nepal.

———. 2021b. Cool Roofs.

———. 2021c. Fenestration.

———. 2021d. Form and Orientation.

———. 2021e. Ground-Source Heat Pump (GSHP) Systems.

———. 2021f. Mobil House, Dhaka, Bangladesh.

———. 2021g. Radiant Cooling Systems.

———. 2021h. Shading.

———. 2021i. Unnati Office, Greater Noida, Uttar Pradesh.

O'Connor, D., J.K.S. Calautit, and B.R. Hughes. 2016. A Review of Heat Recovery Technology for Passive Ventilation Applications. *Renewable and Sustainable Energy Reviews*. 54. pp. 1481–1493.

Oh, E., 2015. o2a's Proposed Tel Aviv University Building Controls Natural Light and Wind for a Sustainable Solution.

Olgun, C.G. et al. 2015. Long-Term Performance of Heat Exchanger Piles. *Acta Geotechnica*. 10 (5). pp. 553–569.

Passivent. 2018. *Roof Ventilation Terminals*.

———. 2021. Night cooling.

Peng, D., X. Zhang, and Y. Yin. 2008. Theoretical Storage Capacity for Solar Air Pretreatment Liquid Collector/Regenerator. *Applied Thermal Engineering*. 28 (11–12). pp. 1259–1266.

Penić, M., N. Vatin, and V. Murgul. 2014. Double Skin Facades in Energy Efficient Design. *Applied Mechanics and Materials*. 680. pp. 534–538.

Pennycook, K. 2009. *The Illustrated Guide to Ventilation: BG 2/2009*, Bracknell, Berkshirem, UK: Building Services Research and Information Association (BSRIA).

Plumer, B. 2022. U.S. Greenhouse Gas Emissions Bounced Back Sharply in 2021.

Richardson, S. 2020. What Are Green Mortgages & How Will They Revolutionise Home Energy Efficiency?

Robatherm. 2014. *Run Around Coils Heat Recovery Systems for High Efficient Heat Recovery*.

Royal Academy of Engineering 2010. *Engineering a Low Carbon Built Environment: The Discipline of Building Engineering Physics*.

Sahlot, M. and S.B. Riffat. 2016. Desiccant Cooling Systems: A Review. *International Journal of Low-Carbon Technologies*. 11 (4). pp. 489–505.

Sarbu, I. and C. Sebarchievici. 2015. *Ground-Source Heat Pumps: Fundamentals, Experiments and Applications*. Elsevier.

———. 2017a. Solar-Assisted Heat Pumps. In I. Sarbu and C. Sebarchievici, eds. *Solar Heating and Cooling Systems – Fundamentals, Experiments and Applications*. Academic Press.

———. 2017b. Solar Thermal-Driven Cooling Systems. In I. Sarbu and C. Sebarchievici, eds. *Solar Heating and Cooling Systems – Fundamentals, Experiments and Applications*. Academic Press. pp. 241–313.

———. 2017c. Solar Water and Space-Heating Systems. In I. Sarbu and C. Sebarchievici, eds. *Solar Heating and Cooling Systems*. Academic Press. pp. 139–206.

Science Learning Hub. 2014. Heat Pumps and Energy Transfer.

Serck. 2019. Heat Pipes.

Sergei, K., C. Shen, and Y. Jiang. 2020. A Review of the Current Work Potential of a Trombe Wall. *Renewable and Sustainable Energy Reviews*. 130.

Shan, M. and B. Hwang. 2018. Green Building Rating Systems: Global Reviews of Practices and Research Efforts. *Sustainable Cities and Society*. 39. pp. 172–180.

Sharon Davis Design. 2019. Bayalpata Hospital.

Siegenthaler, J. 2018. A Look at Air-to-Water Heat Pump Systems.

Society of Building Science Educators. 2012. Carbon Neutral Design Strategies: Reduce Loads / Demand – Shading.

Song, C. et al. 2021. Application and Development of Ground Source Heat Pump Technology in China. *Protection and Control of Modern Power Systems*. 6.

SUNShINE. n.d. SUNShINE.

Sustainabbonds.com. 2017. DBS Gets 'Compelling' Level in Singaporean Green Bond First.

Swiss Agency for Development and Cooperation. 2018. *Thermally Comfortable and Climate-Friendly Affordable Housing in India: The Smart GHAR III Project*.

The Renewable Energy Hub. 2021. Types of Heat Recovery System.

The Skyscraper Center. 2016. Parkview Green FangCaoDi.

Taylor, R. P. et al. 2008. Financing Energy Efficiency: Lessons from Brazil, China, India, and Beyond.

Titon UK. 2021. Vent-Trex Over-Head Extract Fan and Trickle Ventilator.

Transsolar Energietechnik. 2021. Bayalpata Regional Hospital, Achham, Nepal.

Ullah, K.R. et al. 2013. A Review of Solar Thermal Refrigeration and Cooling Methods. *Renewable and Sustainable Energy Reviews*. 24. pp. 499–513.

UN-Habitat. 2018. *Sun Shading Catalogue. Adequate Shading: Sizing Overhangs and Fins*.

United Nations Environment Programme (UNEP). 2020. Seed Capital Assistance Facility.

UNEP. 2021. *2021 Global Status Report for Buildings and Construction: Towards a Zero-emission, Efficient and Resilient Buildings and Construction Sector*. United Nations Environment Programme.

United Nations Framework Convention on Climate Change (UNFCCC). 2021. NDC Registry (interim).

UNFCCC. 2022. About Carbon Pricing.

Urban Systems Design. 2021. Commerzbank HQ Frankfurt, Germany.

UNFCCC. 2022. NDC Registry. United Nations Framework Convention on Climate Change.

United States Agency for International Development. 2014. *HVAC Market Assessment and Transformation Approach for India*.

U.S. Green Building Council (USGBC). 2018a. ArthaLand Century Pacific Tower.

USGBC. 2018b. Unnati.

———. 2019. Mobil House.

Vandervort, D. 2020. How a Heat Pump Works.

VOLKsHouse. 2021. How Does a Passive House Work?

Wang, R.Z., Z.Y. Xu, and T.S. Ge. 2016. Introduction to Solar Heating and Cooling Systems. In R.Z. Wang and T.S. Ge, eds. *Advances in Solar Heating and Cooling*. Elsevier Science. pp. 3–12.

Wettengel, J. 2021. Germany's Carbon Pricing System for Transport and Buildings.

Whole Building Design Guide (WBDG). 2016. Energy Codes and Standards.

World Bank. 2021. Carbon Prices Now Apply to Over a Fifth of Global Greenhouse Gases.

World Green Building Council (WGBC). 2015. Case Study – Construction Industry Council – Zero Carbon Park.

WGBC. 2019. Case Study – Arthaland Century Pacific Tower.

———. 2021. Green Building: Improving the Lives of Billions by Helping to Achieve the UN Sustainable Development Goals.

Whole Building Design Guide. 2016. Windows and Glazing.

Widiatmojo, A. et al. 2019. Ground Source Heat Pump Application in Tropical Countries. In *44th Workshop on Geothermal Reservoir Engineering*, Stanford University.

Williams, J. 2016. Two Ways to Make a Solar Wall. The Earthbound Report.

Wincott, N. 2019. *Low Carbon Heating and Cooling from Groundwater for Commercial and Residential Developments and District Networks*.

World Resources Institute (WRI). 2016. *Accelerating Building Efficiency: Eight Actions for Urban Leaders*.

WRI. 2019. Release: New Research Shows Zero Carbon Buildings Are Possible Where You Might Least Expect Them.

WUNZ:HS. 2012. Evaluation of Warm Up New Zealand: Heat Smart.

WWF. 2021. Rajkot – Efficient Cooling is Key.

Xu, Z.Y. and R.Z. Wang. 2016. Solar-Powered Absorption Cooling Systems. In R.Z. Wang and T.S. Ge, eds. *Advances in Solar Heating and Cooling*. Elsevier Science. pp. 251–298.

Yang, T. and D.J. Clements-Croome. 2013. Natural Ventilation in Built Environment. In V. Loftness and D. Haase, eds. *Sustainable Built Environments*. Springer.

Yasukawa, K. and Y. Uchida. 2018. Space Cooling by Ground Source Heat Pump in Tropical Asia. In *Renewable Geothermal Energy Explorations*. IntechOpen.

ZCP. 2018. ZCP's Design Strategies.

Zhao, K. et al. 2011. Performance of Temperature and Humidity Independent Control Air-Conditioning System in an Office Building. *Energy and Buildings*. 43 (8). pp. 1895–1903.

Handbook on Energy Efficiency in Buildings

As Asia's buildings and building construction sector booms, this handbook provides practical tools and guidance to design, evaluate, and finance energy efficient building projects to help make the sector greener and more resilient. Explaining how energy efficient construction can help reduce costs, lower emissions, and improve affordability, the handbook offers practical guidance and tools, covers key project cycle stages, and delves into sustainable heating and cooling strategies. It details ways to improve procurement, supervise, and rate energy efficiency for the emissions intensive industry while underscoring the need for governments and the private sector to work together to help transition toward a zero-carbon building stock.

About the Asian Development Bank

ADB is committed to achieving a prosperous, inclusive, resilient, and sustainable Asia and the Pacific, while sustaining its efforts to eradicate extreme poverty. Established in 1966, it is owned by 69 members —49 from the region. Its main instruments for helping its developing member countries are policy dialogue, loans, equity investments, guarantees, grants, and technical assistance.

ADB

ASIAN DEVELOPMENT BANK
6 ADB Avenue, Mandaluyong City
1550 Metro Manila, Philippines
www.adb.org